香港原生植物圖鑑

Botanical Illustrated Guide to Hong Kong Native Plants

胡秀英植物標本館科學繪圖系列
《香港原生植物圖鑑》
劉大偉、李敏貞、王曉欣 著
植物繪圖師：王曉欣、李敏貞
編輯、教育顧問：王天行

國際統一書號（ISBN）：978-988-237-270-2

出版：香港中文大學出版社
香港新界沙田．香港中文大學
傳真：+852 2603 7355
電郵：cup@cuhk.edu.hk
網址：cup.cuhk.edu.hk

Shiu-Ying Hu Herbarium Scientific Illustration Series
Botanical Illustrated Guide to Hong Kong Native Plants
By David T. W. Lau, Man-Ching Li and Hiu-Yan Wong
Botanical Illustrators: Hiu-Yan Wong and Man-Ching Li
Editor and Education Consultant: Tin-Hang Wong

ISBN: 978-988-237-270-2

Published by The Chinese University of Hong Kong Press
The Chinese University of Hong Kong
Sha Tin, N.T., Hong Kong
Fax: +852 2603 7355
Email: cup@cuhk.edu.hk
Website: cup.cuhk.edu.hk

胡秀英植物標本館科學繪圖系列
Shiu-Ying Hu Herbarium Scientific Illustration Series

香港原生植物圖鑑
Botanical Illustrated Guide to Hong Kong Native Plants

劉大偉、李敏貞、王曉欣 著
王天行 編

David T. W. Lau, Man-Ching Li and Hiu-Yan Wong
Edited by Tin-Hang Wong

2012 - 2022
10th anniversary
胡秀英植物標本館
SHIU-YING HU HERBARIUM

The Chinese University of Hong Kong Press

目錄
Contents

序言

2012年5月22日，我們敬愛的胡秀英教授與世長辭。從此，植物學界的一顆巨星隕落，損失無可彌補。她對植物的淵博早已聞名；後輩有問題請教，她是從不推搪的，故此她十分忙碌，但也是其助人之道。

在她身體不適的時候，中文大學崇基學院已為她安排了另一起居之所，使她不用攀上樓梯。她的參考書和用品也隨著她移進新居所，除了多年來，為了教學和研究需要經常和同事外出採集，全部經她鑒定的大量本地植物標本，還有和外地交換的珍貴標本。標本本是採自郊野，經過精心的裝裱，連同準確無誤的鑒定後，便成為無價之寶，必須嚴格除蟲、去霉，還須保存在適當的溫度、濕度之下，總之經過一系列的專業處理後，仍留在她辦公室的標本櫃內。

隨著教授的離開，還有多少人認識、感悟到這批標本的價值？關心的人是有的，而且不在少數，尤其是平時和教授在工作上接觸較多的人，但解決辦法卻不易尋求。此事千頭萬緒，牽涉多方面，中大生命科學學院時任院長朱嘉濠教授明白到標本室的重要性和關鍵性，所以一開始便和各相關人員，包括關心標本室的人士，研究成立胡秀英植物標本館的可能。幸好胡教授的人緣甚佳，樂於幫助的人也不少，所以決定分頭進行，以促成此事。

其後，劉啓文博士接到中大大學李賢祉博士來電，被告知此事，李博士與他在植物學範疇上素有聯繫，略知他與伍氏家族成員的關係，所以電告有關情況，希望劉博士可以尋求一些辦法。劉博士原受益於胡教授的指導很深，但礙於日常工作，只可以短時期工作於胡教授的辦公室，其中一項工作是協助她為籌備「香港植物誌計劃」而編寫菊科有關艾納香屬的部分。胡教授是冬青科、錦葵科、菊科和蘭科的專家，負責為《香港植物誌》編寫此四科，工作壓力很大。劉博士常覺得，未能夠更多減輕胡教授在編寫菊科方面的負擔，是一大遺憾。而在鑒定香港蛇菰的過程中，胡教授憑著豐富的經驗和知識，最終證實了香港蛇菰為一個新品種，並得以在權威的植物學報上發表。這些事情在在都説明標本室對基礎研究的重要性。

幸運地，當劉博士和伍絜宜慈善基金會的成員討論這個問題時，他們都認為標本室的工作十分有意義，應該支持。加上在許多其他人的參與下，其中最主要人士如胡教授的尊崇者吳紀嬅女士，經費來源因而得到保障，標本館在原胡教授的辦公室有了一個工作地點。

在胡教授逝世十周年之際，香港中文大學校方決定在39區新科學大樓撥地給予標本館作永久館址，表現出卓越的遠見，本基金會謹致以熱烈祝賀。此舉有助延續胡教授的美名，為後來者展示一個榜樣，希望胡教授在天之靈，能認同我們的做法。大學標本館的目標顧名思義，除了服務社會之外，還應當包括教學在內，涵蓋意義廣泛。

《香港原生植物圖鑑》於標本館成立十周年這時候出版，意義非凡，將科學與藝術兩者結合融入教育當中，傳承了胡秀英教授畢生追求和傳播知識的熱忱。

本會付出微力，能夠參與標本館的工作，覺得無比光榮，同時極希望有其他機構或基金、個人一齊參與這份有意義的工作。

伍絜宜慈善基金會
2022年春

Foreword

On the 22nd of May 2012, our beloved Professor Shiu-Ying Hu left us. A bright star in the botanical field has fallen, causing an irreplaceable loss to us. Her wide-ranging grasp of knowledge in the botanical field is well-known. Enquiries from the younger generation were never denied, so she was kept busy most of the time, but she thoroughly enjoyed this way of life and being of help to others.

When Prof Hu was feeling unwell, Chung Chi College of The Chinese University of Hong Kong assigned a ground-floor accommodation for her, so she did not have to climb stairs to reach her apartment. All her books and belongings were transferred to this accommodation, except the specimens. These specimens, which were collected locally by her and her colleagues over many years for teaching and research purposes, and which were all identified by her—plus many others which were exchanged with outside herbaria—had to stay in the specimen racks in her departmental office. Specimens collected in the wild, after being carefully mounted, then accurately identified, become invaluable assets. A strict set of measures must be followed to ensure that the herbarium specimens are clear of harmful matter, then kept at an appropriate temperature and humidity.

With the departure of Prof Hu, there were initial uncertainties about the herbarium specimens. However, there were still many people who were steadfast about the values of a herbarium, especially those who had worked closely with Prof Hu. Professor Ka-Hou Chu, the then Director of the School of Life Sciences at CUHK, from the beginning realized the value and the important position that the herbarium would hold for the University, and advocated for the establishment of the Shiu-Ying Hu Herbarium. There were many others who did much work in this regard as well. Despite her fame, Professor Hu was kind to many people when she was alive, and the amount of help that people offered in return was widespread.

Dr Eric Y. T. Lee at CUHK, who met Dr Kai-Man Lau on occasions relating to botanical matters, had some idea about Dr Lau's relationship with the Wu family that runs this charitable foundation; he then contacted Dr Lau and talked about support of the Herbarium. It happened that Dr Lau had worked with Prof Hu, though briefly, and benefited from the experience. Due to his regular job, Dr Lau could work only for short periods in Prof Hu's office. He was assisting Prof Hu on the genus Blumea in her work on the family of Compositae (Asteraceae) in the Flora of Hong Kong Project. Prof Hu was an acknowledged expert of Aquifoliaceae, Malvaceae, Compositae (Asteraceae), and Orchidaceae, and she was asked to take on writing up these four families for the Flora of Hong Kong, a heavy responsibility indeed. Dr Lau has often felt regret for not being able to do more on Asteraceaea and so to relieve Prof Hu's heavy load. Another experience relates to the identification and publication of the new species of *Balanophora hongkongensis* K. M. Lau, N. H. Li & S. Y. Hu, which was facilitated with Prof Hu's rich experience and extensive knowledge of the subject. Both these events demonstrated the indispensability of the Herbarium to basic research. Fortunately, when the topic of support was discussed with members of the Wu Jieh Yee Charitable Foundation, the members all felt the work of the Herbarium was very meaningful and should be supported. With the participation of many other donors, among whom we must mention Ms Jenny Ng, an admirer of Prof Hu, the sponsorship was thus assured, and the Herbarium started to function in Prof Hu's original office.

On the 10th anniversary of Prof Hu's passing away, we note with gratification the provisioning of a permanent site for the Herbarium in the new Biomedical Sciences Building of Area 39, showing the University's exceptional vision for the future. It is a remembrance of Prof Hu, carrying with it an exemplary message for excellence, coupled with kindness and generosity. For this we offer our heartfelt congratulations.

The publication of *Botanical Illustrated Guide to Hong Kong Native Plants* is of extraordinary significance, not only in that the year 2022 marks the 10th anniversary of the Herbarium, but also because it integrates science and arts in public education; in this way, the legacy of Prof Hu's life-long passion to pursue and transmit knowledge is being preserved for future generations.

Our contribution is small, but we have an immense feeling of pride and satisfaction in being able to participate in the work of the Herbarium. It is our ardent wish that other organizations, foundations, or individuals may find it worthwhile to participate in the work of the Herbarium in the same way as ours.

Wu Jieh Yee Charitable Foundation
Spring 2022

序言

植物繪圖一直以來是分享資訊和教育的一種溝通媒介，香港中文大學胡秀英植物標本館自2012年成立後，其館長與館員繼續發揮植物繪圖這種作為提供資訊與教育的傳統。在本圖鑑，他們精選了二十幅出色的植物繪圖，細致地呈現出不被大部分人所關注、或慣常被忽略的詳細植物形態。在這種情況下，我們所看到的，是植物在其自然生長環境的形態、植物結構的近距離照片，以及往往被鏡頭所隱蔽、被一般人眼睛所錯過的植物細部，現在卻能以精美的墨線圖所強調和凸顯出來。

這本圖鑑同時亦闡明了畫家需要使用植物樣品、照片與標本館標本，以及採集和在防蟲防腐的環境下保存憑證標本的需要。憑證標本是作為繪圖的永久性紀錄，要是對照片或繪圖中植物的品種存疑，憑證標本可提供資訊，同時可與其接近的品種進行比較，藉此確認其品種的準確性。

本書亦展示了詳盡、細致的繪圖的重要。繪圖者慎密的眼睛往往能注意到被植物學者所遺漏的細節。植物學者在辨識植物品種或鑒定其關係時，可能只專注於植物的某個部分，但是繪圖者則沒有這種偏見，而能均等地觀察到植物的各個部分，因此可以揭示植物學者所完全忽視的，以及微距照片所不能清楚展現的細節。

看著這些美麗的植物繪圖、照片及與之相關的解說，都讓我們期盼香港豐厚、多樣性的植物物種亦能得到同樣的詳細科學記錄。

David E. Boufford
哈佛大學植物標本館高級研究科學家
《中國植物誌》編輯委員
2022年春

Foreword

Botanical illustrations have always been a means to communicate, to share information and to educate. Since its founding in 2012, the curators and staff of the Shiu-Ying Hu Herbarium at The Chinese University of Hong Kong have continued this tradition of providing illustrations and information about plants. In this book they have selected 20 outstanding botanical illustrations to show morphological details that most of us were unaware of, or routinely overlooked. In this case, we are shown the plants as they appear in nature, closeup photographs of the parts of those plants and exquisitely beautiful illustrations that highlight and emphasize details that a camera often obscures and the average human eye misses.

The book also explains the artist's need to use live plants, photos and herbarium specimens, and the need to collect and preserve a voucher specimen in a safe, insect- and fungus-free environment. The voucher specimen serves as a permanent record for the illustrations. If the identity of the plant in the photo or in the illustration is ever questioned, the voucher specimen can be consulted and compared with closely related species to make certain that the identity is correct.

This book also shows the importance of having detailed illustrations. The careful eye of the illustrator often sees details that a botanist misses. A botanist may concentrate only on particular parts of a plant to identify it or to determine its relationships, but the artist has no such bias and sees all part of the plant equally, thereby revealing details that the botanist may overlook entirely and which micrographs do not show clearly.

To see these beautiful illustrations, the photos and the commentary about the plants that accompany them makes us wish that all of the rich and diverse flora of Hong Kong could be treated similarly.

David E. Boufford
Senior Research Scientist, Harvard University Herbaria
Editorial Committee Member of *Flora of China*
Spring 2022

前言

植物科學繪圖在植物分類學上有非常重要的角色，尤其墨線圖可以清晰表達各植物器官的結構，數十年前數碼拍攝仍未普及時，墨線圖所提供的資訊實在功不可沒。植物誌通常都附有植物墨線圖，讓讀者更容易明白各項特徵的描述，加強鑒定的準確性。

中文大學生物系植物標本室自1970年代便開始植物科學繪圖的工作，筆者曾經親眼目睹繪圖工序的細緻，以及繪圖師的耐性和刻苦。要完成一幅繪圖必須採集和鑒定原植物，過程包括在野外尋找結構完整的植物標本，回到實驗室後解剖花和果實，才開始繪圖工序，期間亦需要不斷觀察顯微鏡下的植物特徵。從草稿到鋼筆完成稿，由數天至數星期不等，視乎植物結構的複雜性。再者，不同的植物標本接踵而來，可以讓繪圖師的工作沒休止，長年累月的顯微觀察和手繪動作實令人身心疲累，因此每一幅植物圖都是繪圖師和植物學家的心血結晶，也為植物分類鑒定提供重要的科學資訊，是基礎科學的重要元素。

胡秀英植物標本館（下稱「標本館」）自2012年成立至今，每位同事都竭盡所能履行本館的職務，期望完成標本館的使命。每當我們回憶起胡老師的厚生哲學，並在人生中實踐，盡己所能幫助有需要的人和貢獻社會，都激勵我們注入動力，並儆醒標本館的原始工作目標，包括採集和存護植物標本，作植物分類學的研究，繼而應用在社會有需要的範疇，例如：生態保育、中藥鑒定及園藝設計等。過去十年，我們一直學習和鑽研植物分類，主要工作包括品種鑒別、特徵描述、親緣關係和命名法規——每種研究方向都是無底深潭，雖然艱辛但不斷有新鮮感和挑戰。其中一項鑒定方法便是植物科學繪圖。

筆者自入行已常觀察及使用植物誌內的墨線圖協助判別品種，不過一直對於墨線圖都沒有特別的感覺，只覺得是另類的黑白圖像而已。直至數年前整理館藏的手稿時，發現了一批由教授級畫家——曾孝濂先生和馬平先生繪畫的植物墨線圖，筆者頓時進入了植物的微觀世界一樣，過去從沒有被記錄過的結構都清晰呈現眼前，好像一頭小昆蟲在植物的毛被、鱗片間遊走和冒險一樣。這些震撼人心的畫作，令我回憶起當年馬老師繪圖時的情景。當時他與胡定國先生在野外採集標本，回到標本館時使用殘舊的解剖顯微鏡，把植物標本的各部分，尤其是花和果實作詳細的解剖，在顯微鏡旁即時繪畫鏡中所見的影像，經多次修定和更改後，最後使用墨水筆定型，每個品種可花上半個月的時間，箇中樂趣及辛勞，實不為外人道也。筆者當時也不太明白，只認為相片是快捷方便，彩色呈現結構，乃為最佳之選。其後再次用心觀察曾老師和馬老師的墨線圖，才發現更多相片較難表現的輪廓、深度和質地，都在墨線圖中呈現出來，有助充分了解文字描述和真實結構，並且提供極有力的助證。再者，在細緻的觀察、解剖和繪畫過程中，通常都會記錄到新特徵，為該品種增加鑒定證據。

自數年前，本館開展了植物科學繪圖的項目，包括探討墨線圖的繪畫方法及其衍生的鑒定和教育工作。筆者有幸於三年前邀請曾孝濂老師來港指導，席間一席話更令我銘記於心，曾老師說：「在森林中觀察植物是一件幸福的事情，回到家裏開始畫畫就不能停下來，好像進入了一個平靜安寧的心境。」當時令筆者領會植物學繪畫不但是科學與藝術的結合，也是建立個人修養和安撫心靈的工作。其後，筆者在曾老師及馬老師的驚世工作下，最終明白及深受感動，決心把植物科學繪圖的價值承傳下去。當時適逢王天行先生，他具備相關寶貴經驗，協助本館建立所需的繪圖工具，並指導練習生不斷嘗試並改進繪圖技巧，成功建立了繪圖團隊及工作平台，大量繪圖作品已納入標本館的教學材料，在STEAM教育中發揮重要功能。去年筆者遇見兩位新進的植物學同事，她們在植物繪圖方面極具潛質。經歷本館團隊的共同探討、學習和練習，終於成功繪畫出多種植物的細緻結構和鑒定特徵。筆者特別向本圖鑑的教育顧問王天行和植物繪圖師李敏貞、王曉欣致深切謝意。今年(2022)正是胡秀英植物標本館建館十周年，我們期望此鑒定圖鑑可令讀者更了解植物鑒定的方法及其價值，進一步認識香港原生植物之美。

在此，筆者衷心致謝伍絜宜慈善基金長期支持胡秀英植物標本館的教研及出版工作。是次出版有賴香港中文大學出版社的支持才能成功進行，特別感謝編輯冼懿穎和美術設計吳劍業的努力與專業；我亦十分感謝以下人士協助校對本書(排名不分先後)：王君、葉芷瑜、陳耀文、王顥霖與潘慧德。我們衷心感謝漁農自然護理署香港植物標本室提供重要的憑證標本和鑒定協助。有賴各方單位及同事的支持，本圖鑑才能成功發行，誠希各讀者能透過此書更了解植物分類這門歷史悠久和多姿多采的基礎學科。

劉大偉
胡秀英植物標本館館長
2022年春

Preface

Botanical illustration is a crucial component of plant taxonomy, as pen and ink can clearly illustrate the plant organs. The evidence provided through drawings were especially significant a few decades ago, when digital photography was not commonly applied. Conventionally, most flora would have included line drawings of the plant species, which greatly enhances readers' understanding of the plant description and characteristics for their accurate authentication.

In the 1970s, the herbarium at the Biology Department at The Chinese University of Hong Kong began the work of botanical illustration and documentation. I personally witnessed the sophisticated drawing procedures, which truly reflect the patience and endurance of the botanical illustrators. A series of procedures—from collection to authentication—have to be done for a drawing. That includes specimen collection, flower and fruit dissection, and frequent microscopic investigation. It takes a few days to weeks for the completion of the drawing, depending on the complexity of the plant structures. When many good specimens can be collected in the flowering season, the workload of the botanical illustrators is increased. Long-term usage of a microscope and continuous concentration on drawings is very exhausting, therefore every botanical illustration is a fruitful and scientific product of the artists. Further, each drawing provides strong evidence about the content in plant taxonomy, which is a crucial element in basic science.

Since the establishment of the Shiu-Ying Hu Herbarium ('the Herbarium') in 2012, our colleagues have been fulfilling the Herbarium's mission. Whenever we recall the life-enriching philosophy of Professor Shiu-Ying Hu, who dedicated her life's work to society and the people in need, we are motivated to implement the principle work of the Herbarium. For the past ten years, we have been continuously studying and exploring plant taxonomy, which includes specimen collection and preservation, taxonomic research and its applications in conservation, Chinese medicine authentication, horticulture, and more. All the major categories are immense research areas of species authentication, description, phylogeny and nomenclature. Though every category is challenging, the new knowledge and findings are usually encouraging and refreshing.

One of the authentication methods in taxonomy is using botanical illustration, which is also very supportive for the understanding of plant description. We always inspected the pen and ink drawings of flora in our early stages of learning, but I did not realize their significance until a few years ago when I re-arranged some manuscripts and originals in our Herbarium, and I was shocked by a series of botanical illustrations drawn by two top botanical artists Mr Xiao-Lian Zeng and Mr Ping Ma. Looking at them was like entering into the microcosmic world of the plants, and I was able to observe patterns that had not been documented in any references. I felt like an insect adventuring through the hairs and scales of the plants.

The close-up drawings and sophisticated artistic presentations made me recall Mr Ma and Mr Ting-Kok Woo's work on botanical illustrations. On fieldwork days, they would go to the countryside and collect a lot of specimens. After coming back to the Herbarium, Mr Ma would begin to dissect the flowers and fruit, using a very old style dissecting microscope. Simultaneously, he would carefully record the plant's characteristics from what he saw through the microscope. He repeatedly did the same procedures even for a very tiny plant part, so as to ensure the drawing records were correct before he used an ink pen to finalize the drawing. Usually it took about two weeks to complete a drawing. Most outsiders, including myself, do not realize the importance of botanical illustration, and we usually rely on photographs that are expected to record everything. After years of observation and reviews, I realized that certain outlines, depth of field and surface textures that are inconspicuous in photographs could be better displayed in some botanical illustrations. Such kinds of presentations provide us with significant or new evidence for determining a species, as the dissecting procedures can often reveal new and hidden structures which are taxonomically significant.

Over the past several years, our Herbarium has been working on a botanical illustration project. Some specific drawing skills, along with authentication methods and education projects, were first explored by our education team. Three years ago, we were very honoured to have Mr Xiao-Lian Zeng visit us and provide us with special demonstrations and trainings. One of the thoughts that he shared has been inspiring us for many years. He said: 'It is a blessing to observe plants in a forest. When I get back home and start drawing, I cannot stop. It's like entering a calm and peaceful state of mind.' I was impressed by his words and philosophy, and began to understand the true value of botanical illustration, which is not only an integration of science and art, but also a way to develop a sense of personal spiritual cultivation and to calm the heart and mind. In the following years, I continued to explore and learn from the drawings of Mr Zeng and Mr Ma, and came to deeply feel that their genuine value should be passed on to the next generation. I had the good fortune to meet Mr Tin-Hang Wong, who has relevant experience, and he helped us to establish a team and work platform of botanical art. He also trained up a number of artist trainees and produced a vast amount of artwork for our STEAM projects. In recent years, two young botanists joined our team and really boosted up our capacity, as they have excellent potential to learn and perform. Going through the stages of learning and practicing, a number of well-authenticated and highly precise botanical illustrations were successfully produced. Therefore, I wish to sincerely express my greatest gratitude to our Herbarium colleagues: Mr Tin-Hang Wong (Education Consultant), Ms Man-Ching Li (Botanical Illustrator) and Ms Hiu-Yan Wong (Botanical Illustrator) for their great efforts and contribution. This year is the 10th anniversary of the Shiu-Ying Hu Herbarium, and we would like to take this precious opportunity to publish this illustrated book, not only for readers to understand the methods and value of plant authentication, but also to appreciate the beauty of native plants in Hong Kong.

Last but not least, we would like to express our greatest gratitude to the Wu Jieh Yee Charitable Foundation Limited for their continuous support of our research, education and publication work. The Chinese University of Hong Kong Press is definitely the key to the success of this illustrated book publication, I am grateful to editor Ms Winifred Sin and designer Mr Daniel Ng for their diligence and professionalism. I'd like to acknowledge the following persons for proofreading the book: Yiu-Man Chan, Tsz-Yu Ip, Wai-Tak Poon, Ho-Lam Wang and Kwan Wong. We are also thankful to the Hong Kong Herbarium, AFCD for providing us with the authenticated voucher plant specimens, and all the parties and co-workers' contribution to this illustrated book. Finally, our Herbarium team sincerely hopes that this illustrated book may enhance readers' understanding of plant taxonomy, which is a precious basic science subject with a rich and long history.

David T. W. Lau
Curator, Shiu-Ying Hu Herbarium
Spring 2022

從白簕的科學命名了解植物分類學家工作
——紀念胡秀英教授

「沉醉在自然裏，就想不到苦，只想到樂。」這是胡秀英教授常說的一句話。

與其說研究植物是她的工作，不如說是她生活的一部分習慣和樂趣。植物學家會在野外筆記上記錄採集標本的日期、地點和物種名稱，尤其是胡教授無論身處何時何地都會採集標本，所以憑筆記和標本可以追蹤她的足跡，不管是四川、新英倫還是香港。

胡教授的足跡亦印記在植物學名上。本書記錄的白簕 *Eleutherococcus trifoliatus* (L.) S. Y. Hu，這學名正是她更改的。事緣是 1970 年代，學術界對人蔘的藥用和保健功效感到興趣，同時亦有學者研究五加屬（*Eleutherococcus*）植物的化學成分，以及作為人蔘代用品的成效。然而，*Eleutherococcus* 和 *Acanthopanax* 兩個屬分類和命名混淆。白簕最早由林奈（Carl Linnaeus）命名為 *Zanthoxylum trifoliatum* L.，後來於 1890 年代改為 *Acanthopanax trifoliatus* (L.) Voss，除此以外也有不少異名同物，因此造成溝通上的不便。

胡教授希望以自己的知識貢獻社會，藥用植物和經濟作物皆是她的研究領域，故此便著手調查。在一個沒電腦數據庫和互聯網的年代，胡教授翻查所有相關的文獻和標本，經過一番考證工作，胡教授在 1980 年發表論文，將兩個屬名歸併在 *Eleutherococcus*。胡教授的研究將白簕的學名改為 *Eleutherococcus trifoliatus* (L.) S. Y. Hu。雖然植物學名時有更改，但她的看法至今仍然為《中國植物誌》英文版和《香港植物誌》所採納。這樣的工作，為所有應用研究提供了科學基礎。

植物分類學是一門基礎科學，任何有關生物的研究，都需要定義研究對象的身分。出色的植物分類學家，既要有野外工作的體能和熱情，亦要有查證文獻及標本的耐性。為要將植物分門別類，分類學家不分寒暑翻山涉水，進行田野採集和記錄的活動，同時遊走圖書館和實驗室，釐清物種的命名歷史和發表理據，方能對物種的身分作出整合和更正。

這是一份刻苦而嚴謹的差事，是什麼信念推動胡教授孜孜不倦投身其中？其中之一，是胡教授熱愛世界和人類。人生早年的苦難和基督信仰的薰陶，造就她利用個人專業去造福人群的志向。另一方面，她熱愛和欣賞大自然。植物的美態深奧而微妙，引發她純真的好奇心，牽動她以理性的科學方法探索世界，以致日後有超卓的科研成就。

培養大眾欣賞自然的美，正是科學普及和環境教育的重要元素。昔日，胡教授曾經與馮冬俠、馬平、曾孝濂及華樂庭（Beryl M. Walden）夫人幾位畫家合作。畫家以純熟的技巧，在畫紙上呈現植物的美；胡教授則為科學把關，確保每一筆都是準確的記錄。今天，胡秀英植物標本館的研究人員延續這個理性與情感的組合。在繪圖之前，研究人員會採集新鮮植物來解剖和量度，並用顯微鏡觀察，同時參考大量文獻，把植物的形態和結構了解透徹才動筆繪畫。這些植物樣本亦會被製成臘葉標本，按植物學分類系統收藏在標本櫃內，留待日後翻查及引證之用。這些科學工序，和胡教授過往的手法同出一轍。

胡教授為我們留下豐富的禮物，她創立的標本館、她的著作、她的教導……等等。如果你闔上這本書，想往郊外去感受大自然，想學習植物多元化的結構，想分享給別人……我想，你也承傳了胡教授對人類和大自然的熱愛。

陳耀文
胡秀英植物標本館技術員

In Remembrance of Prof Shiu-Ying Hu: The Naming History of *Eleutherococcus trifoliatus* (L.) S. Y. Hu

'Rejoice in the nature! Sorrow will turn into joy!', Prof Shiu-Ying Hu always used to say.

It was more than a career or duty; botanizing was her habit and joy. She collected plant specimens, recorded their locations and identified them. Her field notes and specimens are indeed scientific records; they are also traces of her pilgrimage.

Her footprints can also be found in scientific names, like *Eleutherococcus trifoliatus* (L.) S. Y. Hu. In the 1970s, there was a growing interest in research on ginseng. Species of *Eleutherococcus* were considered substitutes for ginseng and also drew considerable research interest. However, the nomenclature and taxonomy of *Eleutherococcus* and *Acanthopanax* were unclear and led to much confusion among scientists. With a vision to build a foundation for other scientific disciplines, Prof Hu committed herself to this taxonomic problem.

Taxonomists work in the field and laboratory to seek first-hand data and materials. In libraries and herbaria, they dig out every piece of their predecessors' works to review. There were no computer databases or internet at that time, but Prof Hu managed to uncover every piece of related literature and specimen. The naming history and rationale were reviewed with new findings to compile an up-to-date treatment. In 1980, Prof Hu published an article entitled '*Eleutherococcus* vs. *Acanthopanax*' and combined the genera into *Eleutherococcus*. According to this treatment, *Eleutherococcus trifoliatus* (L.) S. Y. Hu is the only accepted name for a plant which was once called *Zanthoxylum trifoliatum* L. or *Acanthopanax trifoliatus* (L.) Voss. Prof Hu's treatment was adopted in major publications like *Flora of China* and *Flora of Hong Kong*.

Prof Hu's achievement was the result of her diligent work, and was driven by her love for humanity and nature. Despite a tough childhood, her Christian faith conferred on her a big heart to serve others. The beauty of nature inspired her and aroused her curiosity. In response to this sentiment, she explored nature by using scientific methods and shared her knowledge with others. This is why she was called a 'Walking encyclopedia of botany'.

To reveal the diversified plant morphology with scientific accuracy, Prof Hu teamed up with artists and published botanical artworks. Teresa Fung, Ping Ma, Mrs Beryl M.

Walden and Xiao-Lian Zeng visualized the miraculous plant structures in their drawings, while Prof Hu ensured the precision of plant descriptions and measurements.

Today in the Shiu-Ying Hu Herbarium, researchers follow in the footsteps of their beloved mentor. They believe a loving heart for nature is the key to environmental education, so they present the marvels of nature to the public with a blend of Art and Science. Researchers collect and dissect plants for observation before they draw them. Detailed structures are documented with stereomicroscopes and cameras. The plant materials examined are made herbarium specimens and kept in the cabinets according to a classification system. All of these procedures were inherited from Prof Hu.

Scientific achievement was not the only legacy Prof Hu has left. It is hoped that her love for nature can also be passed on to the readers of this book, which is an invitation for people to experience nature in all its wonder. If you close this book and have an impulse to visit the wild, to learn about the plants and to share that with others...then you are indeed passing on the precious gifts that Prof Hu has left for us.

Yiu-Man Chan
Technican, Shiu-Ying Hu Herbarium

References

Deng, Y. F. (2008). Araliaceae. In Q. M. Hu & D. L. Wu (eds.) *Flora of Hong Kong* (Vol. 2., pp. 291–296). Agriculture, Fisheries and Conservation Department, HKSAR.

Hu, S. Y. (1977). *The Genera of Orchidaceae in Hong Kong*. The Chinese University Press.

Hu, S. Y. (1980). *Eleutherococcus* vs *Acanthopanax*. *Journal of Arnold Arboretum*, *61*(1), 107–111. http://www.jstor.org/stable/43782067

Hu, S. Y. (2007). Malvaceae. In Q. M. Hu & D. L. Wu (eds.) *Flora of Hong Kong* (Vol. 1., pp. 215–232). Agriculture, Fisheries and Conservation Department, HKSAR

Hu, S. Y. (2008). Aquifoliaceae. In Q. M. Hu & D. L. Wu (eds.) *Flora of Hong Kong* (Vol. 2., pp. 182–190). Agriculture, Fisheries and Conservation Department, HKSAR.

Hu, S. Y., & Wong, Y. (2009). Asteraceae. In Q. M. Hu & D. L. Wu (eds.) *Flora of Hong Kong* (Vol. 3., pp. 245–328). Agriculture, Fisheries and Conservation Department, HKSAR.

Hu, S. Y., & Kong, S. L. (2011). Orchidaceae. In Q. M. Hu & D. L. Wu (eds.) *Flora of Hong Kong* (Vol. 4., pp. 389–353). Agriculture, Fisheries and Conservation Department, HKSAR.

Walden, B. M., & Hu, S. Y. (1979). *Wild flowers of South China and Hong Kong*. Sino-American Publishing Company.

Walden, B. M., & Hu, S. Y. (1987). *Wild flowers of South China and Hong Kong Part 2*. All Noble Company Ltd.

Xiang, Q. B. [Shang, C. B.] & Lowry II, P. P. (2007). Araliaceae. In Z. Y. Wu, P. H. Raven, & D. Y. Hong (eds.) *Flora of China* (Vol. 13., pp. 435–491). Science Press & Missouri Botanical Garden Press.

關於本圖鑑

為紀念香港中文大學生命科學學院胡秀英植物標本館建館十周年，特別出版《香港原生植物圖鑑》。本圖鑑作為「胡秀英植物標本館科學繪圖系列」的第一冊，以科學繪圖融合植物藝術墨線圖的方式，詳細介紹20種香港原生植物，而仔細的科學繪圖及鑒定級的特徵描述，更可作為鑒定植物品種的參考資料。

本冊圖鑑所收錄的20個本港原生植物品種，是因應其憑證標本是否合乎科學鑒定特徵，以及其標本完整程度作為挑選條件。所有憑證標本的採集，均需配合不同品種的開花或果實成熟的季節，並在合法程序下採集、製作和保存；而其中15種稀有及瀕危植物，更是在漁農自然護理署香港植物標本室符合香港法例下提供。在得到政府部門提供的寶貴協助下，本圖鑑獲得了更準確及珍貴的憑證標本，為讀者提供了難能可貴的品種資料。

本圖鑑的特色及創新

- 彙集20種香港原生植物繪圖，當中包括15種稀有及瀕危植物。
- 四種為藥用相關品種，其中兩種由胡秀英教授命名。
- 體現植物繪圖在科學鑒定上的意義和功能。
- 在保留每個物種所有科學鑒定的特徵情況下，亦具有美術的欣賞性。
- 收錄的植物均根據野外觀察、詳細解剖、憑證標本及標本館的館藏標本考證，由植物學家鑒定及植物繪畫師繪製。
- 收錄的植物繪圖根據多部植物誌編製而成，詳盡記錄該植物的結構及特徵。
- 所有品種以墨線圖形式表達，包括了該品種的典型枝條、葉片、花朵或果實。
- 植物繪圖的結構及特徵描述均配有中英文描述。
- 收錄的植物繪圖附有該品種的在野外、微距、顯微鏡及標本相片。
- 收錄的植物繪圖均附有詳細的參考文獻。
- 本圖鑑可用作植物鑒定、科學參考、科普教育及繪圖藝術欣賞之用。

About This Illustrated Guide

This book, the *Botanical Illustrated Guide to Hong Kong Native Plants*, is being published to commemorate the 10th anniversary of the establishment of the Shiu-Ying Hu Herbarium at The Chinese University of Hong Kong. This guide is the first book in the 'Shiu-Ying Hu Herbarium Scientific Illustration Series', and brings together two methods in one book—scientific illustrations, and pen and ink drawings of plants—to provide a detailed introduction to 20 of Hong Kong's native plant species. These scientific and detailed illustrations and morphological descriptions also provide important reference materials for species authentication of our native plants.

In this guide, 20 native species of Hong Kong have been selected based on the comprehensive results of a study to verify if the voucher specimens conformed with the scientific authentication, which is a process done under challenging conditions. Voucher specimens need to be collected in the right flowering and fruiting seasons according to each species. All specimens being referred to in this guide were collected, held, preserved and stored for legitimate purposes. Since 15 species in the selections are rare and endangered species, the voucher specimens were provided by the Hong Kong Herbarium—officially the Agriculture, Fisheries and Conservation Department. With the invaluable support of the HKSAR Government, this guide provides readers with hard-to-find information on native plants based on precise and precious voucher specimens.

Features and Innovations of this Guide:

- Twenty Hong Kong native plant species have been selected, 15 of which are rare and endangered species.
- Four species are herbal medicinal materials, two of which were named by Prof Shiu-Ying Hu.
- It shows how the meaning and function of botanical illustrations are used in scientific species authentication.
- It maintains all of the scientific information on species authentication, while at the same time providing aesthetic beauty.

- All illustrations are based on observation records, detailed dissecting, research with vouchers and collected specimens of the Shiu-Ying Hu Herbarium. It has been authenticated by professional botanists and illustrated by botanical artists.
- All illustrations are based on referencing multiple flora, recording morphological features and structures in detail.
- All species are illustrated with pen and ink drawings, including their typical branches, leaves, flowers or fruit.
- All illustrated morphological features and structures are labelled with both Chinese and English.
- All illustrations are also accompanied by photos of live plants in their natural environment, microscopic observations and real specimens.
- References are provided for each illustration.
- This guide can be used for plant species authentication, for its scientific references, for science education in schools, and for the artistic appreciation of its illustrations.

本圖鑑的編排
How This Book Is Organised

植物分類系統 Plant Classification System

本圖鑑所收載的裸子植物和被子植物分別採用1990年Kubitzki和2016年被子植物系統發育及演化組 IV (APG IV) 的分類系統排列，文中亦有收錄《中國植物誌》英文版（*Flora of China*）及《香港植物誌》的植物拉丁名稱。

The gymnosperms and angiosperms included in this book comply with Kubitzki and Angiosperm Phylogeny Group IV (APG IV) systems of classification in 1990 and 2016 respectively. The scientific names adopted in the *Flora of China* and the *Flora of Hong Kong* are also documented.

編排結構 The Structure

I. 首頁 Opening Page

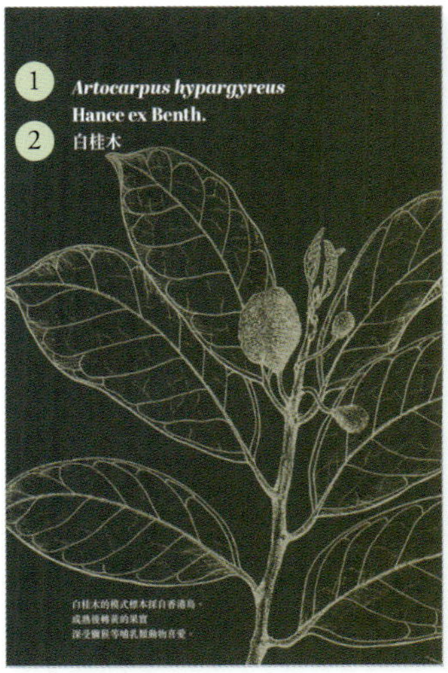

1 學名
Scientific name

根據《國際藻類、真菌及植物命名法規》，標準的植物學名由拉丁文組成：

學名＝屬名＋種加詞＋命名者

The standard scientific names of plants follow the *International Code of Nomenclature for Algae, Fungi, and Plants*, and are constructed from Latin as follows:
Scientific name = Generic name + Specific epithet + Author(s)

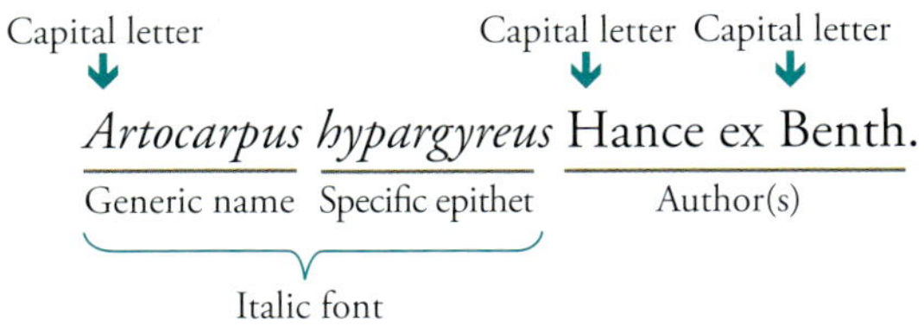

2 中文名稱
Chinese Name

植物的常用中文名稱
The common names of the plants in Chinese

II. 胡秀英教授採集的標本 Specimen Collected by Professor Shiu-Ying Hu

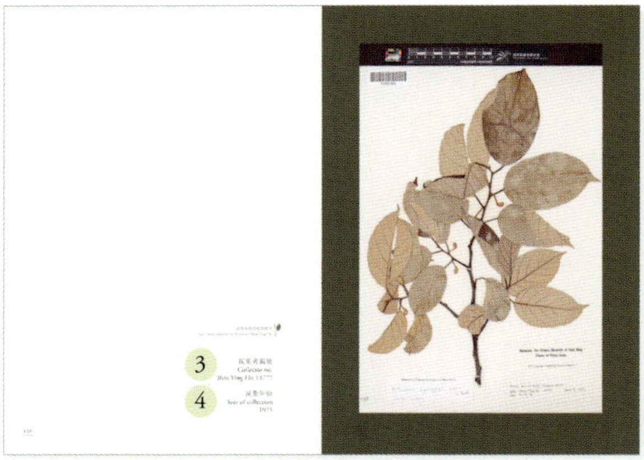

3 採集者編號
Collector no.

4 採集年份
Year of collection

III. 植物簡介 Introduction of Plant Species

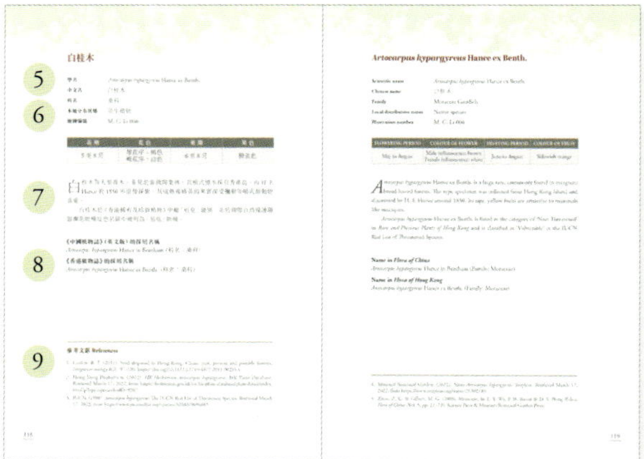

5 植物基本資料
Basic information of plants

6 繪圖編號（根據植物繪圖師的繪畫次序而編定）
Illustration number (complying with the drawing sequence of the botanical illustrators)

7 內文描述
Description

8 在《中國植物誌》與《香港植物誌》的名稱
Names in *Flora of China* and *Flora of Hong Kong*

9 參考文獻
References

IV. 植物繪圖 Botanical Illustration

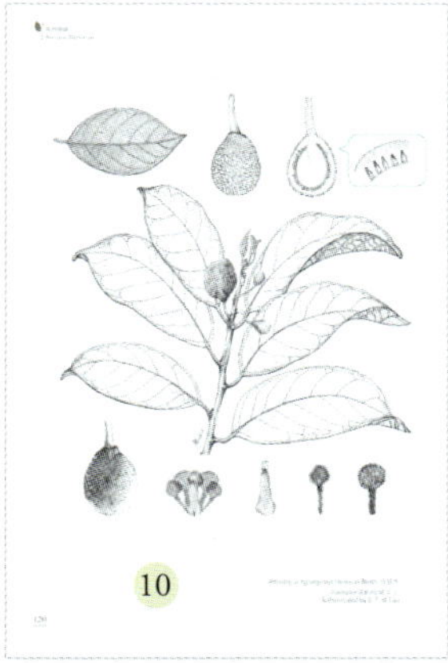

10 展示植物結構

Showing the plant structures

V. 植物繪圖圖解 Botanical Illustration with Annotations

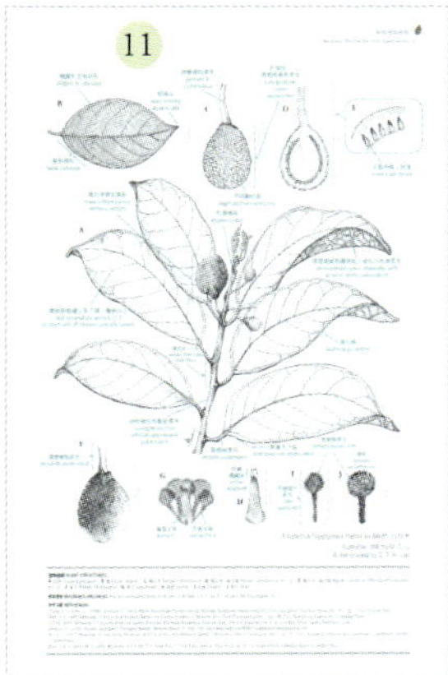

11 展示「植物繪圖圖解」、「植物結構」、「參考標本」及「參考文獻」

Showing 'Botanical Illustration with Annotations', 'Plant Structures', 'Reference Specimens' and 'References'

植物的展示面向 Orientation of plant view

本圖鑑主要的展示面向為「正面觀」和「側面觀」，示意圖如下：

This illustrated book mainly shows the plant's frontal view and lateral view, as indicated here:

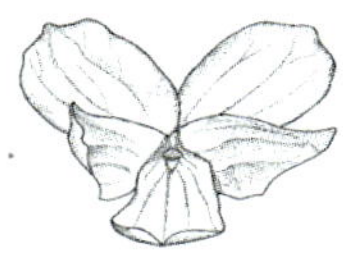

正面觀 Frontal view

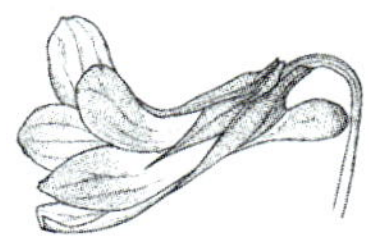

側面觀 Lateral view

植物的解剖面向 Plant dissection

本圖鑑主要的解剖面向為「縱切面」及「橫切面」，以下是垂直或水平剖開雌蕊展示胚珠排列的示意圖：

This illustrated book mainly shows the longitudinal section and cross-section of dissection. The following diagram demonstrates the vertical and horizontal dissection of a pistil to show the arrangement of ovules:

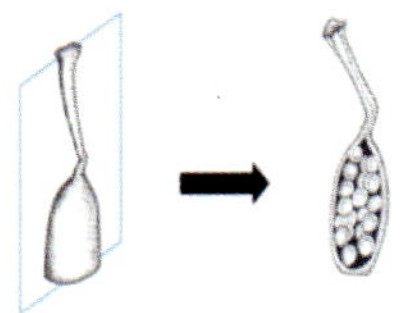

縱切面 Longitudinal section

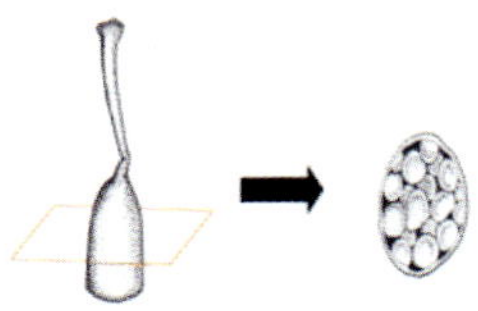

橫切面 Cross-section

VI. 憑證標本 Voucher Specimen

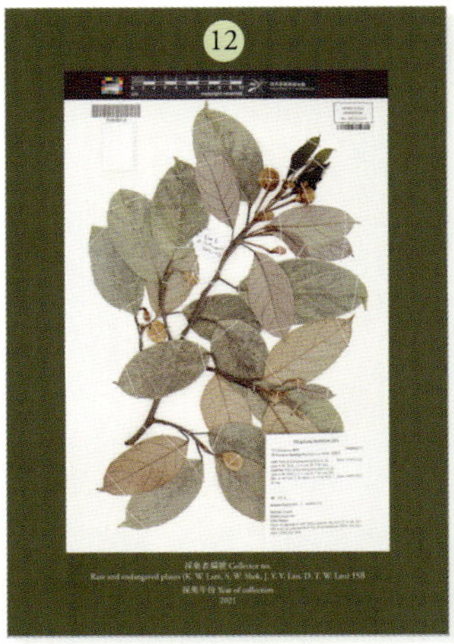

12 展示科學繪圖主要對照的植物標本
Showing the voucher specimen used to draw the scientific illustration

VII. 植物結構與科學繪圖對照 Comparison of Plant Structure to Scientific Illustration

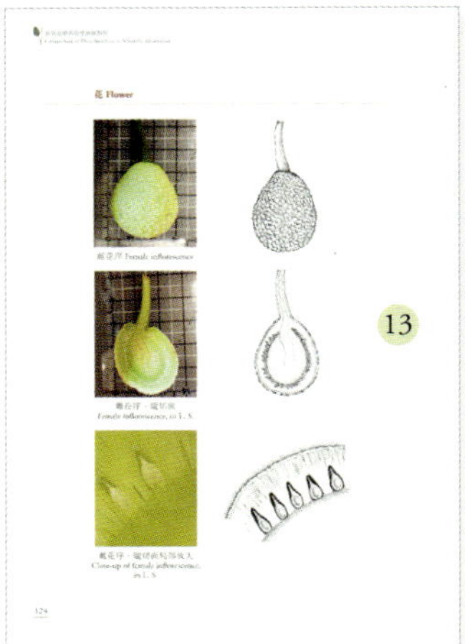

13 展示所引用的植物結構以作科學繪圖之用
Showing the plant structure used for each scientific illustration

李敏貞 Man-Ching Li
胡秀英植物標本館教研助理
Education Assistant, Shiu-Ying Hu Herbarium

如何使用本圖鑑在 STEAM 教育計劃上

STEAM 教育的起源、定義及目標

「STEAM教育」中的STEAM是指科學(Science)、科技(Technology)、工程(Engineering)、美藝(the Arts)及數學(Mathematics)，由於是從英文單字中取出每個詞語的首字母組合而成的簡稱(首字母縮略詞)，因此並不能直接翻譯成「蒸汽」教育。追溯「STEAM教育」的起源，應該是於1986至1996年期間，由美國國家科學基金會使用STEAM的前身 —— SMET為簡稱。當時是一項讓幼兒至初中生，能更早為接受科學與科技培育作準備而制定的題材指引，只有S、M、E及T的元素；2001年，美國國家科學基金會再次改變排列次序成STEM，STEM隨時間發展及得以推廣。2010年以後，為配合青少年發展創意和解難的能力，以及讓他們有更多機會在現實生活中運用所學，便加入了美藝元素補充STEM的不足而成為S、T、E、A、M教育，現已在國際間的教育工作者中得到普及。

「STEAM教育」的目標是教導學習者擁有「創新的思維」和「思辨的能力」，並透過運用S、T、E、A、M的元素去解決生活中的難題，是一個跨學科、具體的、理性思考及解難的思維模式。這種思維模式不僅適用於中、小學教育上，在大學專科教育、成人持續進修以至日常生活中，以STEAM思維模式去觀察、了解以及解決問題，都可以產生更有效率及更佳的結果。

本圖鑑的讀者定位

本圖鑑記載了20種本港原生植物，其中15為稀有及瀕危植物，當中繪畫圖像時所參考的樣本均是在符合香港法例下，由漁農自然護理署香港植物標本室提供，以如此難得及珍貴的資源，來作憑證及根據製作的圖鑑，實屬難能可貴。

- 喜歡認識植物分類或植物品種的讀者：

  本圖鑑提供了非常科學及深入的圖解植物辨認特徵，能夠非常有效地輔助鑒定所載的20種植物品種。

- 喜歡繪畫自然事物，或特別喜歡植物作為繪畫主題的讀者：

  本圖鑑提供了20個品種植物的繪圖，特寫部分更是非常寶貴的參考。

- 從事教育工作的讀者：

  運用本圖鑑中的多項植物資料及繪圖，以植物學主題配合美術元素，在學校進行以認識植物多樣性為主題的STEAM教育學習活動。本單元將以範例形式去展示相關的運用方法。

- 對STEAM教育或科普有興趣的讀者：

  從本單元中可了解到如何以STEAM思維模式，去了解如何判斷不同植物品種的獨特之處。

本圖鑑的內容如何成為 STEAM 教育活動內的教學資源

本圖鑑所載的20種香港原生植物的辨認圖版，均經嚴謹的觀察憑證標本，並以真實的記錄相片或微距拍攝，引證繪圖內的植物結構、比例及特徵。這些結構的繪圖，也可作為學習植物分類上專有名詞的範例，對認識葉形、葉序、花序、雄蕊、雌蕊及子房等花的結構，提供了具體的真實例子作為參考，也是學習植物結構特徵的良好教材。

在圖版以外，還載有多項科學資訊，包括植物品種的大小、花、果、葉的種類、枝條的外表及顏色，還有開花結果的時期等。這些資料都是通過反覆驗證及準確的科學記錄，使用者在運用這些資料時可作為該品種的資料範本，與其他來源的資料作對比參照。

這些圖像及資訊作為學習活動的資源，可從以下方式配合STEAM教育：

圖像 / 資訊	用法	STEAM 活動
植物品種主圖（花枝圖或果枝圖）	比較真實植物和繪圖中的結構特徵，以辨認出真實植物的品種名稱	讓學習者以本圖鑑所收錄的植物相片與黑白植物繪圖進行配對，尋找出相同植物品種
植物部件圖（葉片、花朵）	認識不同葉片、花朵結構及形態	簡單繪畫基本外型，從而加深對該類結構的認識及記憶
	可以配合課堂科目內容，例如花的結構，認識雄蕊、雌蕊、子房等，重要的生物結構	除認識科學定義外，還可從結構及形態推測花粉的傳播方式
植物資料（開花時期、結果時期）	認識不同品種的植物有不同的開花期，及由花發育至果實在不同品種的植物中所需時間的長短	統計校內不同植物的開花時期，與本圖鑑的品種開花時間作比較。比較不同植物由花發展為果實所需時間等
植物部件圖（果實、種子）	認識不同果實結構及形態	推測種子傳播方式
	比較食用果實與大自然的果實形態	以蔬菜及水果的果實形態，比較其他品種的果實

（接下頁）

圖像 / 資訊	用法	STEAM活動
多張不同品種的微距植物結構相片	學習如何觀察重要的植物微觀結構	以本圖鑑的繪圖或微距相片為參考，在沒有一體式的顯微拍攝工具情況協助下，設計如何拍攝或記錄微觀植物結構
本圖鑑記載植物的受保護情況	認識本港部分受保護的植物品種及原因	了解本港監察受保護植物的方法，討論並嘗試以STEAM思維改良相關的方法並進行實驗

當以上不同的STEAM活動以不同的深淺程度、進行方式及持續類型來組合活動時，便可產生多種以植物學為主體的校本STEAM教育活動。為了讓讀者更易了解如何運用本圖鑑在STEAM活動上，以下兩個教育活動例子將可具體說明。

項目類型	所需活動時間
小型課堂植物分類STEAM工作坊（workshop）	短：連續二至三課節
季度校本植物生態主題STEAM教育活動（campaign）	中：橫跨三至四個月內不連續的四至八課節

1. 小型課堂植物分類STEAM工作坊教學流程範例（適用於小學）

主題：鑒別本書所載的植物品種
學習課題：認識生物多樣性 / 香港常見植物 / 植物基本結構

流程	活動	STEAM元素	時間
導入	• 展示多張（十張以上）不同的植物圖片（包括本港常見和本圖鑑相同的品種）但不提供品種名稱	Science 科學觀察	5分鐘
引起學習動機活動1	• 向學生提出以遊戲形式（例如印有圖像的紙卡）以黑白墨線圖與真實植物圖片配對植物名稱 • 挑戰成為正確配對最多的一組 • 以三至六名學生為一組進行活動	Science 科學觀察	15分鐘
學生回饋	• 由成功配對最多品種的組別，分享他們如何成功配對品種的技巧	Science 科學比較	5分鐘
講解	• 導師可引導學生正確地以黑白墨線圖中的植物特徵，如葉片、花及果的形態特徵來作比較的理據，判斷是否相同品種 • 指出正確配對答案	Science 科學比較	10分鐘

（接下頁）

流程	活動	STEAM 元素	時間
活動2	• 統計活動1所顯示的植物品種在校園內 / 班中同學曾經見過的數量 • 雖然統計數量與品種的常見程度未必一致，導師應指出活動1內的常見與稀有及瀕危的植物品種	Mathematics 數學統計	5分鐘
活動3	• 以電腦或流動資訊科技工具，找出一種常見和一種稀有及瀕危的植物簡介及圖片，學生可以直接拍攝黑白墨線圖或彩色植物圖片，以圖片取代關鍵詞在搜尋引擎進行搜尋。	Technology 應用資訊科技	10分鐘
活動4	• 以搜尋得來的一個植物品種的葉片、花朵或果實進行繪畫	Art 美藝繪畫	20分鐘
活動5	• 讓學生繪畫活動2中記錄數量最多的並見於校園內的植物結構，例如葉片及花朵 • 比較學生的畫作是否能作為分辨植物品種的工具	Art 美藝繪畫	20分鐘
講解	• 指出科學墨線圖的使用，是為了記錄植物的外型，其相關資訊能有助他人辨認植物品種 • 指出傳統照相機及數碼相機用於拍攝植物時的方便 • 指出墨線圖細緻、簡單的線條，能令研究人員更清楚了解被描繪的品種的結構細節 • 指出繪圖是文字以外更客觀、更具體的一種有效性描述和記錄	Engineering 工具應用與工程設計	10分鐘
回饋及討論	• 讓學生分享活動5中繪畫植物時，如何有效地讓其他同學了解圖中的形態特徵 • 討論在數碼年代繪圖和攝影如何協助辨認植物品種	Art and Sciences 美術與科學的平衡	5分鐘
結語	• 透過觀察專業的科學植物繪圖，較容易學習不同植物的形態 • 認識了有些植物是比較常見，有些植物卻非常罕有 • 認識了植物的多樣性		5分鐘

2. 季度校本植物生態主題STEAM教育活動範例（適用於中學）

主題：校園開花植物的花朵特徵及開花統計
學習課題：認識開花植物 / 花朵的基本結構

時期	活動	STEAM 元素	課節
前期準備1（可由學生或教師準備）	• 先以本圖鑑的品種資料簡介中的「科屬」，嘗試找出校園中相同科屬的植物	Science 資料比較	一節課中的20分鐘
前期準備2（可由學生或教師準備）	• 統計在前期準備1中找出的植物在校園內的位置及數目	Mathematics 數學統計	一節課中的20分鐘
資料搜集	• 以電腦或流動資訊科技工具，找出同科屬植物的花朵及開花月份資料	Technology 應用資訊科技	一節課
專題活動	• 以小組或個別學員專題討論，及投票形式選出一至兩個品種作為校園植物開花記錄統計的目標 • 選出品種的理據應與能否配合花期作出統計，及複核植物的數目是否能得出科學統計的效果	Mathematics 數學統計	一節或兩節課
實踐及統計	• 在該等選出品種的開花月份，記錄開花日子及每株植物的花朵數目 • 開花的日數或許會較所搜集的資訊為長或短，也一併記錄	Mathematics 數學統計	在小憩、午飯或課外時間；於選定品種的開花時期，維持每天在相同時段、每星期作最少兩次記錄
增強知識	• 學習花的結構	Science 觀察	一節課
解決難題 1	• 配合導師講解讓學員設計 / 運用工具（例如長枝剪）採集花朵	Engineering 工程設計及應用	一節課
解決難題 2	• 配合導師講解讓學員設計 / 運用工具（例如顯微鏡）觀察和記錄花朵裏的微小結構，以準備花朵結構解剖	Engineering及Technology 科技應用與工程設計	一節課

（接下頁）

時期	活動	STEAM 元素	課節
實踐與記錄	• 以校園內各品種的花朵進行結構解剖 • 根據樣本繪畫花朵，包括完整的花朵以及其經解剖的細微結構	Art 美藝繪畫	兩節課
證實	• 比較本書中同科屬的微小結構部分是否一致	Science 科學比較	一節課
回饋及討論	• 讓學生分享從調查中的同科屬花朵的花期，以及花的結構是否一致 • 討論繪製植物圖版在現今的數碼年代是否重要	Engineering + Technology + Science	一節課
總結	• 總結學生的發現、經驗及科學記錄的實行方法：由資料搜集、篩選對象、設計工具、實際操作以及證實假設及求證 • 指出一些傳統發展的科學記錄方法其價值與功能。例如在現今數碼年代，仍可透過植物繪圖讓使用者更清晰、更容易地學習植物的形態		最後一節課

為不同程度的學習者調適

基於學習者不同程度的學習階段，教育工作者應按學習者的能力調適活動內容的深度，從學習主體植物的品種挑選、所需記錄的科學特徵、繪畫植物的構圖及特徵所需的技巧，以至運用科技和工具的操作技巧等，都是STEAM教育活動中根據學生的不同程度與潛能因材施教，可作出調適的部分。

以下是其中一些不同學習階段學習調適活動的例子：

活動	年級	調適部分	合適資訊
以圖版鑒別植物品種	6–9歲 學習階段一 KS1	植物品種的名稱、特徵數目	• 植物品種名稱 • 生長形態 • 葉片形狀 • 花朵顏色 • 果實顏色

（接下頁）

活動	年級	調適部分	合適資訊
以圖版鑒別植物品種	10–12歲 學習階段二 KS2	植物品種的名稱、特徵數目	• 葉序 • 葉的種類 • 果實種類
	12–15歲 學習階段三 KS3	植物品種的名稱、特徵數目、生長環境	• 枝條形態 • 生長環境 • 分布狀況
	16歲以上 學習階段三 KS3	植物品種的名稱、特徵數目、生長環境、部分科學專業資訊	• 科屬名稱 • 花序

活動	年級	調適部分	合適資訊
植物結構繪圖	6–9歲 學習階段一 KS1	植物品種的名稱、明顯外型	• 葉片外型
	10–12歲 學習階段二 KS2	植物品種的名稱、明顯外型、部分結構種類名稱	• 複葉外型及不同種類 • 花的基本結構、果實外型
	12–15歲 學習階段三 KS3	植物品種的名稱、特徵數目、部分結構種類名稱、結構功能	• 葉尖、葉基形狀 • 枝條，花序的外型 • 果實內外形態
	16歲以上 學習階段三 KS4	植物品種的名稱、特徵數目、部分結構種類名稱、結構功能、部分科學專業資訊	• 花序以及微觀部分

最後期望本單元能為教育工作者提供多一種STEAM教育活動的題材，令STEAM活動多一些自然科學的元素。

王天行
胡秀英植物標本館教育經理

How to Use This Guide for STEAM Education

The origin, definition, and objective of STEAM education

The word 'STEAM' in 'STEAM Education' is an acronym of Science, Technology, Engineering, the Arts and Mathematics. STEAM education's origins developed in the US in the 1980s and 1990s by the National Science Foundation, which used the acronym, SMET (Science, Mathematics, Engineering, and Technology). These were guidelines that helped US educators shape their curriculum to better prepare K-12 students in these four subjects, but at that time, it only included S, M, E and T. Later, in 2001, The National Science Foundation changed the acronym to STEM. Since 2010s, it has been developed and targeted towards helping young people to develop effective creative and critical thinking skills. Also, with the aim of facilitating the youth to have more opportunities to apply their knowledge and skills in the real world, elements of Arts education have been added to STEM, turning it into 'STEAM'. Now, STEAM education is a very popular topic for educators throughout the world.

'STEAM education' aims to help learners develop 'creative thinking' and 'critical thinking', with the help of the elements of S, T, E, A and M to solve real-world problems. It is a mode of thinking which is cross-subject, concrete, logical and used for problem-solving. This mode of thinking is not only suitable for primary and secondary school education, but also good for professional training in university, continuous education for adults and even in daily life. Observation, analysis and solving problems in STEAM ways will always have more efficient and better results.

Usage of this illustrated guide

This guide recorded 20 native Hong Kong plant species, 15 of which are rare and endangered species. All of those voucher specimens and referencing materials for the illustrations were transferred by AFCD and legitimate under Hong Kong law. Therefore, the illustrations and records in this book are very precious and special collections.

This book is intended for,

- Botany learners and plant lovers:

  This guide provides scientific and detailed illustrations of a plant's morphological characteristics, which can help to efficiently identify the 20 recorded species.

- Sketchers and painters who love drawing nature and plants:

  This guide provides pen and ink drawings of 20 species. The close-ups are very rare and precious references for drawing.

- Educators:

  Educators can make use of the scientific information and detailed illustrations of plants in this guide, incorporating the Art elements in teaching, which can provide very fruitful STEAM education activities. Suggested education activities and workshop examples will be covered in this chapter.

- Those who are interested in STEAM or Botany and Science:

  This chapter will introduce how to identify different plant species in a STEAM way.

How the content of this guide may be used in STEAM activities

All of the recorded 20 species illustrations in this guide were based on rigorous observation of voucher specimens, as well as the comparison of photo records of live specimens or comparisons of macroscopic photos of dissected parts to confirm the detailed structures, scales and characteristics of the illustrations. Therefore, these delicate illustrations can be used as a kind of glossary of botanical scientific terminology. They are concrete and realistic references and good learning materials for studying the structures of leaf shapes, leaf arrangements, inflorescences, stamen, pistils and ovaries, and so on.

Besides the illustrations of these 20 plants, much scientific information is provided, including the typical dimensions of the plants, leaf types, flower types, fruit types, branches surfaces, and their colours. The flowering and fruiting periods have been also included. Since all of this information has been authenticated with scientific records, users can make use of them as a datasheet template for comparing different data sources of information.

Specifically, these illustrations and information may be used in STEAM education activities as follows:

Illustration or information	Usage	STEAM activities
Plant portrait illustrations (branches with flowers or fruits)	To compare with live plants and identify the species by the illustrated morphological characteristics provided in this book	Prepare some live plant photos which are listed in this book and let learners find out the species through comparing plant photos with pen and ink illustrations
Plant part illustrations (leaves, flowers)	To learn different types of leaves and flowers by their appearance and structures	Learners can try to illustrate the simple shapes of the leaves and flowers, and speculate on the types of pollination of different sample flowers
	Use as teaching materials of the 'basic structure of a flower' to learn the important structures of the anther, pistil and ovary, etc.	Let learners draw a simple outlined shape of different structures inside the flower. The exercise should enhance the observation and understanding of the detailed structures
Plant specific information e.g. flowering period and fruiting period	To learn about different plant species that have different flowering and fruiting periods, as well as different development times from flower to fruit in different species	Survey different flowering periods of plants on the school campus, comparing with the species recorded in this book. Compare the fruit development times required for different species
Plant part illustrations, e.g. fruits and seeds	To learn different types of fruits and seeds	Let learners draw simple shapes of the fruit or seeds and try to identify different types of seed dispersal of the samples
	To compare fruit structures collected in the wild with agricultural foods	To learn the fruit structures by observing different types of agricultural fruits
Macroscopic photos of dissected plant parts	To learn how to observe specific macroscopic morphological characteristics of different plant species	Use the illustrations and macroscopic photos as a sample to design how to take a photo or record macroscopic structures without an all-in-one microscope camera system
The information of plant rarity or endangered status	To learn what the endangered plant species are and the reasons for the vulnerable species	To learn how local authorities protect the vulnerable species and try to discuss or improve the methodology in STEAM ways

Many botany STEAM education programmes for a school-based curriculum can be created, by combining the above STEAM activities, which are of different difficulty level adjusted to suit student's varied learning level, with various teaching methods and types of activities. To let you understand more how this could look in a real situation, below are examples of two typical STEAM education models:

Activity Type	Duration
Small-scale basic plant diversity STEAM workshop	Short. Requires two or three lessons
Seasonal (three to four months) school-based plant investigation STEAM campaign	Medium. Involves four to eight non-continuous lessons held over three to four months

1. Small-scale basic plant diversity STEAM workshop lesson plan example (suitable for primary school students)

Topic: Identify plants based on species in this book
Lesson objectives: Biodiversity / Hong Kong common trees / Basic botany morphological characteristics

Lesson flow	Activities	STEAM elements	Duration
Introduction	• Present multiple live photos of plants (including the species in this book and local common plants) without species names	Science (observe)	5 mins.
Activity 1	• Ask students to play a matching game (e.g. using photos pasted on cards) of different plant species. The students should match the real colour photos with the pen and ink illustrations • Arrange the activity as a group competition to find the most correct matched numbers • Three to six students to a group	Science (observe)	15 mins.
Student feedback	• Let the winners share their skills on how to match the species successfully	Science (compare)	5 mins.
Explanation	• The teacher guides the students to compare the morphological characteristics, such as the shapes of leaves, flowers and fruit in the illustration, with the colour photos, and point our the correct anwers	Science (observe and compare)	10 mins.

(Continued on next page)

Lesson flow	Activities	STEAM elements	Duration
Activity 2	• Take statistics about the numbers of observed plant species (presented in Activity 1) on the school campus or by the students • Though the survey may not reflect the rarity of the species, the teacher should point out the common and rare species involved in Activity 1	Mathematics (statistics)	5 mins.
Activity 3	• Ask students to use a tablet computer to search for the information and photos of one common plant species and one rare and endangered plant species • Students may search for the information by taking photos of the illustrations or colour plant photos by using the 'search images' function of a search engine	Technology (apply IT technology)	10 mins.
Activity 4	• Ask students to draw a picture of a leaf, flower or fruit of the species being searched for in Activity 3	Art (illustration)	20 mins.
Activity 5	• Ask students to draw a leaf, flower or fruit of the most recorded species in Activity 2 • Contrast the drawings of the students to see if they can serve the purpose of identifying different plant species	Art (illustration)	20 mins.
Explanation	Point out the following: • 'Pen and ink' scientific illustrations are a good tool for recording plant morphological characteristics and helping others to identify the species • Traditional cameras and digital cameras can be used as recording tools and are much easier than illustration • The simplified and solid lines of 'pen and ink' scientific illustrations, neverthless, clearly illustrate the plant's structure, make it easier for the botanist to learn the species' details • Besides the text in the illustration itself, doing illustrations is an objective and specific way to effectively describe and record the plants	Engineering (tools design and application)	10 mins.
Feedback and discussion	• Let students share how they can make the morphological characteristics easier to be recognized in their own illustrations • Discuss how illustrations and digital photos help to identify plant species	Arts and sciences	5 mins.

(Continued on next page)

Lesson flow	Activities	STEAM elements	Duration
Conclusion	• Observing a professional botanical illustration will help to learn morphological characteristics more clearly and easily • Students learn that some plant species are common and some are under many different types of threat and need to be protected • Students get to experience the biodiversity of plants		5 mins.

2. Seasonal school-based plant investigation STEAM campaign lesson plan example (suitable for secondary school students)

Topic: The survey and characteristics of campus flowering plants
Learning objective: To learn the diversity of flowering plants and the structures of flowers

Period	Activities	STEAM elements	Lesson times
Preparation 1 (by teacher or student)	• Look for some plants growing on the campus that belong to the families of species recorded in this book	Science (compare)	20 minutes of one lesson
Preparation 2 (by teacher or student)	• Take records of the location and numbers of the plants found in Preparation 1	Mathematics (statistics and recording of locations)	20 minutes of one lesson
Information search	• Search for information on the species recorded in Preparation 2 using a mobile device • The students may need to compare the real plants with photos provided by the information they find online • The information required is about the flowering season of the species and the morphological characteristics of the flowers	Technology (information search) Science (compare)	One lesson period
Investigation study	• Students present their findings and discuss to select the individual plants to use for this campaign activity • Their rationale for deciding should include which flowering period is suitable for the survey and numbers of individual plants for scientific statistics	Science and mathematics	One or two lessons

(Continued on next page)

Period	Activities	STEAM elements	Lesson times
Apply and take the survey	• Take surveys of the days and numbers of the flowering period of the selected individual plants • The days of flowering may be shorter or longer than the information provided. The differences should be also recorded	Mathematics (statistics)	The records can be taken during recess, lunchtime or after school. Records should be taken at least two days a week during the flowering period.
Enhance knowledge	• Learn the typical flower structures • Compare the difference of the structures of the species which is the same family of the surveyed plants listed in this book with the typical flower structures	Science (observe and compare)	One lesson
Problem-solving 1	• Guide students to design or apply tools (e.g. Tree Pole Pruner) to collect flower specimens	Engineering (tool design and application)	One lesson
Problem-solving 2	• Guide students to design and apply tools (e.g. microscopes) to observe and record the macroscopic structures of the collected flowers	Engineering (tool design and application)	One lesson
Apply and record	• Dissect collected flowers • Illustrate the structures, including the whole flower and macroscopic structures	Art (sketching)	Two lessons
Verify	• Compare the collected flower structures with the structures of the same family species listed in this book	Science (compare)	One lesson
Feedback and discussion	• Let students share the survey results of the flowering period and flower structure are the same with the same family of species • Discuss the importance of botany illustration in the modern digital ages	Engineering + Technology + Science	One lesson
Conclusion	• Conclude the findings and experience the students have gained and how scientific records can be made • Explain the importance of traditional scientific records, and how they are still very valuable and important, such as scientific illustrations which help to learn morphological characteristics more clearly and easily		The last lesson

Adjustments for different levels of learners

Due to the different levels of students' learning stages, educators should adjust the activities' content and depth according to the learners' abilities. Selecting plant species, recording their morphological characteristics, composing and drawing botanical illustrations, and the skills required to operate the tools and IT equipment are all STEAM elements that can be fine-tuned to suit different levels or talents of students.

Examples of how to adjust the different learning stages of the content are listed below:

Activities	Ages / Stages / Levels	Adjustment elements	Suitable content types
Identify plants with botanical illustrations	6–9 years old Key stage 1	Plant species names (common names); number of morphological characters	Common names, growing habit, leaf shape, flower colour, fruit colour
	10–12 years old Key stage 2		Leaf arrangement, leaf type, fruit type
	12–15 years old Key stage 3	Plant species names (common names); number of morphological characters; growing habitat	Branch character, growing habitat, distribution
	16 years old or above Key stage 4	Plant species names (common names); number of morphological characters; scientific terminologies	Family name, inflorescences

Activities	Ages / Stages / Levels	Adjustment elements	Suitable content types
Botanical illustration training	6–9 years old Key stage 1	Plant species names (common names); the whole shape of the part	Leaf shape
	10–12 years old Key stage 2	Plant species names (common names); the whole shape of the part; selected structure terms or names	Compound or single leaf, and types of leaf, typical flower structures and fruit shape
	12–15 years old Key stage 3	Plant species names (common names); the whole shape of the part; more structured terms and functions	Leaf tips and leaf base; branch shapes and surfaces, a whole inflorescence's shape but not the type of inflorescence, the structure inside and outside of the fruit
	16 years old or above Key stage 4	Plant species names (common names); the whole shape of the part; selected professional scientific information	Inflorescence types, ovary positions, microscopic structures

Lastly, I hope this chapter can provide some new ideas and topics for STEAM education. More traditional science elements related to STEAM activities will appear in the future.

Tin-Hang Wong
Education Manager, Shiu-Ying Hu Herbarium

References

Bassachs, M., Cañabate, D., Nogué, L., Serra, T., Bubnys, R., & Colomer, J. (2020). Fostering Critical Reflection in Primary Education through STEAM Approaches. *Education Sciences, 10*(12), 384. https://doi.org/10.3390/educsci10120384

Curriculum Development Council. (2014). *Basic education curriculum guide.* Education Bureau, The Government of the HKSAR.

Curriculum Development Council. (2015). *Promotion of STEM education—unleashing potential in innovation.* Education Bureau, The Government of the HKSAR.

Directorate for Education, & Human Resources. (1996). *Shaping the future: Perspectives on undergraduate education in science, mathematics, engineering, and technology* (Vol. 2). National Science Foundation, United States government.

Education Bureau, the Government of the HKSAR (2020 June 29). *To sustain, deepen and focus on learning to learn (Primary 1 – 6).* Curriculum Development, Education Bureau. Retrieved March 1, 2022 from http://www.edb.gov.hk/en/curriculum-development/doc-reports/guide-basic-edu-curriculum/index.html

Harris, A., & de Bruin L. R. (2018). L.R. Secondary school creativity, teacher practice and STEAM education: An international study. *Journal of Educational Change, 19.* 153–179. https://doi.org/10.1007/s10833-017-9311-2

Liao, C. (2016) From Interdisciplinary to Transdisciplinary: An Arts-Integrated Approach to STEAM Education, *Art Education, 69*(6). 44-49. https://doi.org/10.1080/00043125.2016.1224873

The Government of the HKSAR. (2015). *Policy address: VII. Youth education and development.* 2015 Policy Address - Press Releases. Retrieved March 1, 2022 from http://www.policyaddress.gov.hk/2015/eng/p150.html

Wojnowski, B. & Pea C. (2014). *Models and approaches to STEM professional development.* NSTA Press.

二十種原生植物圖鑑

Botanical Illustrated Guide to 20 Hong Kong Native Plants

劉大偉、李敏貞、王曉欣

By David T. W. Lau, Man-Ching Li & Hiu-Yan Wong

Alsophila spinulosa (Wall. ex Hook.) R. M. Tryon
刺桫欏

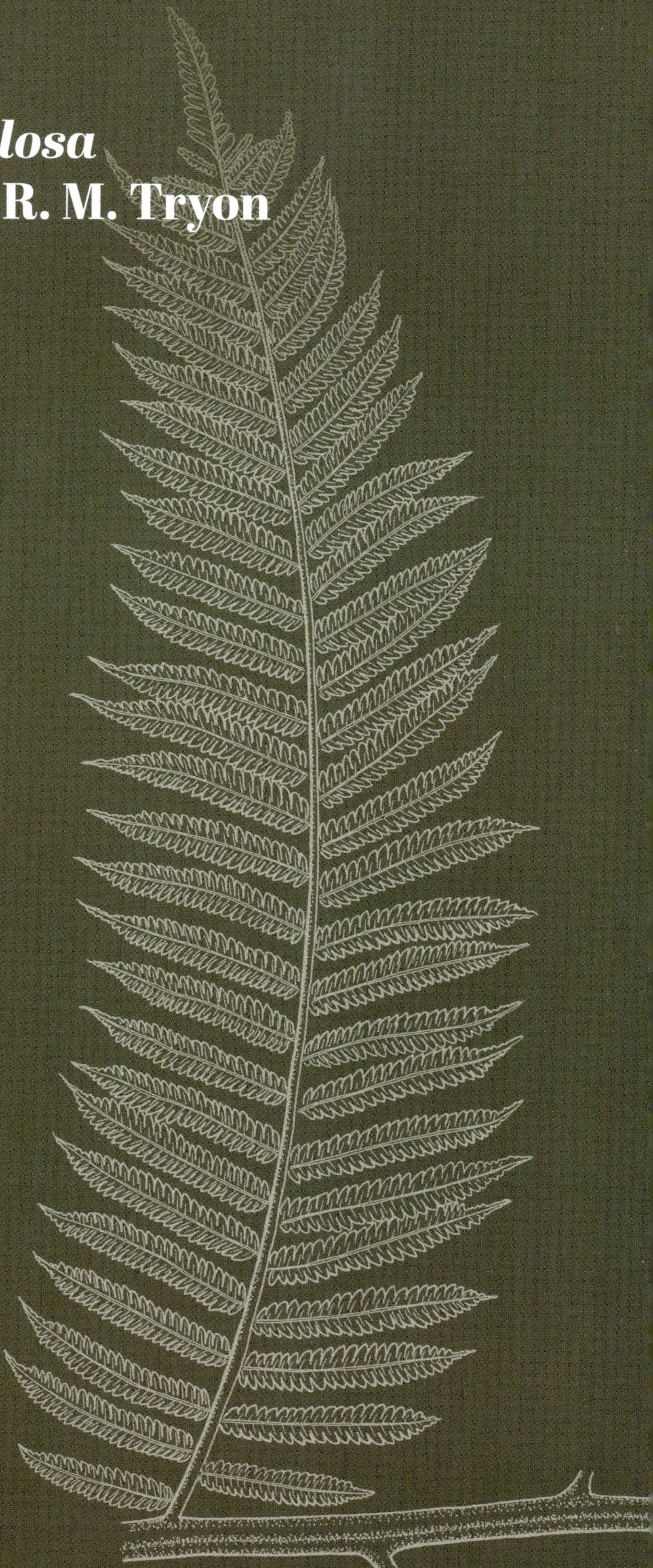

桫欏科植物是活化石，
曾在晚三疊紀至侏羅紀時期
廣泛分布，與恐龍並存。

採集者編號
Collector no.
Shiu Ying Hu 11978

採集年份
Year of collection
1972

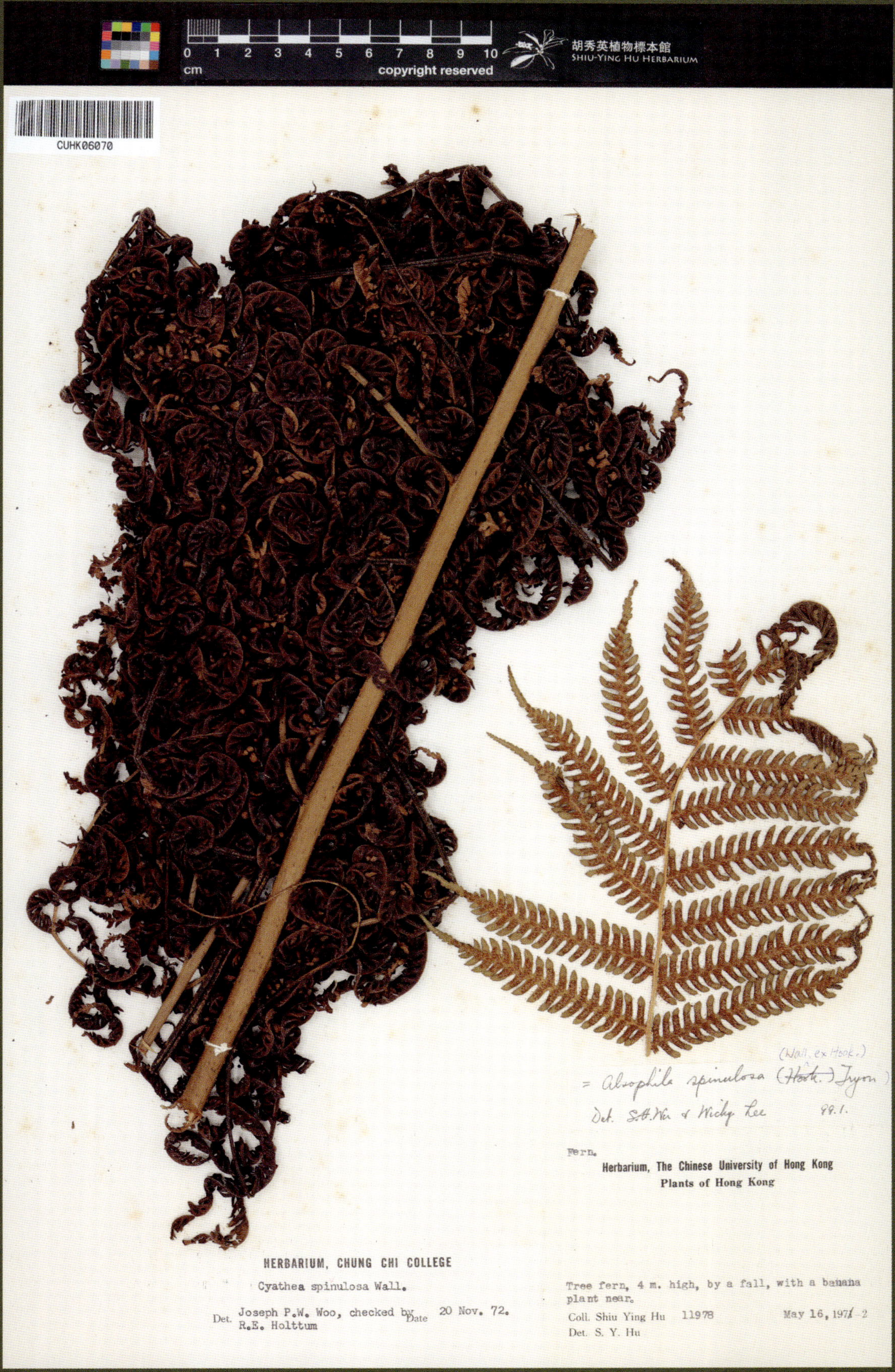
0 1 2 3 4 5 6 7 8 9 10
cm
copyright reserved
胡秀英植物標本館
SHIU-YING HU HERBARIUM
CUHK06070
(Wall. ex Hook.)
= Alsophila spinulosa (Hook.) Tryon
Det. S.A. Wu & Wicky Lee 99.1.
Fern.
Herbarium, The Chinese University of Hong Kong
Plants of Hong Kong
HERBARIUM, CHUNG CHI COLLEGE
Cyathea spinulosa Wall.
Det. Joseph P.W. Woo, checked by R.E. Holttum
Date 20 Nov. 72.
Tree fern, 4 m. high, by a fall, with a banana plant near.
Coll. Shiu Ying Hu 11978
May 16, 1971-2
Det. S. Y. Hu

刺桫欏

學名	*Alsophila spinulosa* (Wall. ex Hook.) R. M. Tryon
中文名	刺桫欏
科名	桫欏科
本地分布狀態	原生植物
繪圖編號	H. Y. Wong 010

花期	花色	果期	果色
不適用	不適用	不適用	不適用

刺桫欏為大形喬木狀蕨類，擁有筆直的莖幹，可見於山谷和溪邊。桫欏科植物是活化石，曾在晚三疊紀至侏羅紀時期廣泛分布，與恐龍並存。它們對於植物演化、古代氣候及地質環境等研究有重要價值。

刺桫欏於《香港稀有及珍貴植物》中屬「易危」，而在《中國植物紅皮書》被列為「漸危」。

刺桫欏列於《瀕危野生動植物種國際貿易公約》附錄II；亦受香港法例第96A章《林務規例》及第 586 章《保護瀕危動植物物種條例》規管。

《中國植物誌》(英文版) 的採用名稱

Alsophila spinulosa (Wallich ex Hooker) R. M. Tryon (科名：桫欏科)

《香港植物誌》的採用名稱

Alsophila spinulosa (Wall. ex Hook.) R. M. Tryon (科名：桫欏科)

參考文獻 References

1. Hong Kong Herbarium. (2021a). *HK Herbarium-Alsophila spinulosa*. HK Plant Database. Retrieved December 20, 2021, from https://www.herbarium.gov.hk/en/hk-plant-database/plant-detail/index.html?pType=species&oID=8694
2. Hong Kong Herbarium. (2021b). *Alsophila spinulosa.* Rare and Precious Plants of Hong Kong (Online Version). Retrieved December 20, 2021, from https://www.herbarium.gov.hk/en/publications/books/book2/text/alsophila-spinulosa/index.html
3. Missouri Botanical Garden. (2021). *Name-Alsophila spinulosa.* Tropicos. Retrieved December 13, 2021, from https://www.tropicos.org/name/26623167

Alsophila spinulosa (Wall. ex Hook.) R. M. Tryon

Scientific name	*Alsophila spinulosa* (Wall. ex Hook.) R. M. Tryon
Chinese name	刺桫欏
Family	Cyatheaceae Kaulf.
Local distribution status	Native species
Illustration number	H. Y. Wong 010

FLOWERING PERIOD	COLOUR OF FLOWER	FRUITING PERIOD	COLOUR OF FRUIT
N/A	N/A	N/A	N/A

Alsophila spinulosa (Wall. ex Hook.) R. M. Tryon is a large tree fern with a straight trunk. It can be found in valleys and by streams.

Cyatheaceae are a family of living fossils; they were widely distributed during the Late Triassic and Jurassic periods, and co-existed with dinosaurs. They are important to the study of plant evolution, paleoclimatology, paleogeography, etc.

Alsophila spinulosa (Wall. ex Hook.) R. M. Tryon is classified as 'Vulnerable' in *Rare and Precious Plants of Hong Kong* and is listed as 'Vulnerable' in the *China Plant Red Data Book*. It is also listed in CITES Appendix II. Lastly, it is regulated under Hong Kong's Forestry Regulations (Cap. 96A) and the Protection of Endangered Species of Animals and Plants Ordinance (Cap. 586).

Name in *Flora of China*
Alsophila spinulosa (Wallich ex Hooker) R. M. Tryon (Family: Cyatheaceae)

Name in *Flora of Hong Kong*
Alsophila spinulosa (Wall. ex Hook.) R. M. Tryon (Family: Cyatheaceae)

4. The CITES Secretariat. (n.d.). *Cyathea spinulosa*. Checklist of CITES Species. Retrieved December 13, 2021, from https://checklist.cites.org/#/en
5. Zhang, X. C., & Nishida, H. (2013). Cyatheaceae. In Z. Y. Wu, P. H. Raven & D. Y. Hong (Eds.), *Flora of China* (Vol. 2–3, pp. 134–138). Science Press & Missouri Botanical Garden Press.
6. 彭霄鵬 (2015)。〈蕨類植物活化石 ——"樹蕨"桫欏利用及保護的探討與展望〉。《化石》，2，69–71。

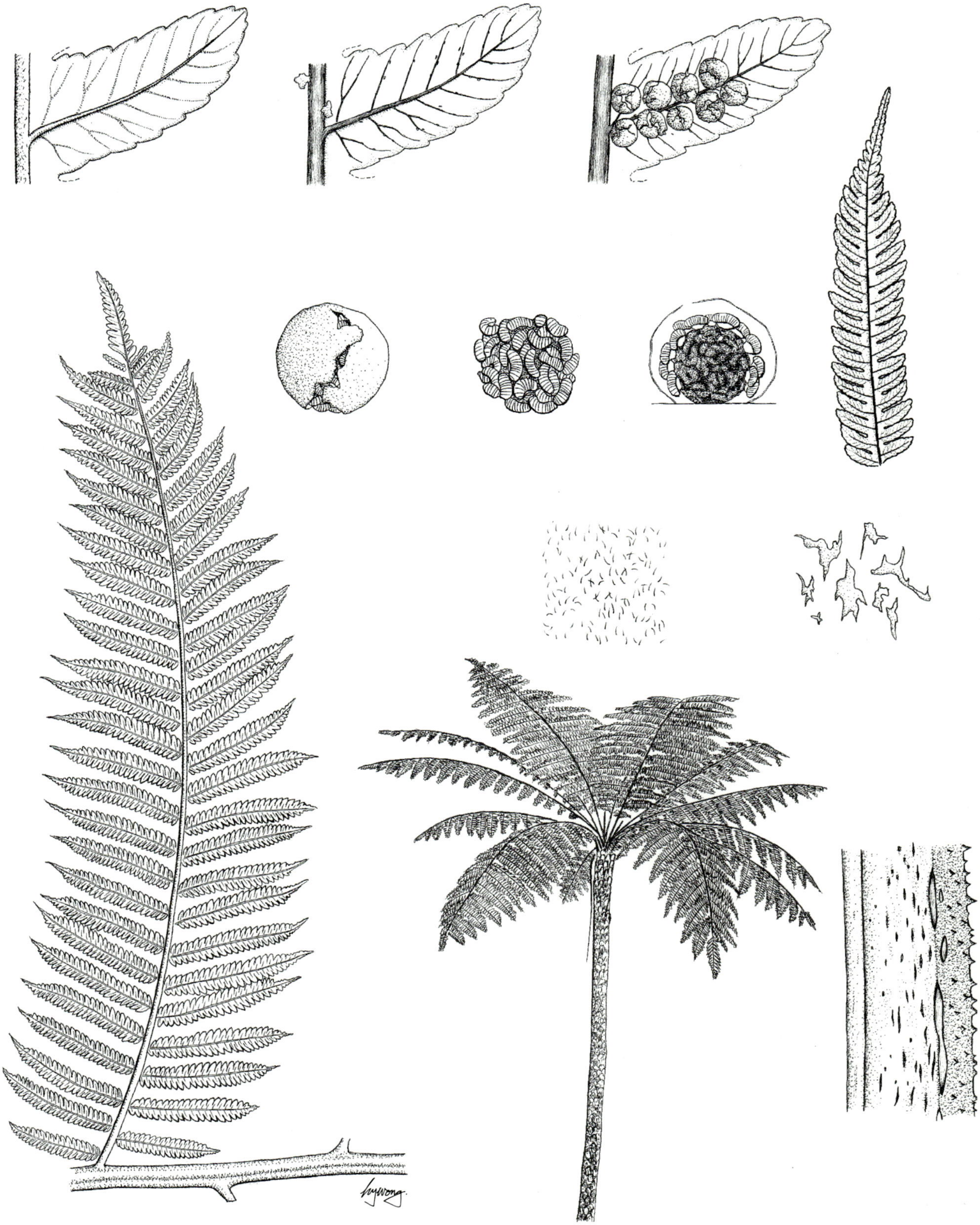

Alsophila spinulosa (Wall. ex Hook.) R. M. Tryon 刺桫欏

Illustration 010 by H. Y. Wong
Authenticated by D. T. W. Lau

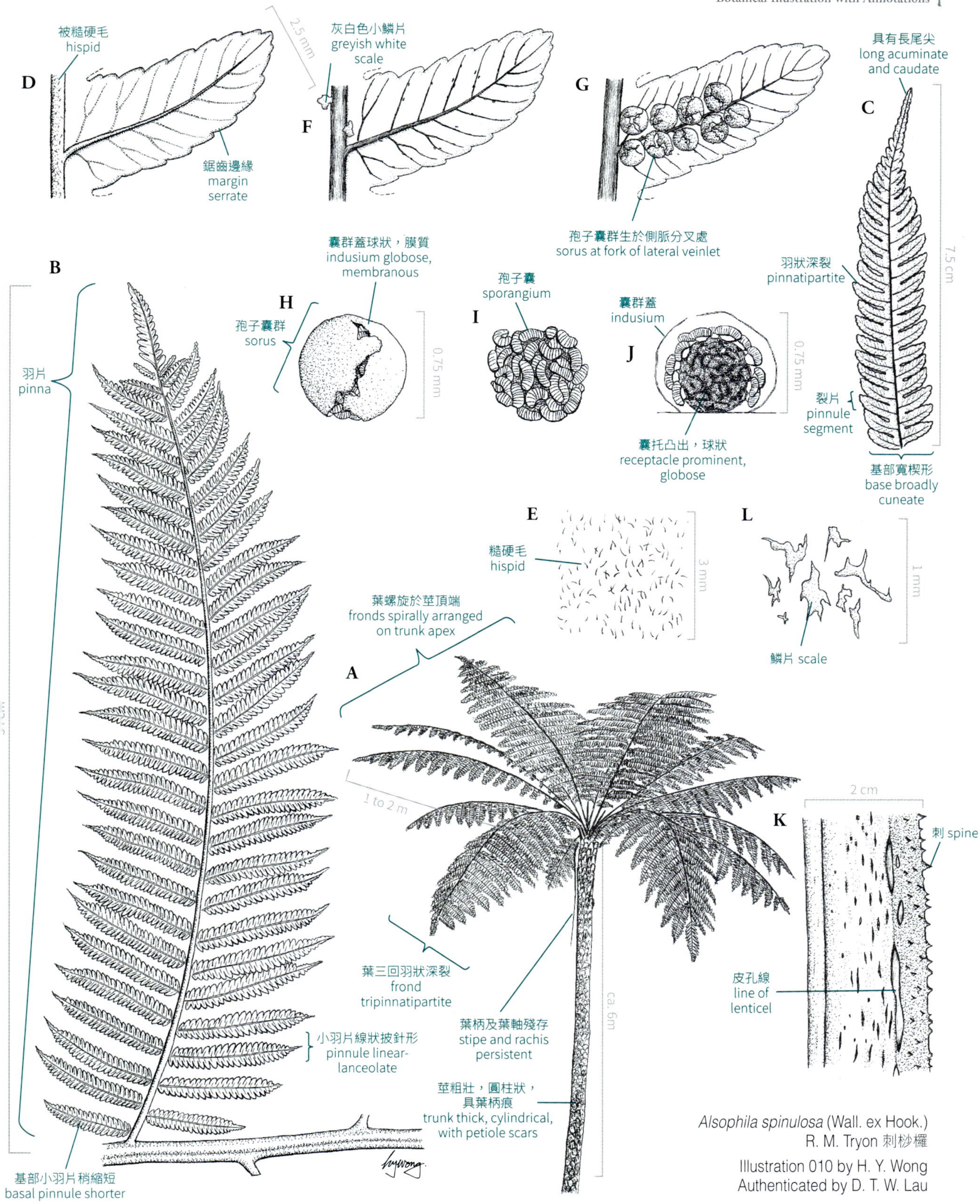

Alsophila spinulosa (Wall. ex Hook.)
R. M. Tryon 刺桫欏

Illustration 010 by H. Y. Wong
Authenticated by D. T. W. Lau

植物結構 PLANT STRUCTURES:
A. 葉與莖 Fronds and trunk **B.** 羽片，正面 Pinna, adaxial **C.** 小羽片，正面 Pinnule, adaxial **D.** 裂片，正面 Pinnule segment, adaxial **E.** 小羽片中脈上的被毛 Hairs on the midrib of pinnule **F.** 裂片，背面 Pinnule segment, abaxial **G.** 有孢子囊群的裂片，背面 Pinnule segment, with sori, abaxial **H.** 孢子囊群，正面觀 Sorus, frontal view **I.** 孢子囊群解剖圖，展示孢子囊，正面觀 Dissected sorus, showing sporangia, frontal view **J.** 孢子囊群解剖圖，展示囊托，側面觀 Dissected sorus, showing receptacle, lateral view **K.** 葉柄，側面觀 Stipe, lateral view **L.** 葉柄上的鱗片 Scales on stipe

參考標本 REFERENCE SPECIMENS: Rare and endangered plants (D. T. W. Lau, K. W. Lam, S. W. Shek) 29B; 嚴岳鴻 YYH11613a (CSH0123426); Shiu Ying Hu 11978

參考文獻 REFERENCES:

Dong, S. Y., & Chen, Z. C. (2017). Cyatheaceae. In Fairy Lake Botanical Garden, Shenzhen & Chinese Academy of Sciences (Eds.), *Flora of Shenzhen* (Vol. 1, pp. 82–85). China Forestry Publishing House.
Hong Kong Herbarium. (2021). *Alsophila spinulosa*. Rare and Precious Plants of Hong Kong (Online Version). Retrieved December 13, 2021, from https://www.herbarium.gov.hk/en/publications/books/book2/text/alsophila-spinulosa/index.html
Li, T. J., Zhou, J. C., & Wu, Z. H. (2003). Cyatheaceae. In T. J. Li, J. C. Zhou & Z. H. Wu (Eds.), *The Flora of Hong Kong-Pteridophyta* (pp. 103–111). Kadoorie Farm and Botanic Garden.
Stevens, P. F. (2021). Cyatheales. Angiosperm Phylogeny Website. Retrieved December 12, 2021, from http://www.mobot.org/MOBOT/research/APweb/welcome.html
Zhang, X. (2004). Cyatheaceae. In Delectis Florae Reipublicae Popularis Sinicae Agendae Academiae Sinicae Edita (Ed.), *Flora Reipublicae Popularis Sinicae* (Vol. 6(3), pp. 249–274). Science Press.
Zhang, X. C., & Nishida, H. (2013). Cyatheaceae. In Z. Y. Wu, P. H. Raven & D. Y. Hong (Eds.), *Flora of China* (Vol. 2–3, pp. 134–138). Science Press & Missouri Botanical Garden Press.

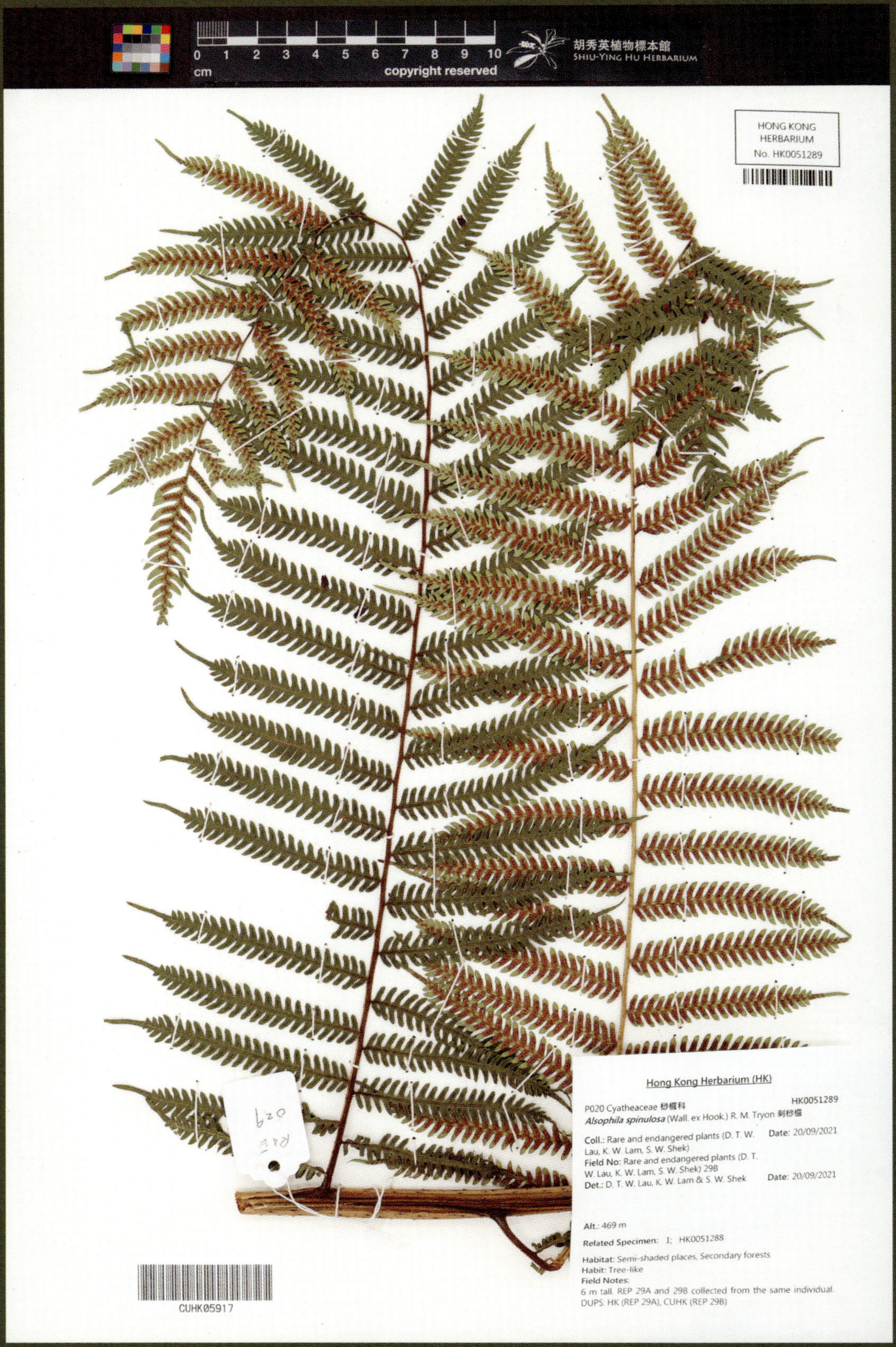

採集者編號 Collector no.
Rare and endangered plants (D. T. W. Lau, K. W. Lam, S. W. Shek) 29B

採集年份 Year of collection
2021

外觀 Overview

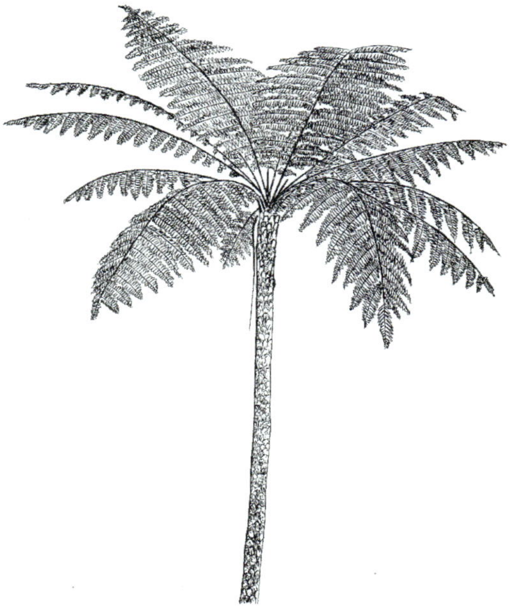

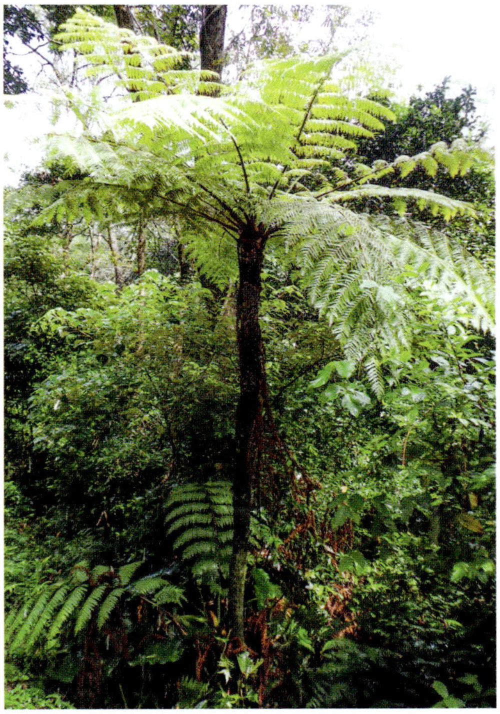

葉與莖 Fronds and trunk

葉 Leaf

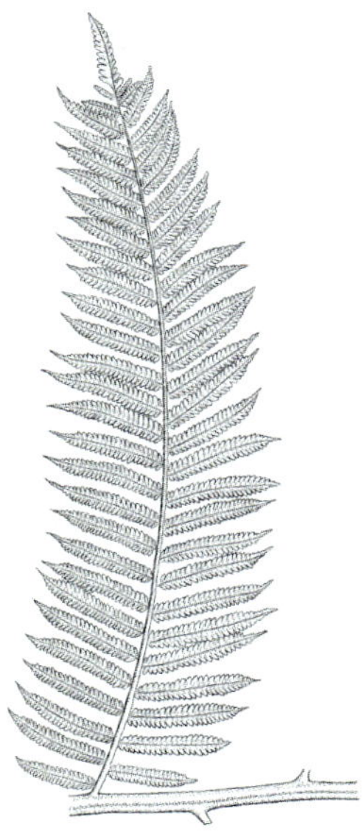

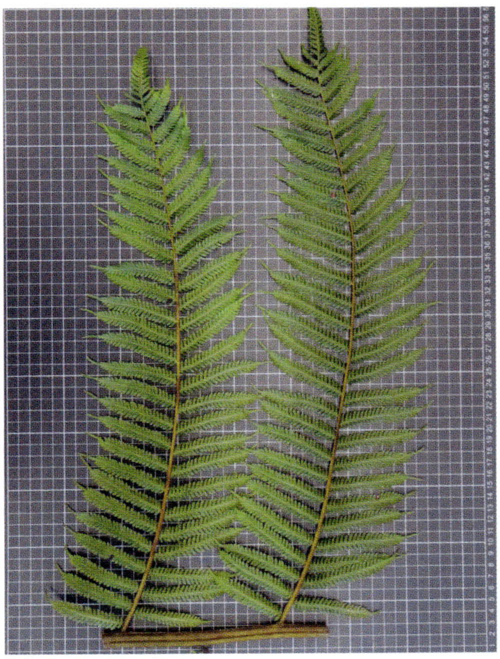

羽片，正面 Pinnae, adaxial

葉 Leaf

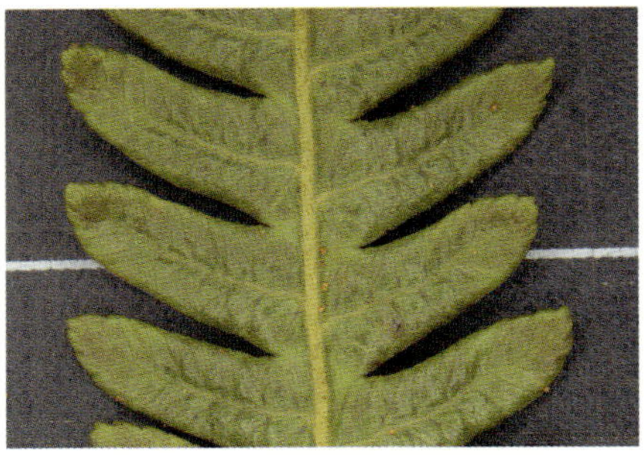

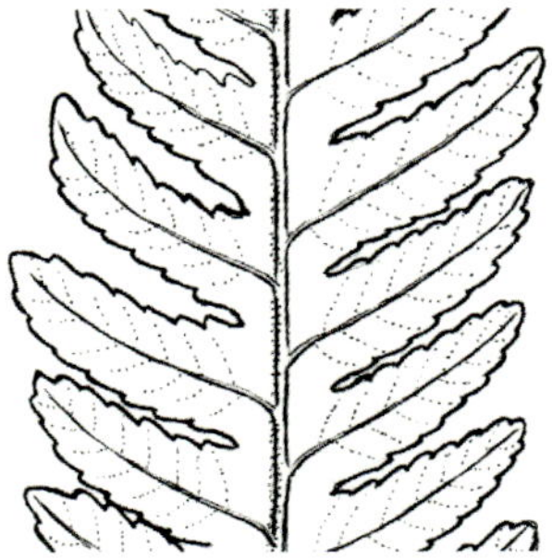

裂片，正面 Pinnule segments, adaxial

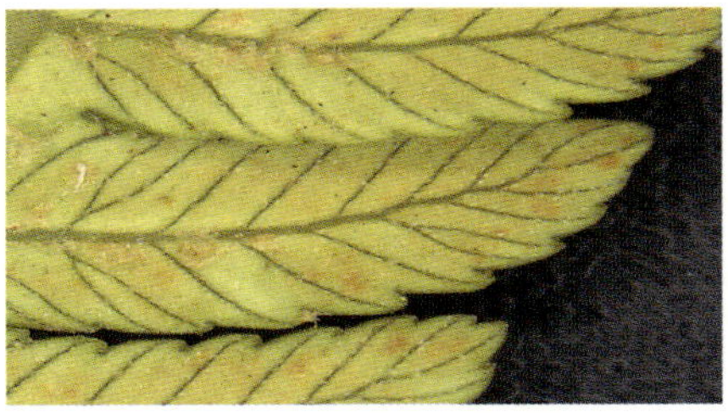

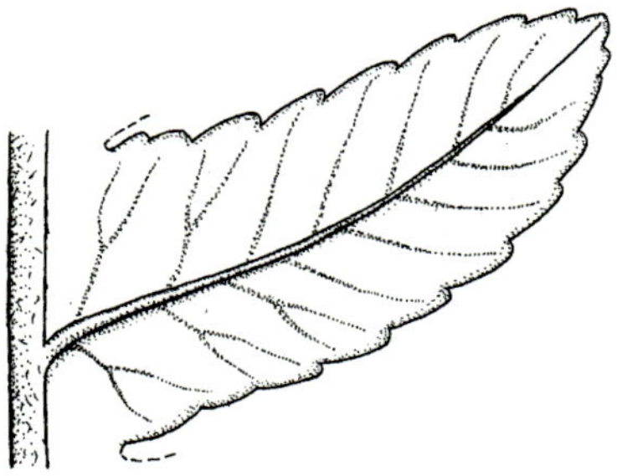

裂片，背面 Pinnule segments, abaxial

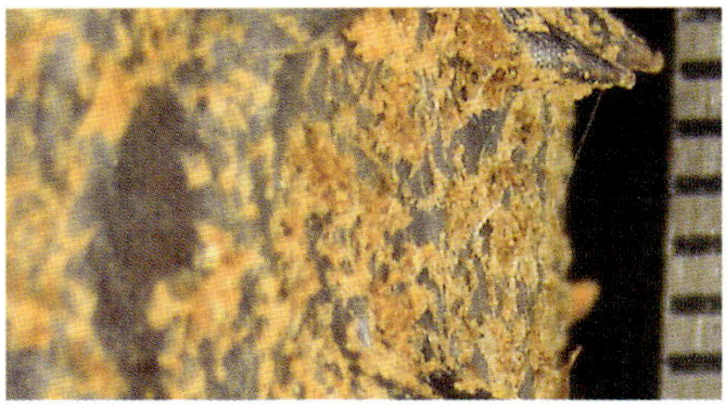

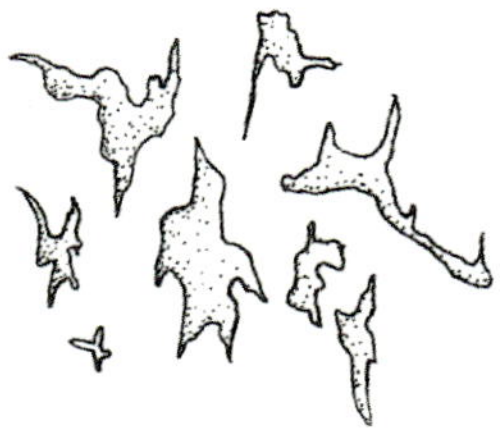

葉柄上的鱗片 Scales on stipe

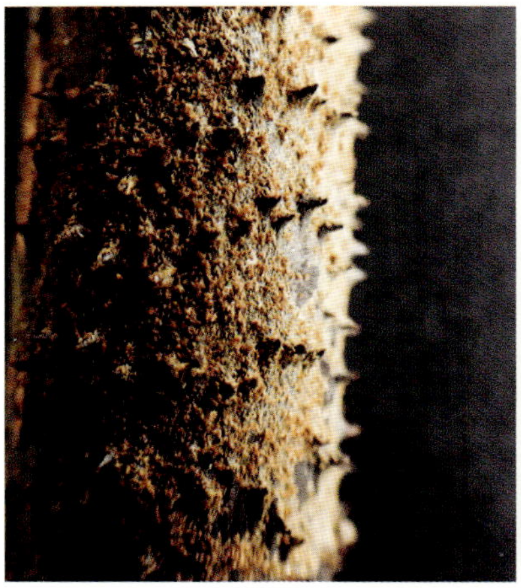

葉柄上的刺 Spines on stipe

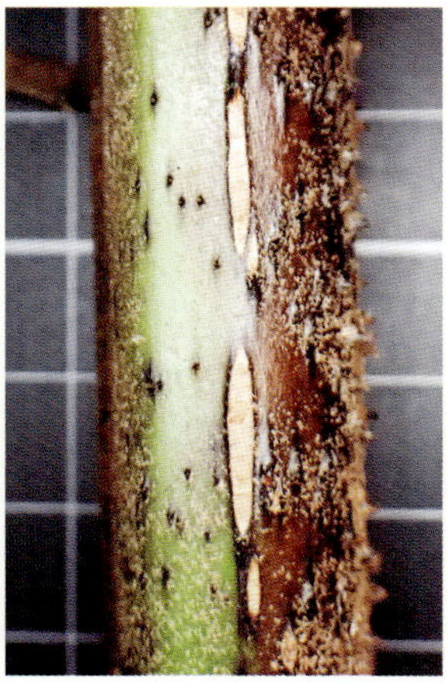

皮孔線 Line of lenticel

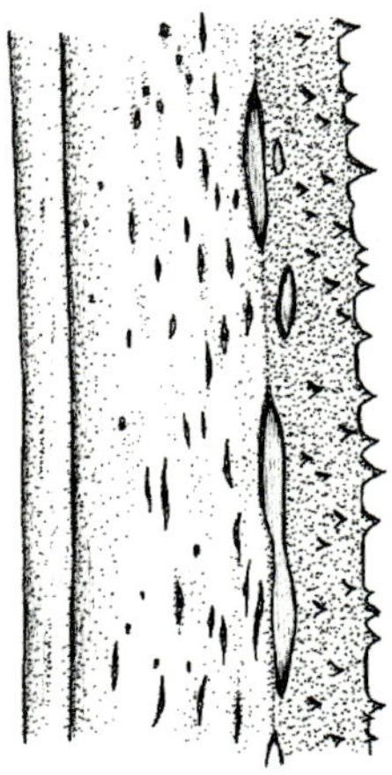

孢子葉結構 Sporophyll Structures

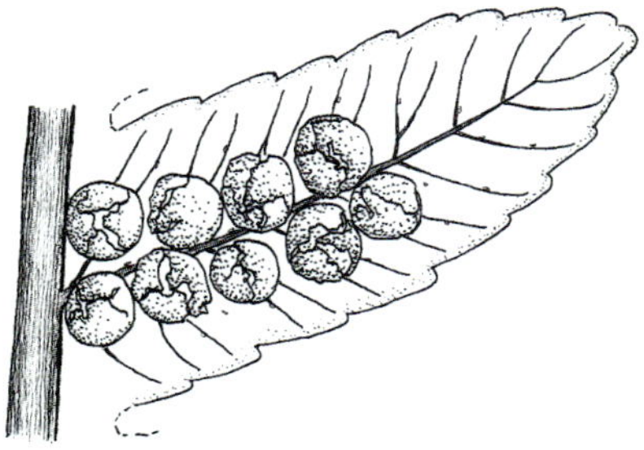
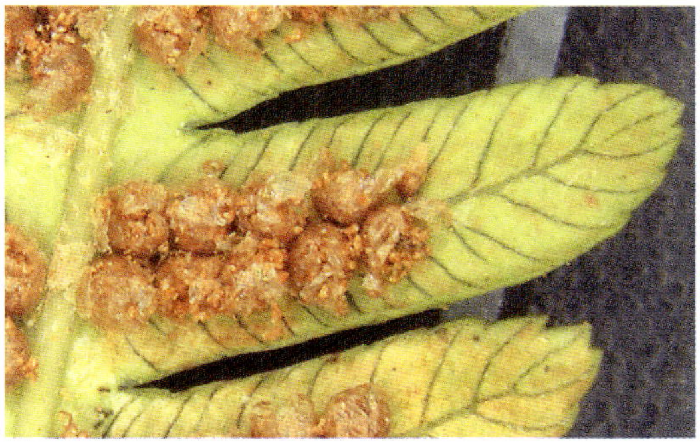

有孢子囊群的裂片，背面
Pinnule segments, with sori, abaxial

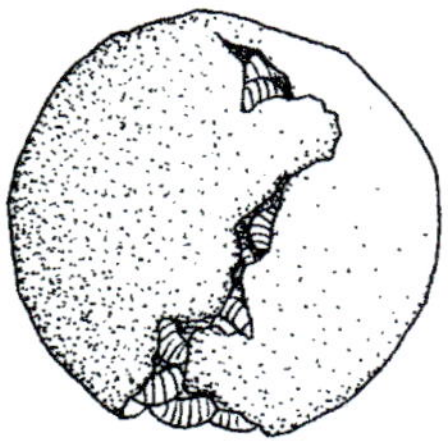
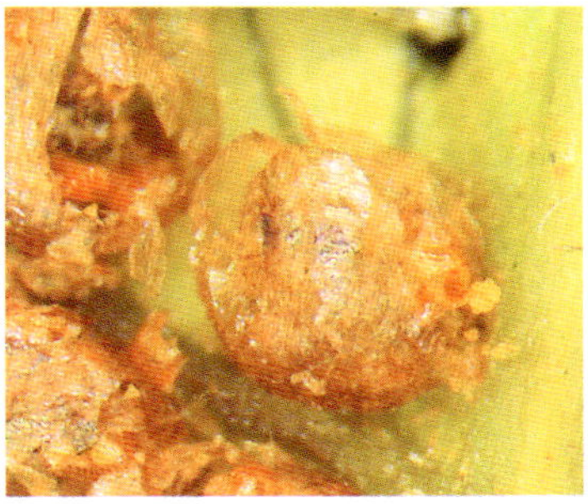

孢子囊群，正面觀
Sorus, frontal view

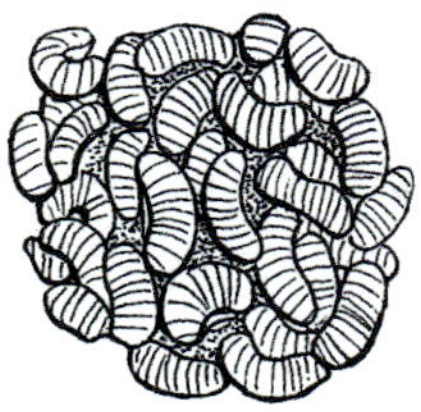
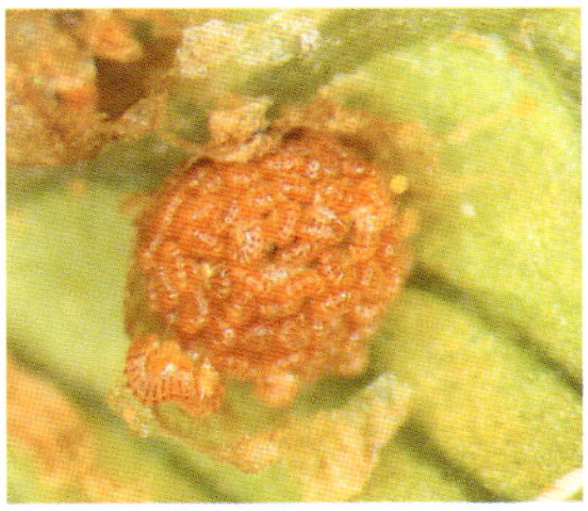

孢子囊群解剖圖，
展示孢子囊，正面觀
Dissected sorus,
showing sporangia,
frontal view

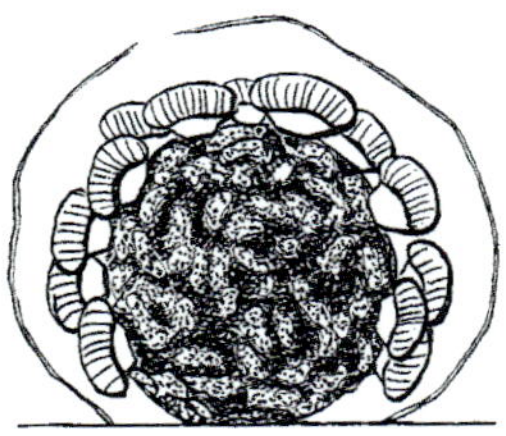
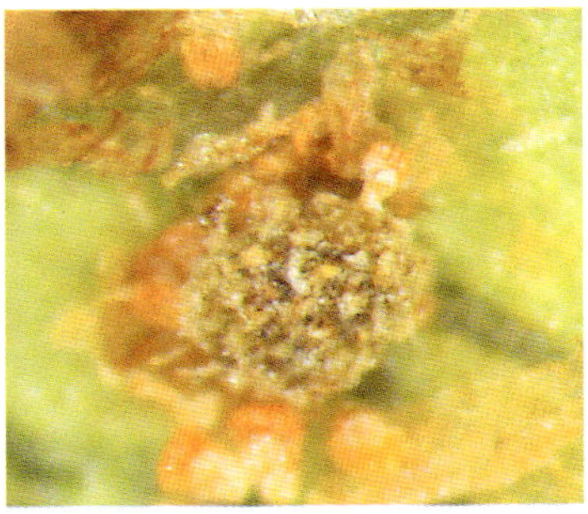

孢子囊群，
展示囊群蓋，正面觀
Dissected sorus,
showing
receptacle,
frontal view

Illicium dunnianum Tutcher
紅花八角

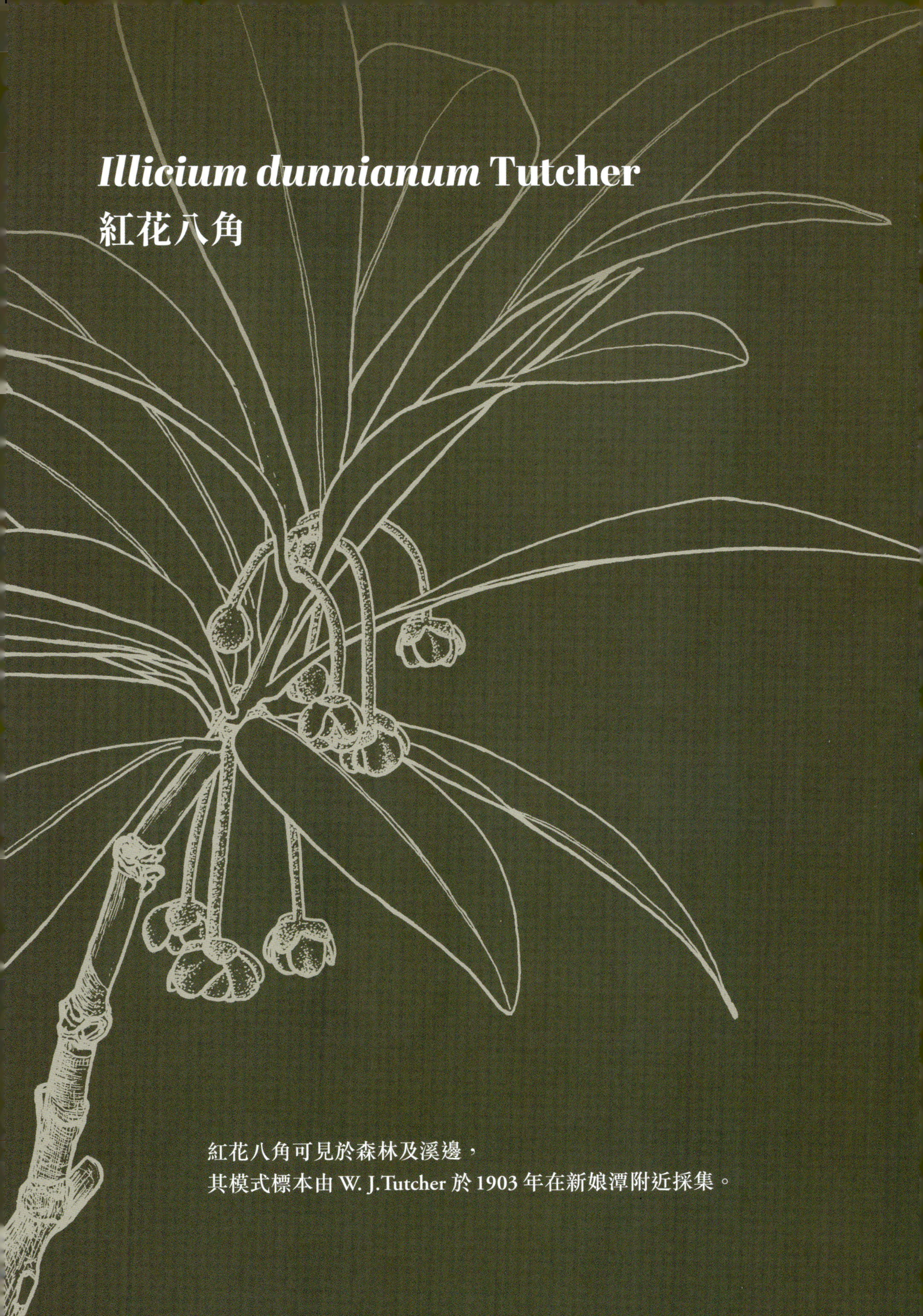

紅花八角可見於森林及溪邊，
其模式標本由 W. J.Tutcher 於 1903 年在新娘潭附近採集。

採集者編號
Collector no.
Shiu Ying Hu 13027

採集年份
Year of collection
1973

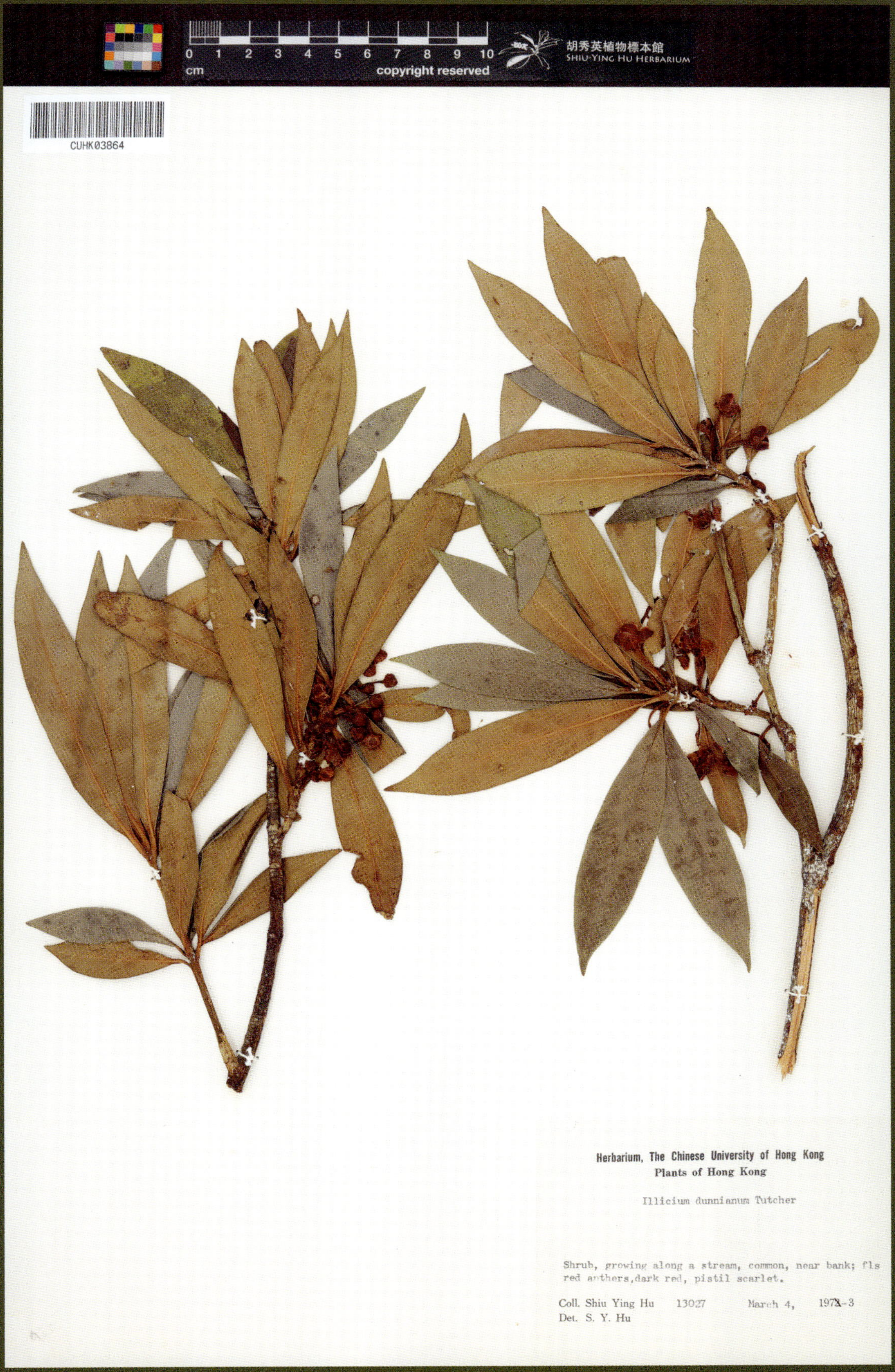
0 1 2 3 4 5 6 7 8 9 10
cm
copyright reserved
胡秀英植物標本館
SHIU-YING HU HERBARIUM
CUHK03864
Herbarium, The Chinese University of Hong Kong
Plants of Hong Kong
Illicium dunnianum Tutcher
Shrub, growing along a stream, common, near bank; fls red anthers,dark red, pistil scarlet.
Coll. Shiu Ying Hu 13027 March 4, 1972-3
Det. S. Y. Hu

紅花八角

學名	*Illicium dunnianum* Tutcher
中文名	紅花八角
科名	五味子科
本地分布狀態	原生植物
繪圖編號	H. Y. Wong 003

花期	花色	果期	果色
3至7月，10至11月	紅色	7至10月	褐色

紅花八角常為 1 至 2 米高的灌木，可見於森林及溪邊，其模式標本由 W. J. Tutcher 於 1903 年在新娘潭附近採集。

八角屬的植物果實皆為聚合果，由數個至十多個蓇葖果組成，單輪排列，呈星狀。

紅花八角無毒並能入藥，具抗炎、止痛等功效，而其他個別的八角屬植物則有毒。

紅花八角受香港法例第 96A章《林務規例》規管。

《中國植物誌》(英文版) 的採用名稱

Illicium dunnianum Tutcher (科名：八角科)

《香港植物誌》的採用名稱

Illicium dunnianum Tutcher (科名：八角科)

參考文獻 References

1. Hong Kong Herbarium. (2021a). *HK Herbarium*-Illiciaceae. HK Plant Database. Retrieved April 10, 2021, from https://www.herbarium.gov.hk/tc/hk-plant-database/plant-detail/index.html?pType=family&oID=468
2. Hong Kong Herbarium. (2021b). *HK Herbarium-Illicium dunnianum*. HK Plant Database. Retrieved April 10, 2021, from https://www.herbarium.gov.hk/tc/hk-plant-database/plant-detail/index.html?pType=species&oID=5513
3. Liu, Y. N., Su, X. H., Huo, C. H., Zhang, X. P., Shi, Q. W., & Gu, Y. C. (2009). Chemical Constituents of Plants from the Genus *Illicium. Chemistry & Biodiversity 6* (7), 963–989. https://doi.org/10.1002/cbdv.200700433

Illicium dunnianum Tutcher

Scientific name	*Illicium dunnianum* Tutcher
Chinese name	紅花八角
Family	Schisandraceae Blume
Local distribution status	Native species
Illustration number	H. Y. Wong 003

FLOWERING PERIOD	COLOUR OF FLOWER	FRUITING PERIOD	COLOUR OF FRUIT
March to July, October to November	Red	July to October	Brown

Illicium dunnianum Tutcher is usually a 1 to 2 m tall shrub, which can be found in forests and by streams. Its type specimen was collected near Bride's Pool by W. J. Tutcher in 1903.

All species in *Illicium* have star-shaped aggregate fruits, formed by a few to more than 10 follicles, arranging in 1 whorl.

Illicium dunnianum Tutcher is a non-toxic medicine, being both anti-inflammatory and pain relieving. However, some species in *Illicium* are toxic.

Illicium dunnianum Tutcher is regulated under Hong Kong's Forestry Regulations (Cap. 96A).

Name in *Flora of China*
Illicium dunnianum Tutcher (Family: Illiciaceae)

Name in *Flora of Hong Kong*
Illicium dunnianum Tutcher (Family: Illiciaceae)

4. Missouri Botanical Garden. (2021). *Name-Illicium dunnianum.* Tropicos. Retrieved April 15, 2021, from https://www.tropicos.org/name/50002067
5. Xia, N. H., & Saunders, R. M. K. (2008). Illiciaceae. In Z. Y. Wu, P. H. Raven & D. Y. Hong (Eds.), *Flora of China* (Vol. 7, pp. 32–38). Science Press & Missouri Botanical Garden Press.
6. 張連富、林開中、楊武德、魏俊蓮、楊桂華、呂興文（1995）。〈紅花八角原植物及易混品的鑒別〉。《中國中藥雜誌》，20(12)，717–718。

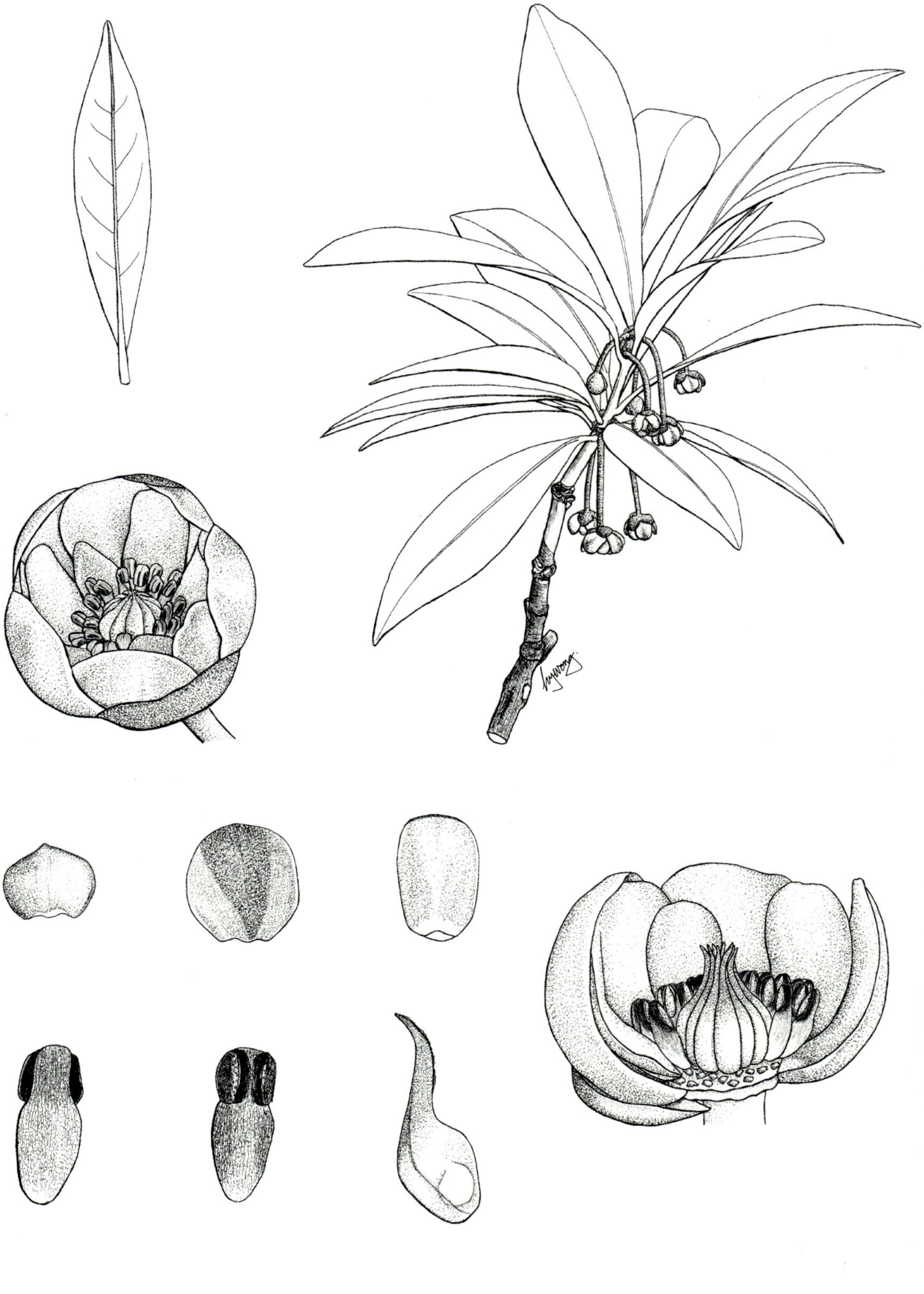

Illicium dunnianum Tutcher 紅花八角
Illustration 003 by H. Y. Wong
Authenticated by D. T. W. Lau

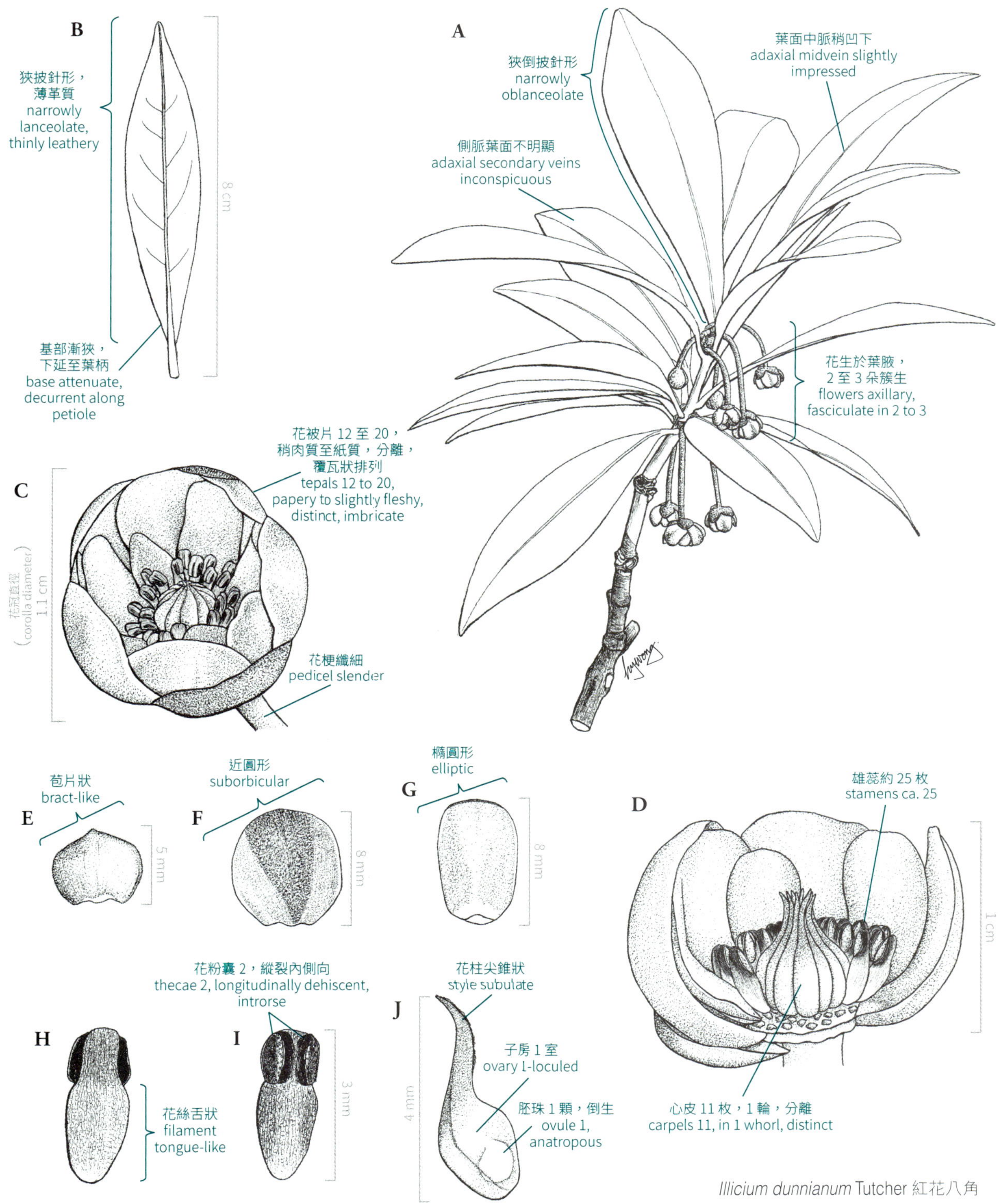

Illicium dunnianum Tutcher 紅花八角

Illustration 003 by H. Y. Wong
Authenticated by D. T. W. Lau

植物結構 PLANT STRUCTURES:
A. 花枝 Flowering branch **B.** 葉底 Leaf, abaxial **C.** 花，正面觀 Flower, frontal view **D.** 花解剖圖，展示花被片、多條雄蕊及雌蕊 Dissected flower, showing tepals, many stamens and gynoecium
E. 外輪花被片 Outer tepal **F.** 中層花被片 Middle layer tepal **G.** 內輪花被片 Inner tepal **H.** 雄蕊，背軸面 Stamen, abaxial **I.** 雄蕊，向軸面 Stamen, adaxial **J.** 雌蕊 Pistil

參考標本 REFERENCE SPECIMENS: Rare and endangered plants (J. Y. Y. Lau, D. T. W. Lau, K. W. Lam, S. W. Shek) 5B; Hu & But 20962; Shiu Ying Hu 13027

參考文獻 REFERENCES:

Chang, B. N. (1996). Illicieae. In Delectis Florae Reipublicae Popularis Sinicae Agendae Academiae Sinicae Edita (Ed.), *Flora Reipublicae Popularis Sinicae* (Vol. 30(1), pp. 198–231). Science Press.
Stevens, P. F. (2021). Austrobaileyales. Angiosperm Phylogeny Website. Retrieved December 20, 2021, from http://www.mobot.org/MOBOT/research/APweb/welcome.html
Wu, R. F. (1991). Illiciaceae. In South China Botanical Garden, the Chinese Academy of Sciences (Ed.), *Flora of Guangdong* (Vol. 2, pp. 1–6). Guangdong Science and Technology Press.
Xia, N. H. (2007). Illiciaceae. In Hong Kong Herbarium, AFCD & South China Botanical Garden, CAS (Eds.), *Flora of Hong Kong* (Vol. 1, pp. 65–66). Agriculture, Fisheries and Conservation Department, HKSAR Government.
Xia, N. H., & Saunders, R. M. K. (2008). Illiciaceae. In Z. Y. Wu, P. H. Raven & D. Y. Hong (Eds.), *Flora of China* (Vol. 7, pp. 32–38). Science Press & Missouri Botanical Garden Press.
Zhou, R. Z., & Wang, Y. L. (2017). Illiciaceae. In Fairy Lake Botanical Garden, Shenzhen & Chinese Academy of Sciences (Eds.), *Flora of Shenzhen* (Vol. 1, pp. 354–355). China Forestry Publishing House.

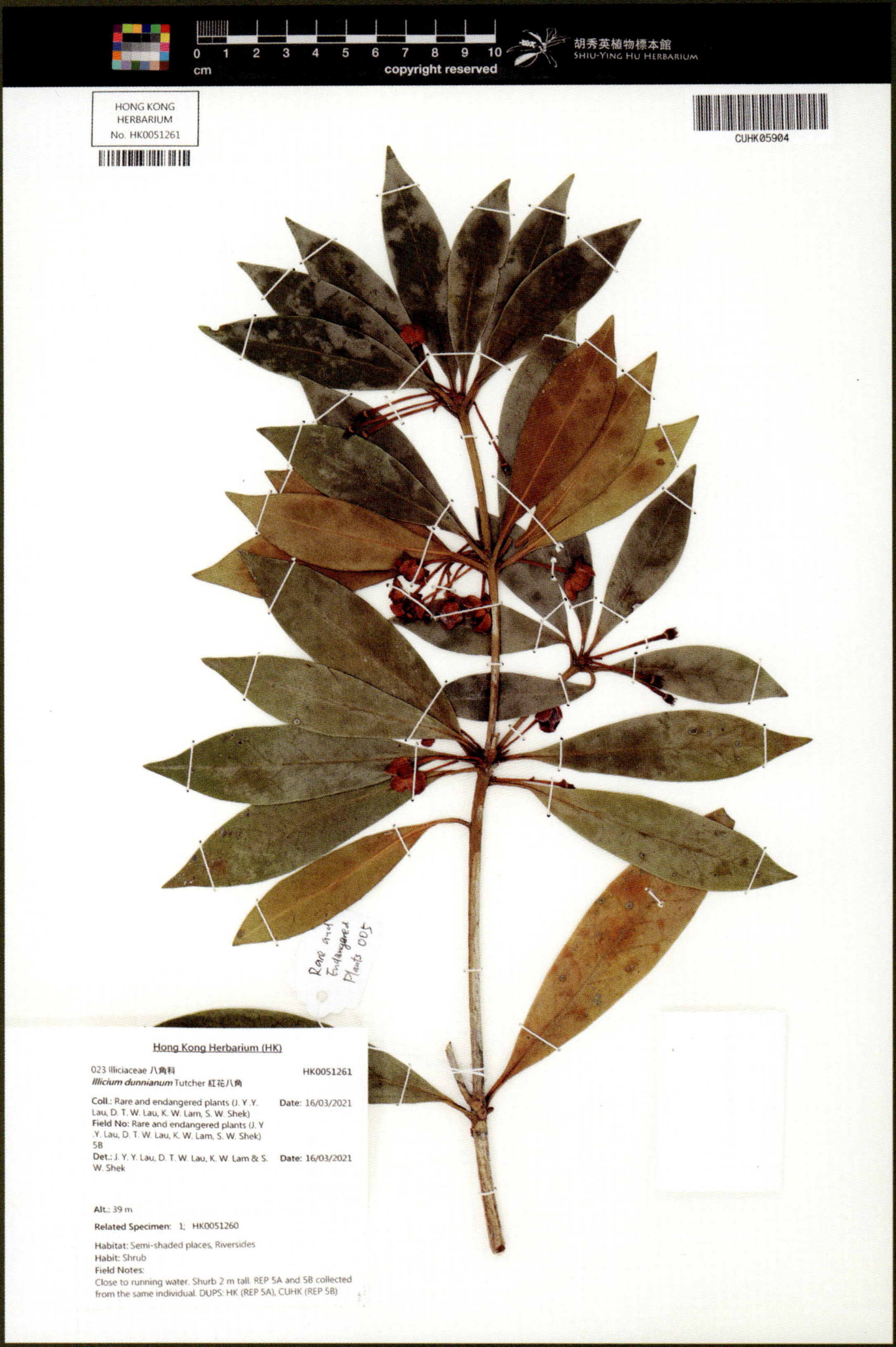

採集者編號 Collector no.

Rare and endangered plants (J. Y. Y. Lau, D. T. W. Lau, K. W. Lam, S. W. Shek) 5B

採集年份 Year of collection

2021

外觀 Overview

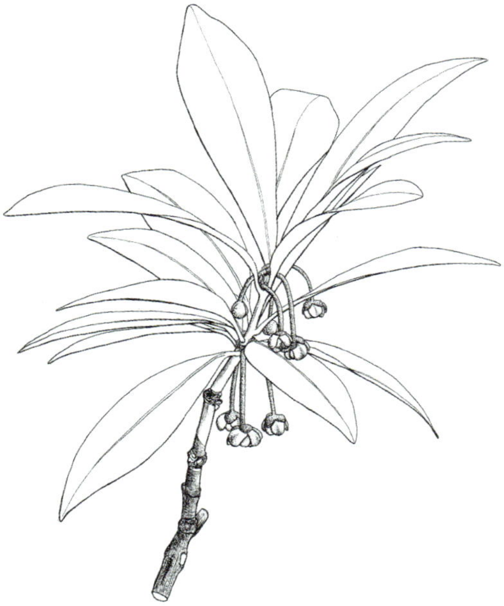

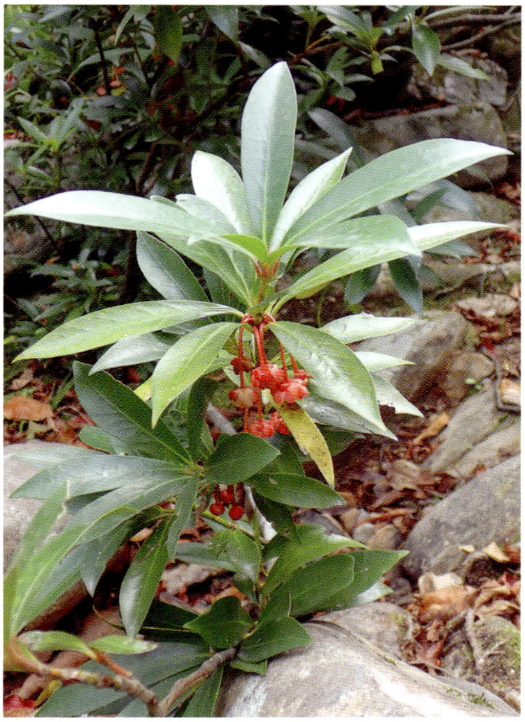

花枝 Flowering branch

葉 Leaf

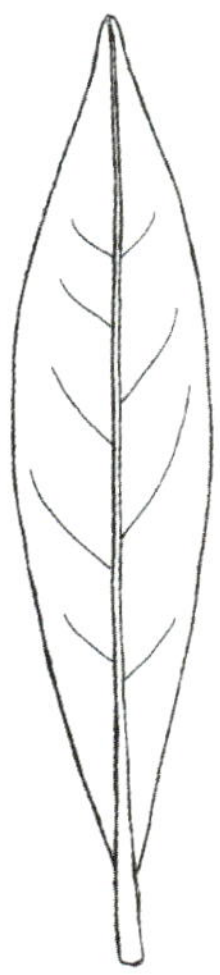

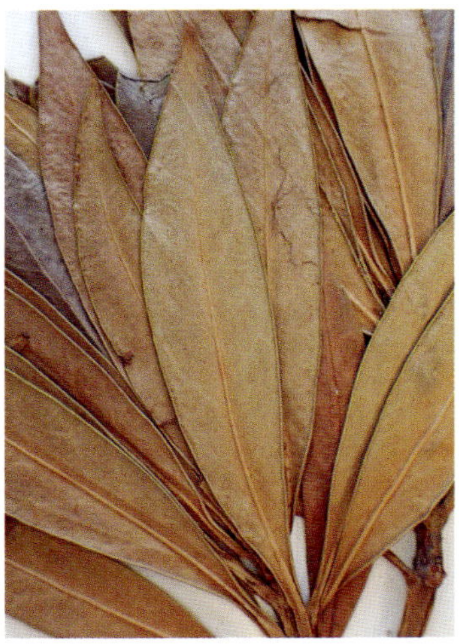

葉底 Leaves, abaxial

花 Flower

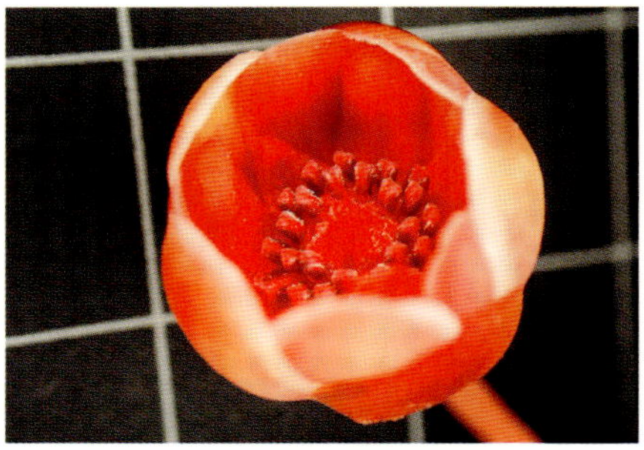
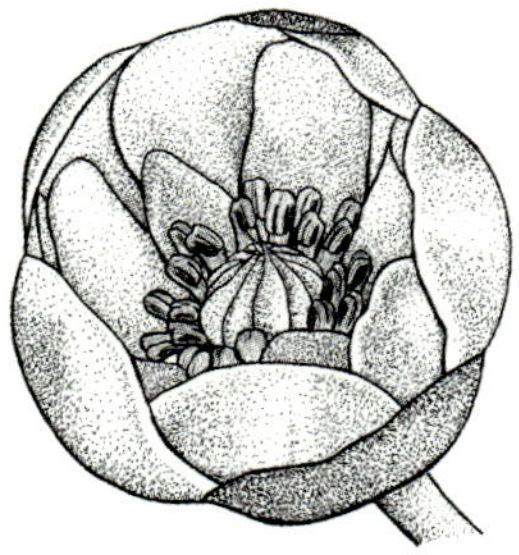

花，正面觀 Flower, frontal view

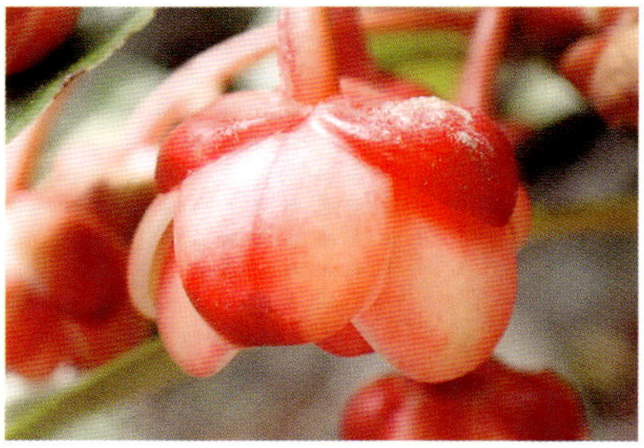
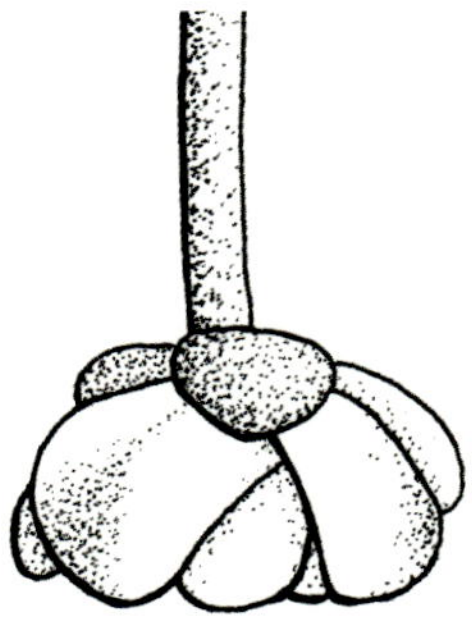

花，側面觀 Flower, lateral view

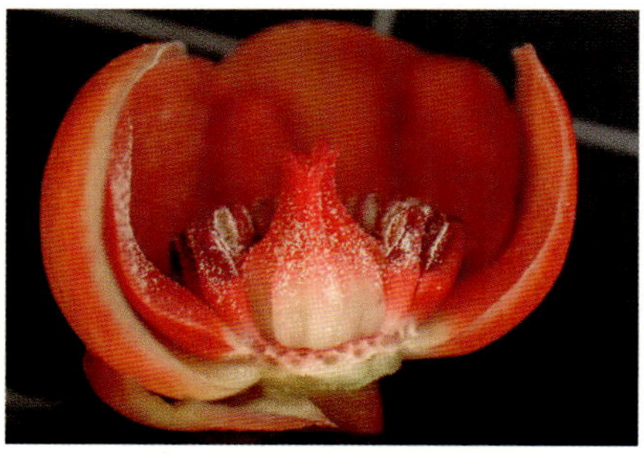
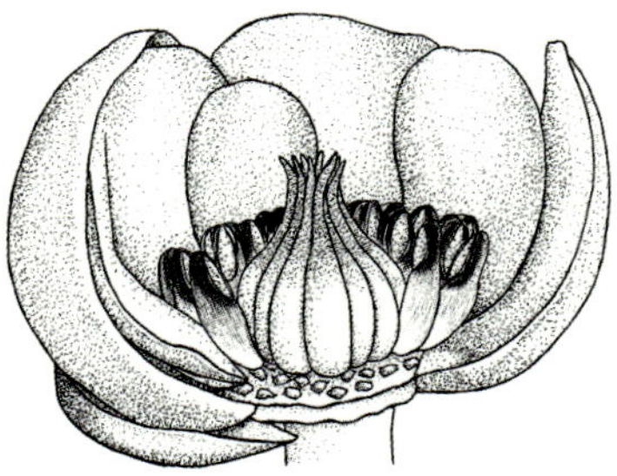

花解剖圖，展示花被片、
多條雄蕊及雌蕊
Dissected flower, showing tepals, many stamens and gynoecium

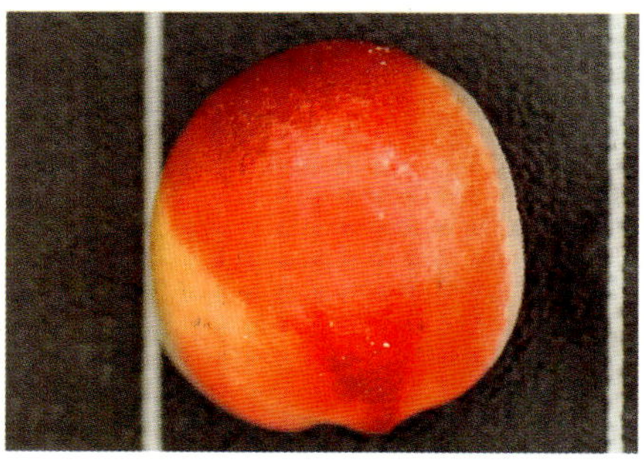
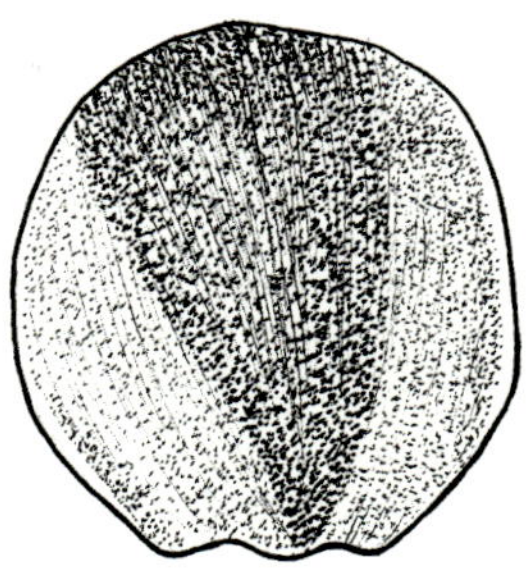

外輪花被片 Outer tepal

花 Flower

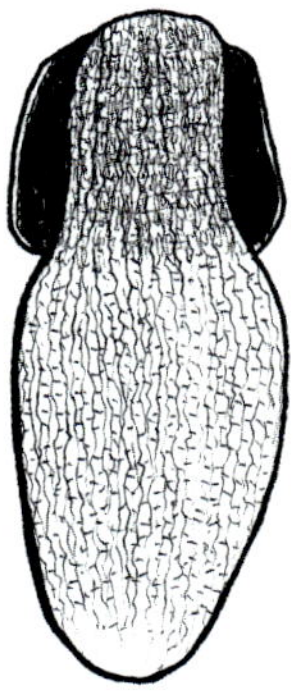
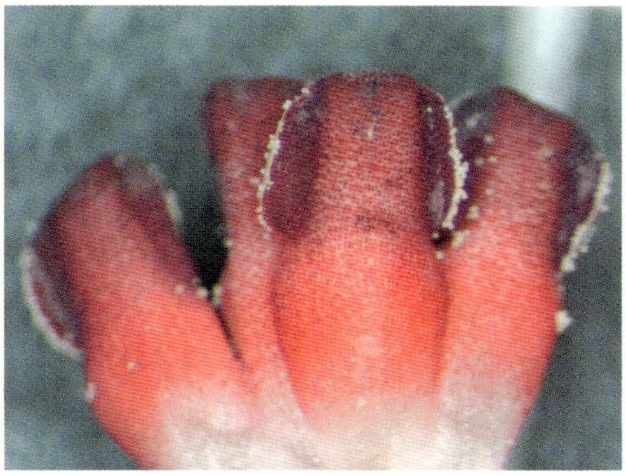

雄蕊，背軸面 Stamens, abaxial

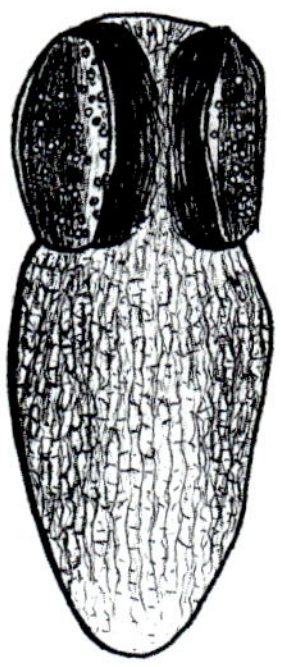
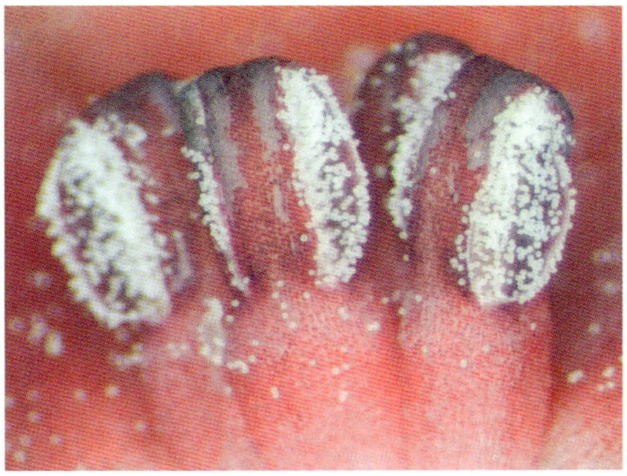

雄蕊，向軸面 Stamens, adaxial

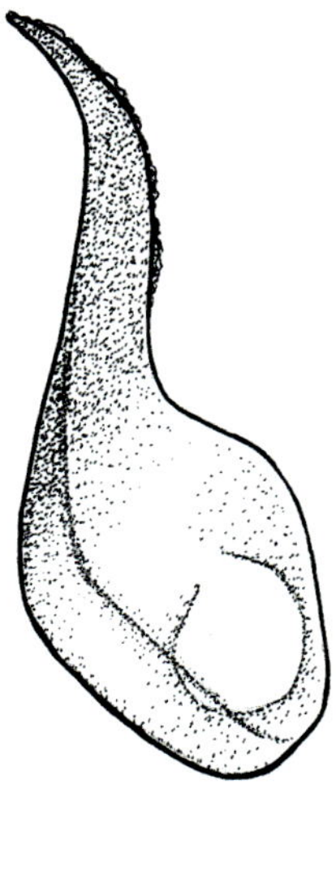
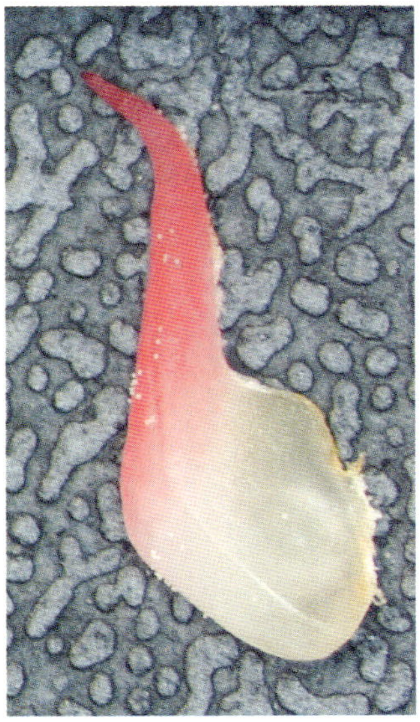

雌蕊 Pistil

Michelia chapensis Dandy
樂昌含笑

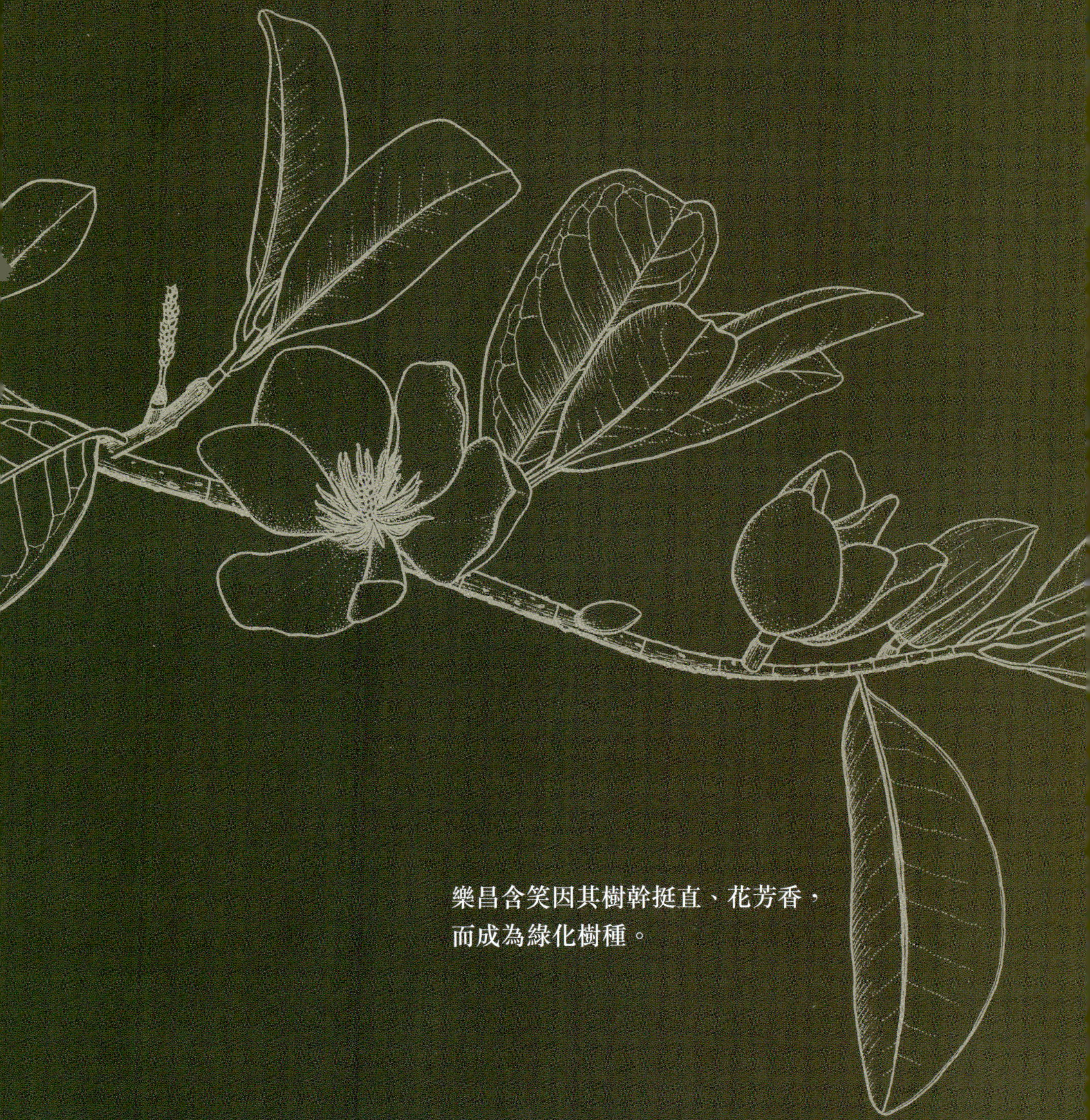

樂昌含笑因其樹幹挺直、花芳香，
而成為綠化樹種。

樂昌含笑

學名	*Michelia chapensis* Dandy
中文名	樂昌含笑
科名	木蘭科
本地分布狀態	原生植物
繪圖編號	H. Y. Wong 002

花期	花色	果期	果色
3至4月	淡黃色	8至9月	未有記錄

樂昌含笑是可長至30米的喬木，常見於森林。在中國，樂昌含笑除了是夾板及紙漿的原料外，亦因其樹幹挺直、花芳香，而成為綠化樹種。

樂昌含笑於《香港稀有及珍貴植物》中屬「無危」，在國際自然保護聯盟瀕危物種紅色名錄中亦被列為「無危」。

樂昌含笑受香港的《林務規例》(第96A章) 規管。

《中國植物誌》(英文版) 的採用名稱

Michelia chapensis Dandy (科名：木蘭科)

《香港植物誌》的採用名稱

Michelia chapensis Dandy (科名：木蘭科)

參考文獻 References

1. Hong Kong Herbarium. (2021). *HK Herbarium-Michelia chapensis*. HK Plant Database. Retrieved March 21, 2021, from https://www.herbarium.gov.hk/en/hk-plant-database/plant-detail/index.html?pType=species&oID=3435
2. Hu, D. H., Wei, R. P., Zhang, Z. X., Deng, F. F., Xu, J. X., Liu, Y. Z., Wang, A. J., & Deng, R. B. (2006). Study on Sowing Nursery of Provenance and Family for *Michelia chapensis* Dandy. *Guangdong Forestry Science And Technology 22*(3), 1–7. https://doi.org/10.3969/j.issn.1006-4427.2006.03.001
3. IUCN. (2014). *Magnolia chapensis*. The IUCN Red List of Threatened Species. Retrieved March 21, 2021, from https://www.iucnredlist.org/species/191491/1985501

Michelia chapensis Dandy

Scientific name	*Michelia chapensis* Dandy
Chinese name	樂昌含笑
Family	Magnoliaceae Juss.
Local distribution status	Native species
Illustration number	H. Y. Wong 002

FLOWERING PERIOD	COLOUR OF FLOWER	FRUITING PERIOD	COLOUR OF FRUIT
March to April	Pale yellow	August to September	Not recorded

Michelia chapensis Dandy is a tree, growing up to 30 m. It can be found in forests. In mainland China, *Michelia chapensis* Dandy is not only used as plywood and pulp, but is also used in greening due to its straight trunk and fragrant flowers.

Michelia chapensis Dandy is classified as 'Least Concern' in *Rare and Precious Plants of Hong Kong* and is listed as 'Least Concern' in the IUCN Red List of Threatened Species.

It is regulated under Hong Kong's Forestry Regulations (Cap. 96A).

Name in *Flora of China*
Michelia chapensis Dandy (Family: Magnoliaceae)

Name in *Flora of Hong Kong*
Michelia chapensis Dandy (Family: Magnoliaceae)

4. Missouri Botanical Garden. (2021). *Name-Michelia chapensis.* Tropicos. Retrieved March 20, 2021, from https://www.tropicos.org/name/19300505
5. Xia, N. H., Liu, Y. H., & Nooteboom, H. P. (2008). Magnoliaceae. In Z. Y. Wu, P. H. Raven & D. Y. Hong (Eds.), *Flora of China* (Vol. 7, pp. 48–91). Science Press & Missouri Botanical Garden Press.

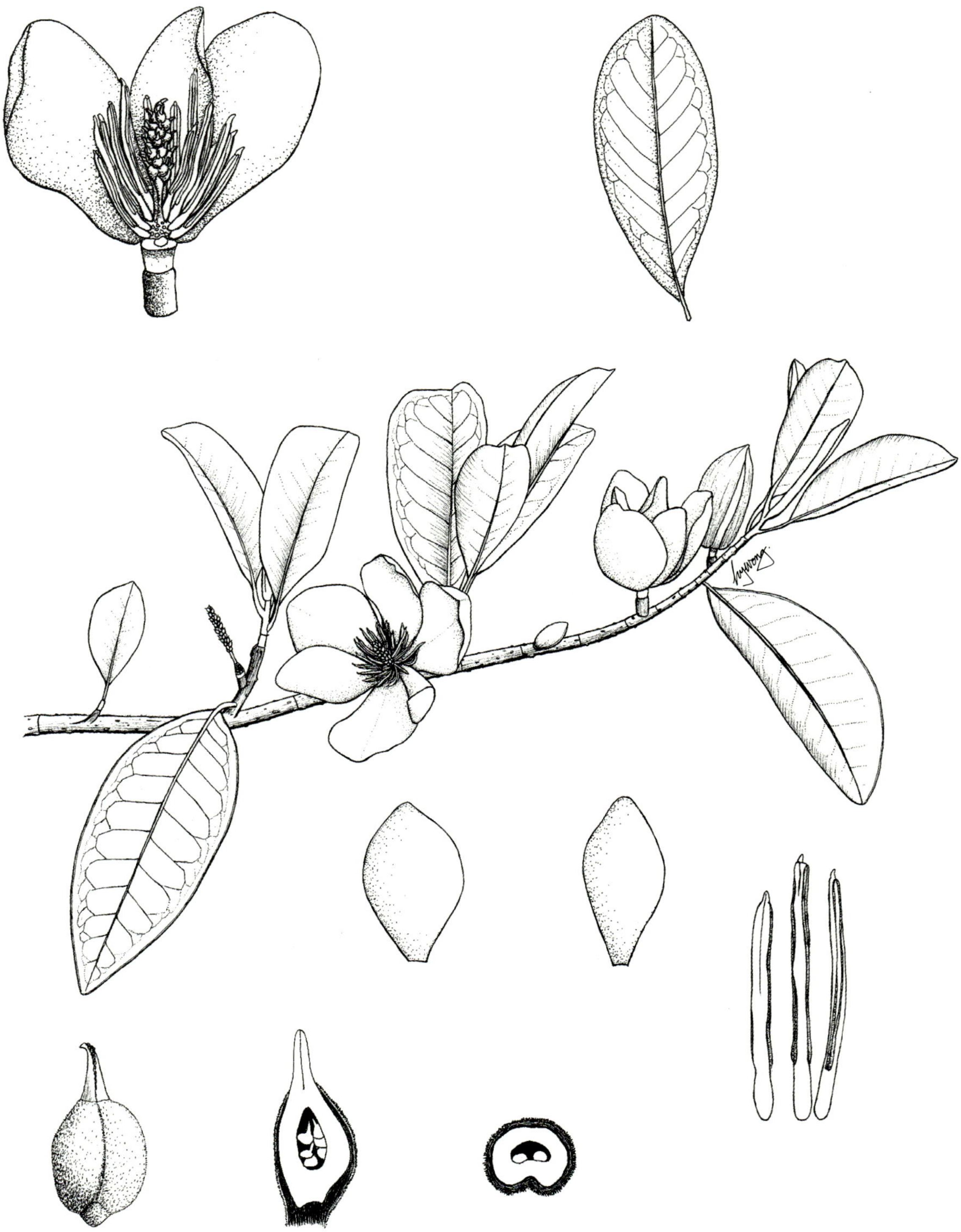

Michelia chapensis Dandy 樂昌含笑
Illustration 002 by H. Y. Wong
Authentication by D. T. W. Lau

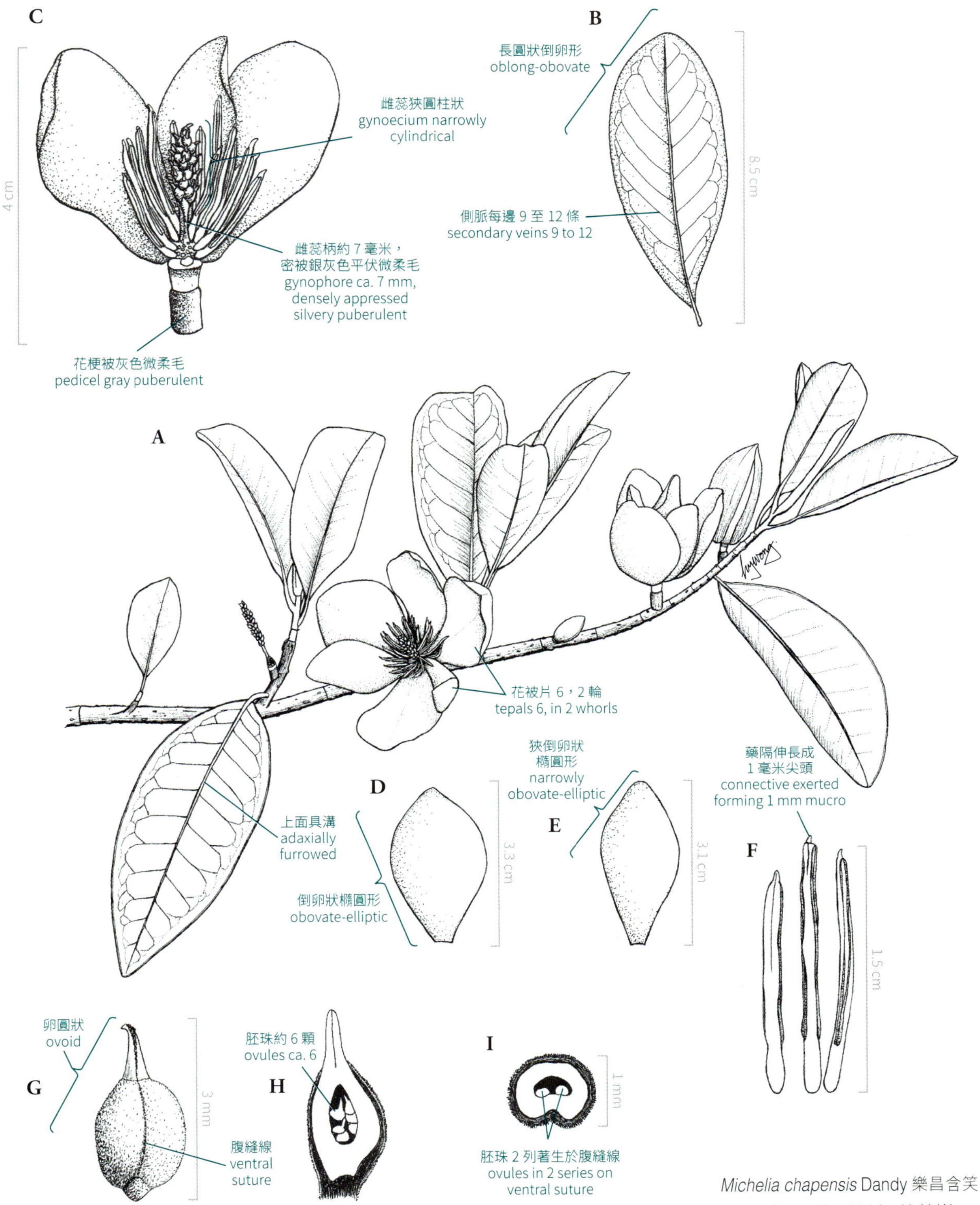

Michelia chapensis Dandy 樂昌含笑

Illustration 002 by H. Y. Wong
Authentication by D. T. W. Lau

植物結構 PLANT STRUCTURES:
A. 花枝 Flowering branch **B.** 葉底 Leaf, abaxial **C.** 花解剖圖，展示 3 片花被片、多條雄蕊及雌蕊 Dissected flower, showing 3 tepals, many stamens and gynoecium **D.** 外輪花被片 Outer tepal **E.** 內輪花被片 Inner tepal **F.** 雄蕊 Stamens **G.** 雌蕊 Pistil **H.** 子房，縱切面 Ovary, in L. S. **I.** 子房，橫切面 Ovary, in C. S.

參考標本 REFERENCE SPECIMENS: Rare and endangered plants (J. Y. Y. Lau, D. T. W. Lau, K. W. Lam, S. W. Shek) 2B; 周仁章 023260 (SZG00009170)

參考文獻 REFERENCES:
Law, Y. W. (1987). Magnoliaceae. In South China Botanical Garden, the Chinese Academy of Sciences (Ed.), *Flora of Guangdong* (Vol. 1, pp. 1–21). Guangdong Science and Technology Press.
Law, Y. W., Wu, Y. F., & Chang, B. N. (1996). Magnoliaceae. In Delectis Florae Reipublicae Popularis Sinicae Agendae Academiae Sinicae Edita (Ed.), *Flora Reipublicae Popularis Sinicae* (Vol. 30(1), pp. 82–269). Science Press.
Stevens, P. F. (2021). Magnoliales. Angiosperm Phylogeny Website. Retrieved March 20, 2021, from http://www.mobot.org/MOBOT/research/APweb/welcome.html
Wang, Y. L., & Zhou, R. Z. (2017). Magnoliaceae. In Fairy Lake Botanical Garden, Shenzhen & Chinese Academy of Sciences (Eds.), *Flora of Shenzhen* (Vol. 1, pp. 266–283). China Forestry Publishing House.
Xia, N. H. 2007. Magnoliaceae. In Hong Kong Herbarium, AFCD & South China Botanical Garden, CAS (Eds.), *Flora of Hong Kong* (Vol. 1, pp. 25–30). Agriculture, Fisheries and Conservation Department, HKSAR Government.
Xia, N. H., Liu, Y. H., & Nooteboom, H. P. (2008). Magnoliaceae. In Z. Y. Wu, P. H. Raven & D. Y. Hong (Eds.), *Flora of China* (Vol. 7, pp. 48–91). Science Press & Missouri Botanical Garden Press.

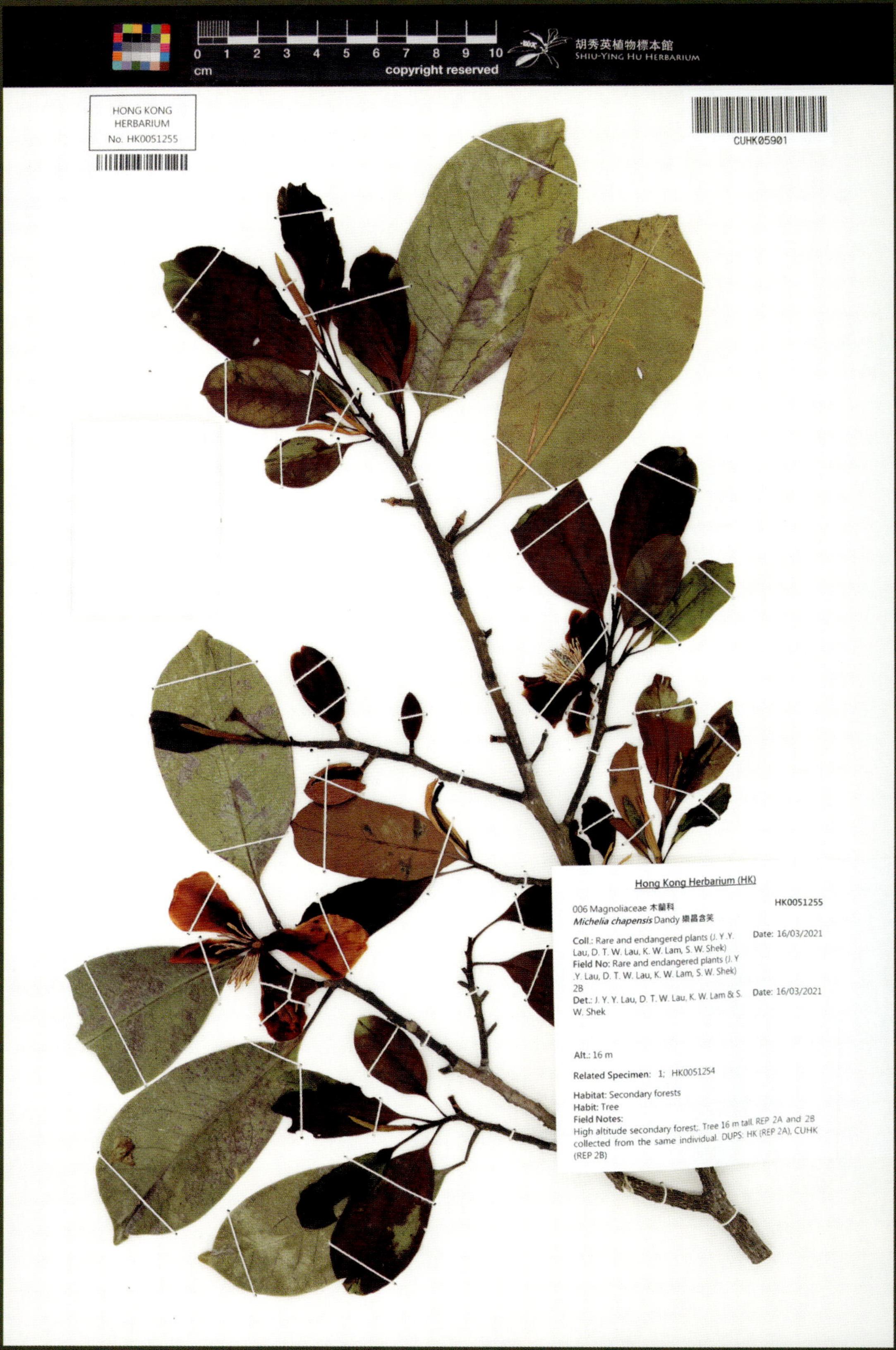

採集者編號 Collector no.

Rare and endangered plants (J. Y. Y. Lau, D. T. W. Lau, K. W. Lam, S. W. Shek) 2B

採集年份 Year of collection

2021

外觀 Overview

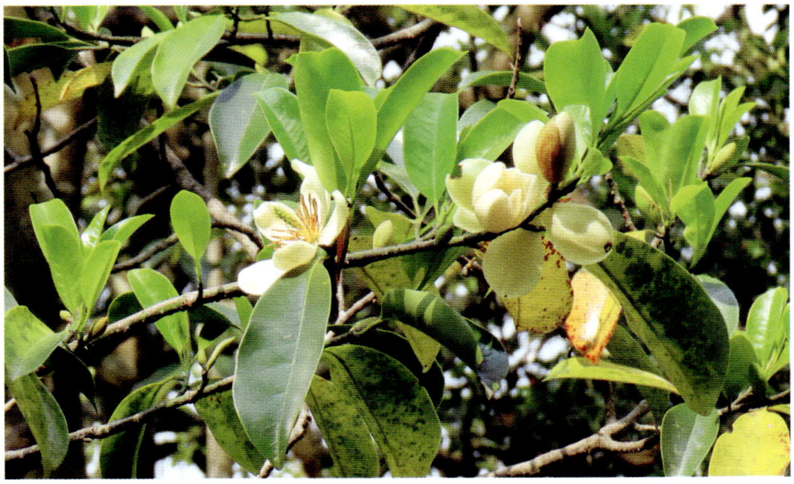

花枝 Flowering branch

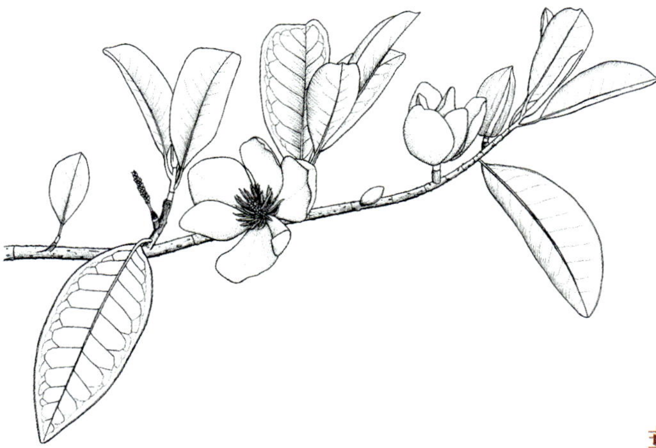

葉 Leaf

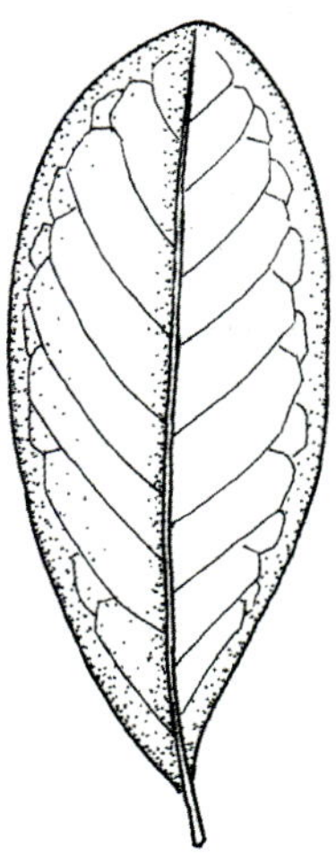

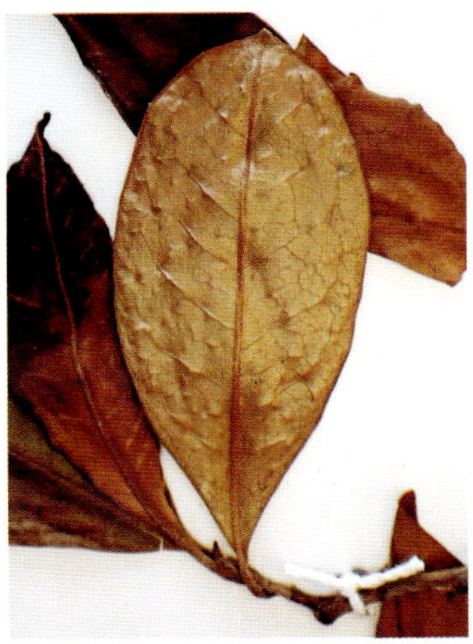

葉底 Leaf, abaxial

花 Flower

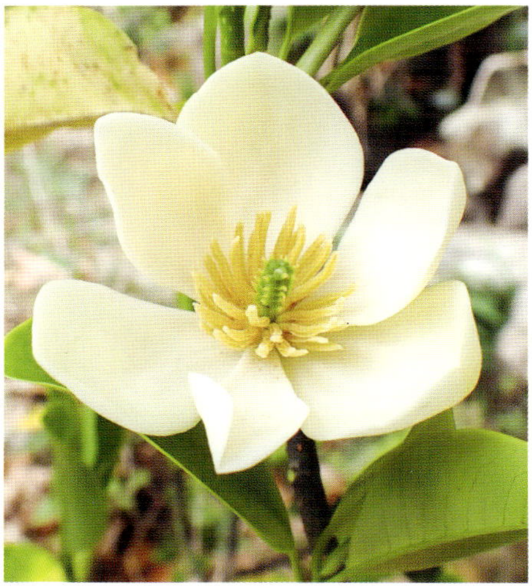
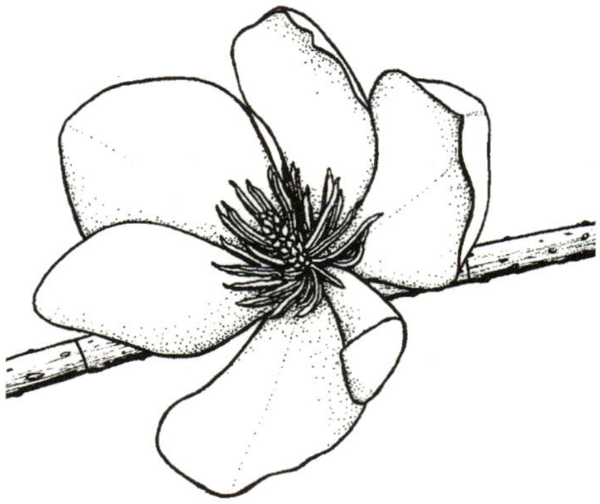

花 Flower

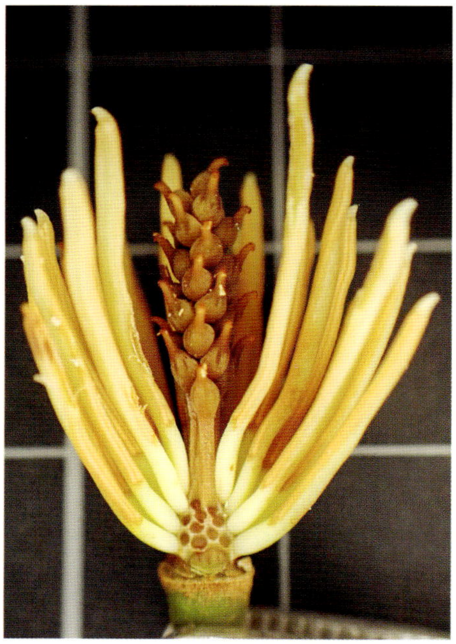
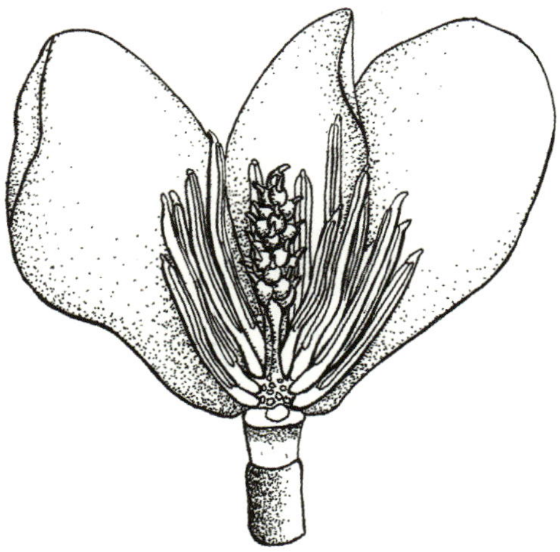

花解剖圖，展示多條雄蕊及雌蕊
Dissected flower, showing many stamens and gynoecium

花 Flower

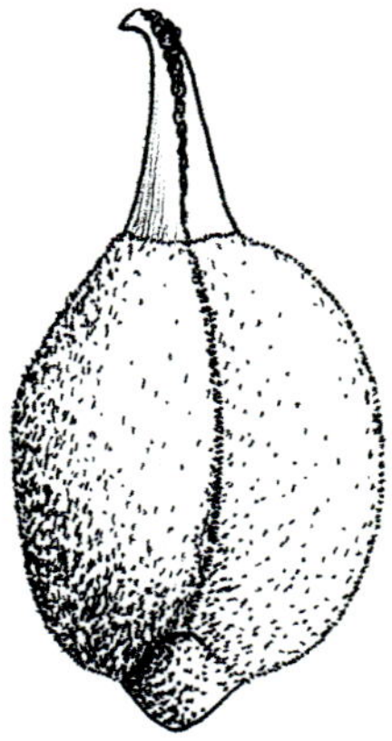

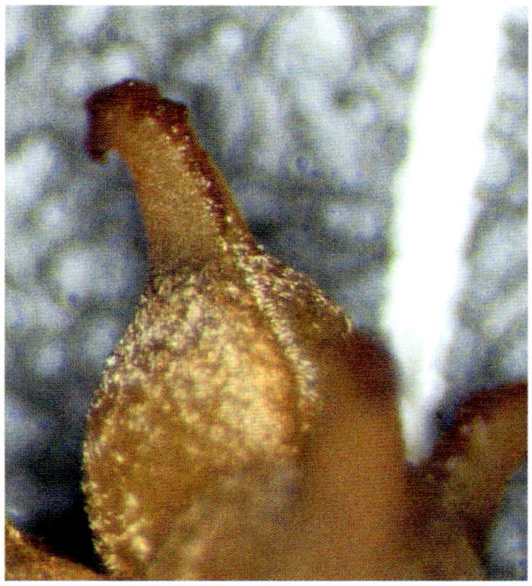

雌蕊 Pistil

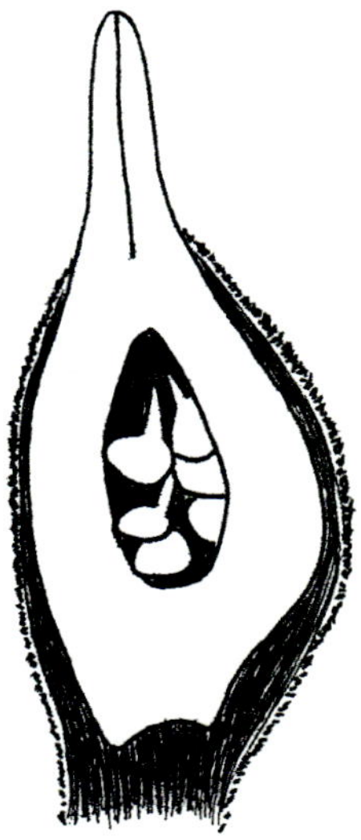

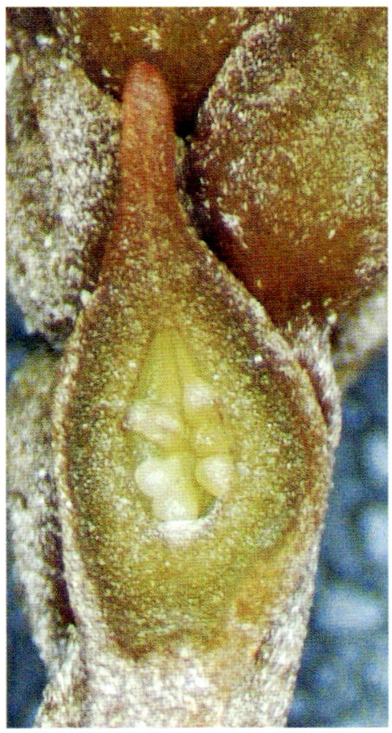

雌蕊，縱切面 Pistil, in L. S.

Artabotrys hongkongensis Hance
香港鷹爪花

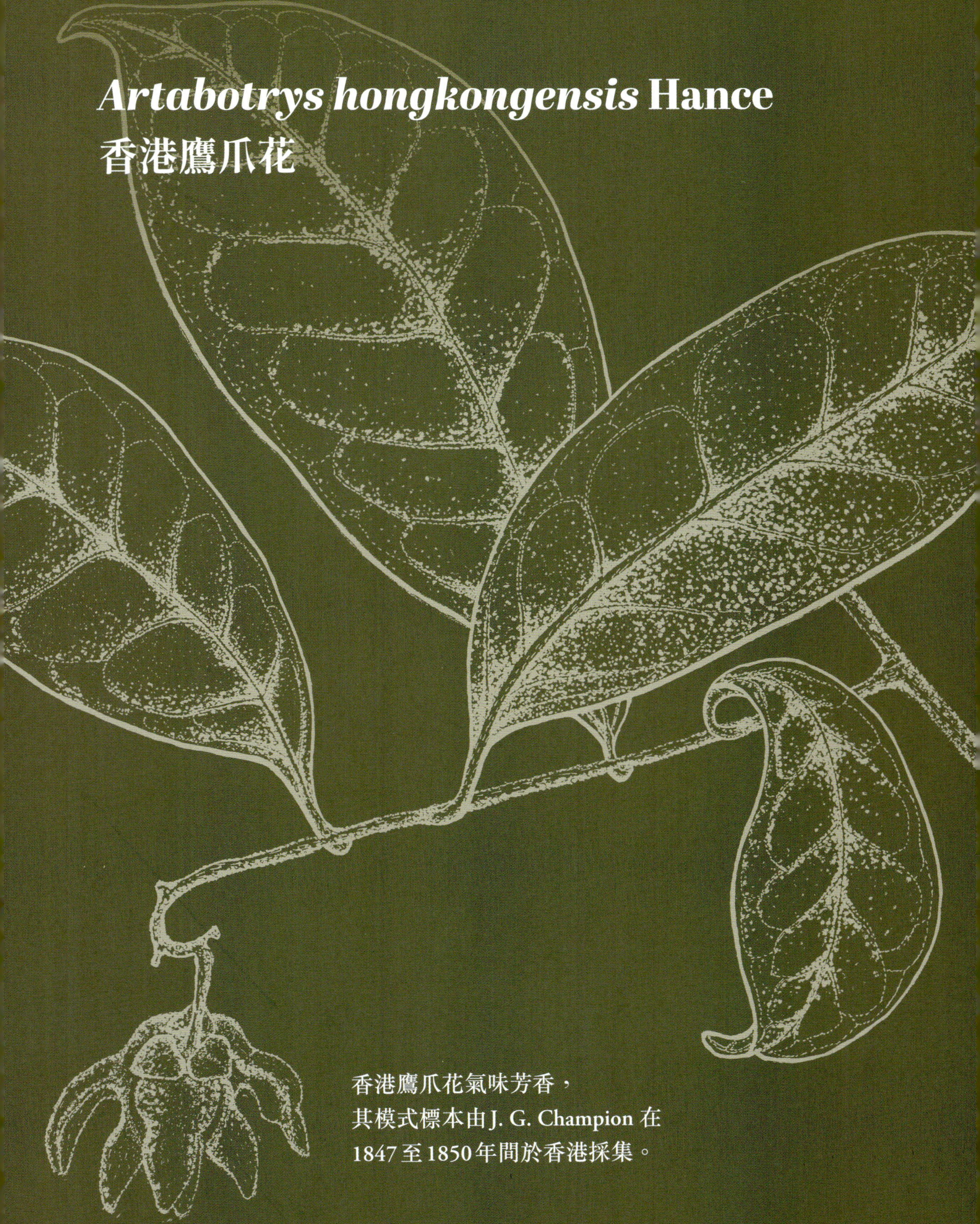

香港鷹爪花氣味芳香，
其模式標本由J. G. Champion 在
1847至1850年間於香港採集。

胡秀英教授採集的標本
Specimen Collected by Professor Shiu-Ying Hu

採集者編號
Collector no.
Hu & But 20985

採集年份
Year of collection
1992

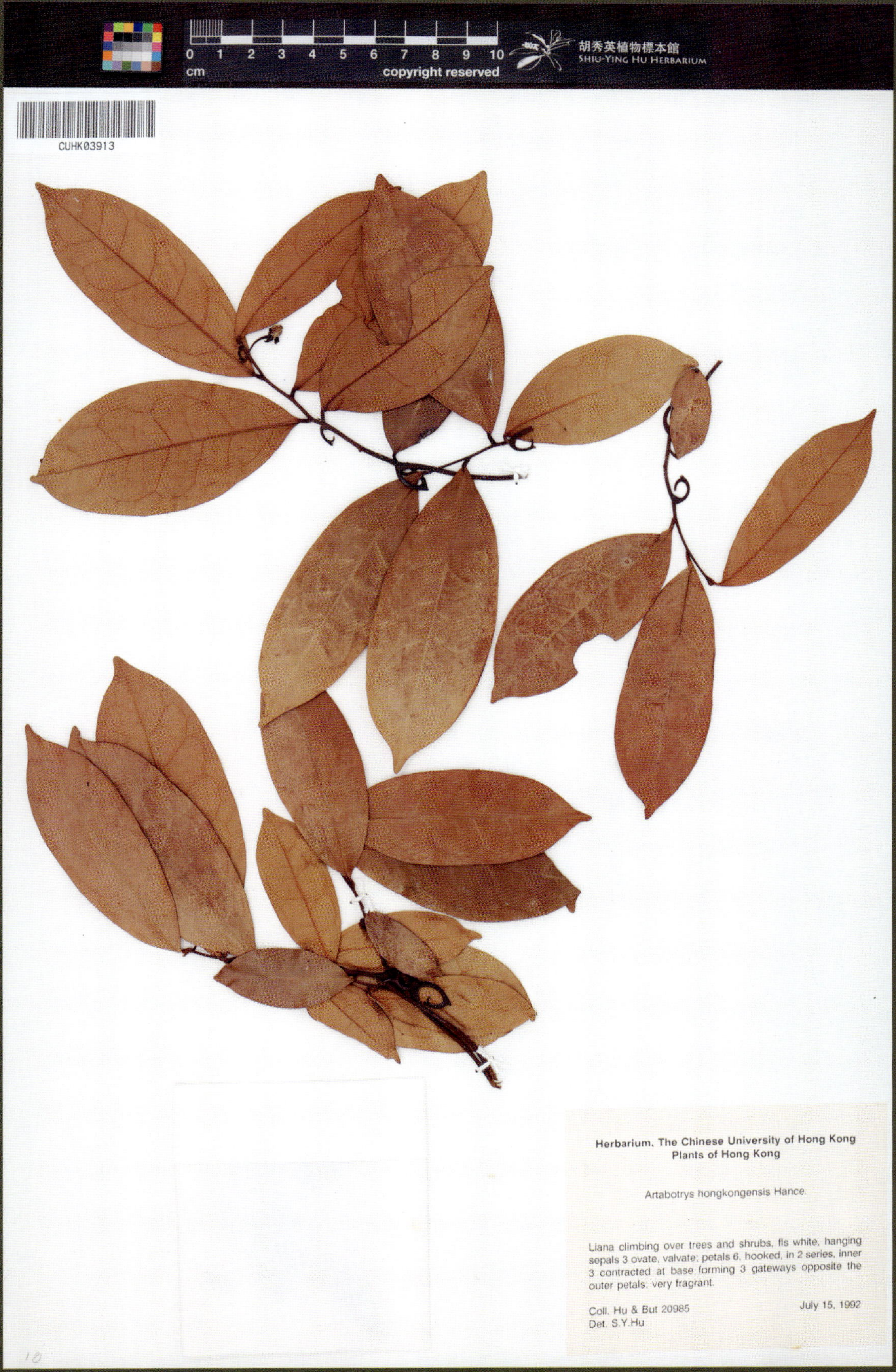
0 1 2 3 4 5 6 7 8 9 10
cm
copyright reserved
胡秀英植物標本館
SHIU-YING HU HERBARIUM
CUHK03913
Herbarium, The Chinese University of Hong Kong
Plants of Hong Kong
Artabotrys hongkongensis Hance
Liana climbing over trees and shrubs, fls white, hanging sepals 3 ovate, valvate; petals 6, hooked, in 2 series, inner 3 contracted at base forming 3 gateways opposite the outer petals; very fragrant.
Coll. Hu & But 20985
July 15, 1992
Det. S.Y.Hu
10

香港鷹爪花

學名 *Artabotrys hongkongensis* Hance
中文名 香港鷹爪花
科名 番荔枝科
本地分布狀態 原生植物
繪圖編號 M. C. Li 007

花期	花色	果期	果色
5至8月	黃色	8至翌年2月	枯乾時黑色

香港鷹爪花是其中一種以香港命名的攀援灌木，其模式標本由 J. G. Champion 在 1847 至 1850 年間於香港銅鑼灣採集。

香港鷹爪花氣味芳香，同科的依蘭（*Cananga odorata* (Lam.) Hook. f. & Thomson）所萃取的精油，是製造香水的原材料。

香港鷹爪花於《香港稀有及珍貴植物》中屬「無危」級別。

《中國植物誌》（英文版）的採用名稱
Artabotrys hongkongensis Hance（科名：番荔枝科）

《香港植物誌》的採用名稱
Artabotrys hongkongensis Hance（科名：番荔枝科）

參考文獻 References

1. Ekundayo, O. (1989). A review of the volatiles of the Annonaceae. *Journal of Essential Oil Research 1*(5), 223–245. https://doi.org/10.1080/10412905.1989.9697789
2. Hong Kong Herbarium. (2022). *HK Herbarium-Artabotrys hongkongensis*. HK Plant Database. Retrieved March 17, 2022, from https://herbarium.gov.hk/en/hk-plant-database/plant-detail/index.html?pType=species&oID=3727
3. Li, B. T., & Gilbert, M. G. (2011). Annonaceae. In Z. Y. Wu, P. H. Raven & D. Y. Hong (Eds.), *Flora of China* (Vol. 19, pp. 672–713). Science Press & Missouri Botanical Garden Press.

Artabotrys hongkongensis Hance

Scientific name	*Artabotrys hongkongensis* Hance
Chinese name	香港鷹爪花
Family	Annonaceae Juss.
Local distribution status	Native species
Illustration number	M. C. Li 007

FLOWERING PERIOD	COLOUR OF FLOWER	FRUITING PERIOD	COLOUR OF FRUIT
May to August	Yellow	August to following February	Black when dry

Artabotrys hongkongensis Hance is one of the scandent shrubs named after Hong Kong. The type specimen was collected by J. G. Champion between 1847 and 1850 in Causeway Bay, Hong Kong.

Artabotrys hongkongensis Hance can produce pleasant smell. Essential oils are usually extracted from *Cananga odorata* (Lam.) Hook. f. & Thomson of the same family and are used as the raw ingredients to produce perfume.

Artabotrys hongkongensis Hance is listed as 'Least Concern' in *Rare and Precious Plants of Hong Kong.*

Name in *Flora of China*
Artabotrys hongkongensis Hance (Family: Annonaceae)

Name in *Flora of Hong Kong*
Artabotrys hongkongensis Hance (Family: Annonaceae)

4. Missouri Botanical Garden. (2022a). *Name-Artabotrys hongkongensis.* Tropicos. Retrieved March 17, 2022, from https://www.tropicos.org/name/1601649
5. Missouri Botanical Garden. (2022b). *Name-Cananga odorata*. Tropicos. Retrieved March 17, 2022, from https://www.tropicos.org/name/1600664

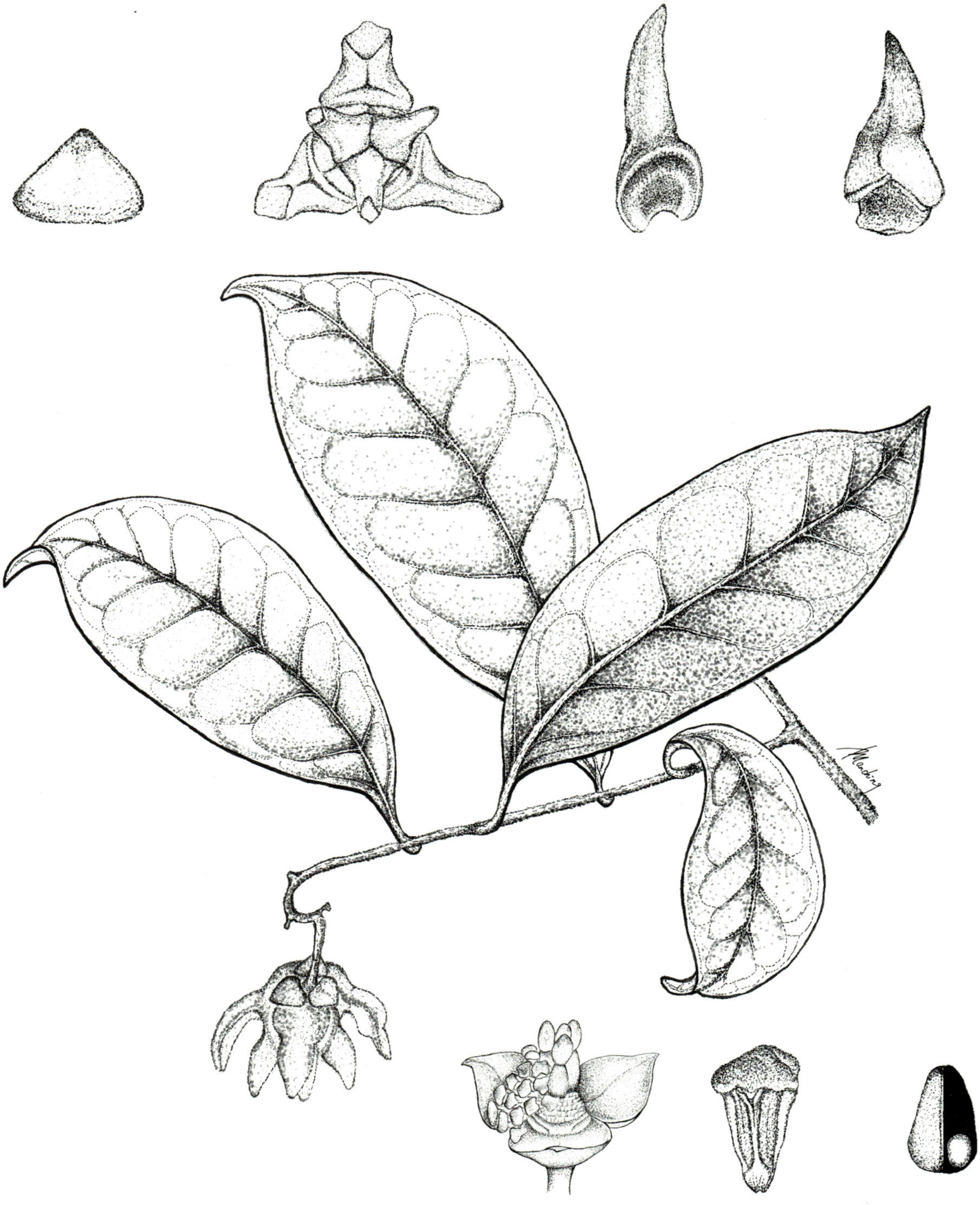

Artabotrys hongkongensis Hance 香港鷹爪花
Illustration 007 by M. C. Li
Authenticated by D. T. W. Lau

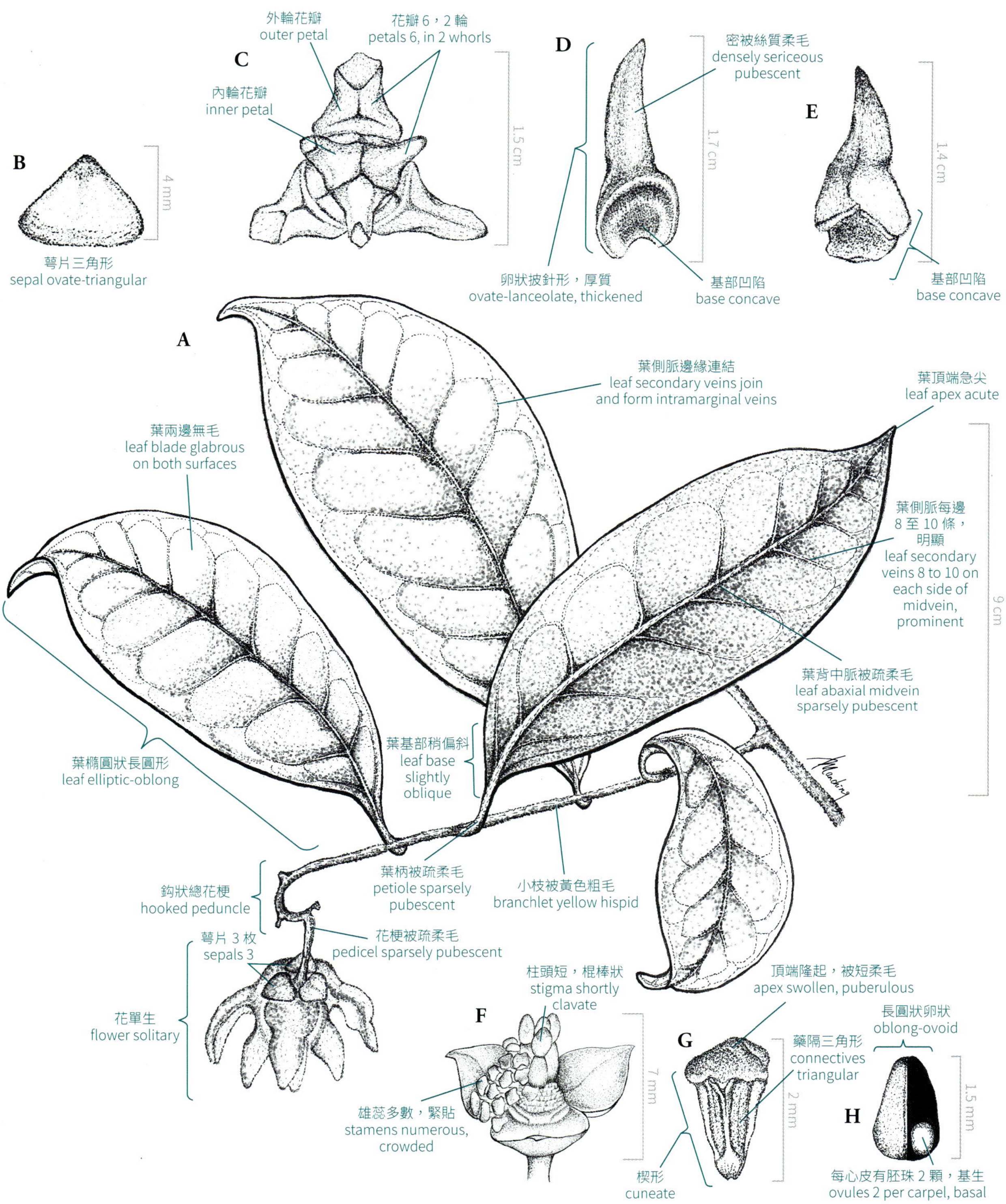

Artabotrys hongkongensis Hance 香港鷹爪花

Illustration 007 by M. C. Li

Authenticated by D. T. W. Lau

植物結構 PLANT STRUCTURES:

A. 花枝 Flowering branch **B.** 萼片 Sepal **C.** 花，正面觀 Flower, frontal view **D.** 外輪花瓣 Outer petal **E.** 內輪花瓣 Inner petal **F.** 花解剖圖，展示數枚雄蕊及雌蕊 Dissected flower, showing some stamens and gynoecium **G.** 雄蕊 Stamen **H.** 心皮 Carpel

參考標本 REFERENCE SPECIMENS: Rare and endangered plants (D. T. W. Lau, J. Y. Y. Lau, S. W. Shek, K. W. Lam) 11B; Hu & But 20985; K. T. Lam 056

參考文獻 REFERENCES:

Chen, F. H. (1991). Annonaceae. In South China Botanical Garden, the Chinese Academy of Sciences (Ed.), *Flora of Guangdong* (Vol. 2, pp. 7–40). Guangdong Science and Technology Press.

Li, B. T., & Gilbert, M. G. (2011). Annonaceae. In Z. Y. Wu, P. H. Raven & D. Y. Hong (Eds.), *Flora of China* (Vol. 19, pp. 672–713). Science Press & Missouri Botanical Garden Press.

Li, Z. K., & Li, B. T. (2017). Annonaceae. In Fairy Lake Botanical Garden, Shenzhen & Chinese Academy of Sciences (Eds.), *Flora of Shenzhen* (Vol. 1, pp. 284–297). China Forestry Publishing House.

Stevens, P. F. (2022). Magnoliales. Angiosperm Phylogeny Website. Retrieved March 17, 2022, from http://www.mobot.org/MOBOT/research/APweb/welcome.html

Tsiang, Y., & Li, P. T. (1979). Annonaceae. In Delectis Florae Reipublicae Popularis Sinicae Agendae Academiae Sinicae Edita (Ed.), *Flora Reipublicae Popularis Sinicae* (Vol. 30(2), pp. 10–175). Science Press.

Wang, R. J., & Saunders, R. M. K. (2007). Annonaceae. In Hong Kong Herbarium, AFCD & South China Botanical Garden, CAS (Eds.), *Flora of Hong Kong* (Vol. 1, pp. 30–36). Agriculture, Fisheries and Conservation Department, HKSAR Government.

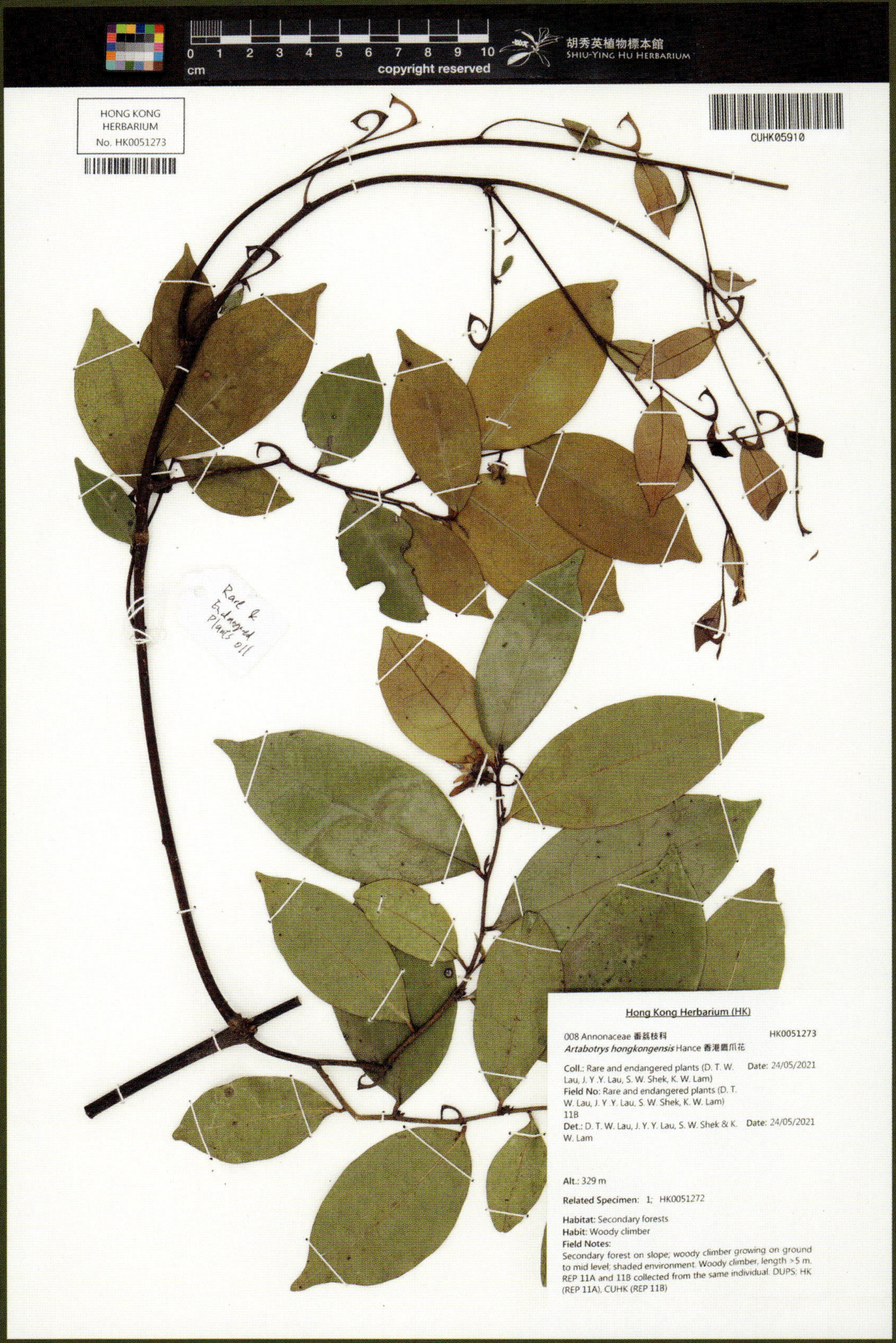

採集者編號 Collector no.

Rare and endangered plants (D. T. W. Lau, J. Y. Y. Lau, S. W. Shek, K. W. Lam) 11B

採集年份 Year of collection

2021

外觀 Overview

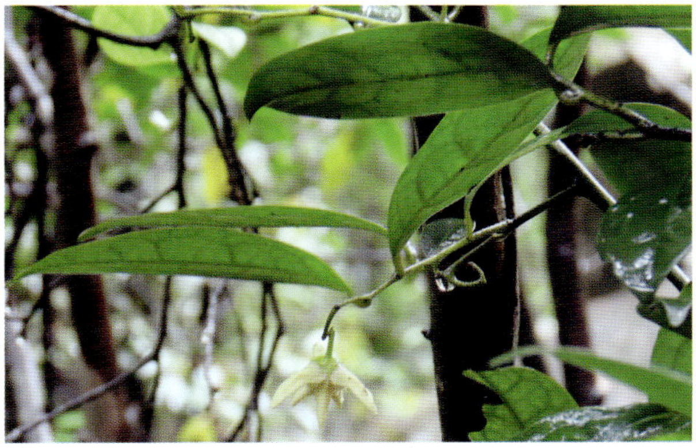

花枝 Flowering branch

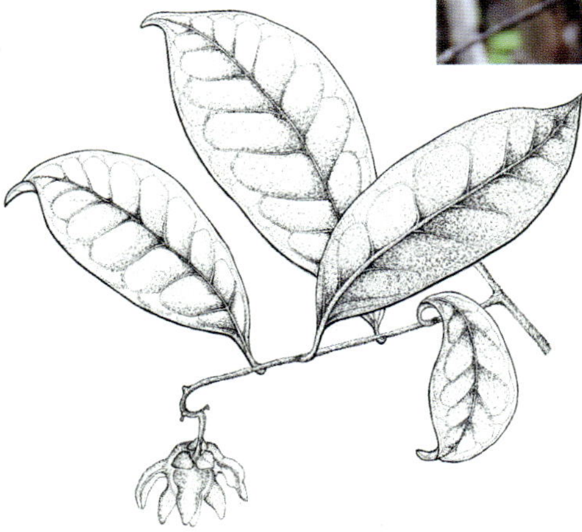

葉 Leaf

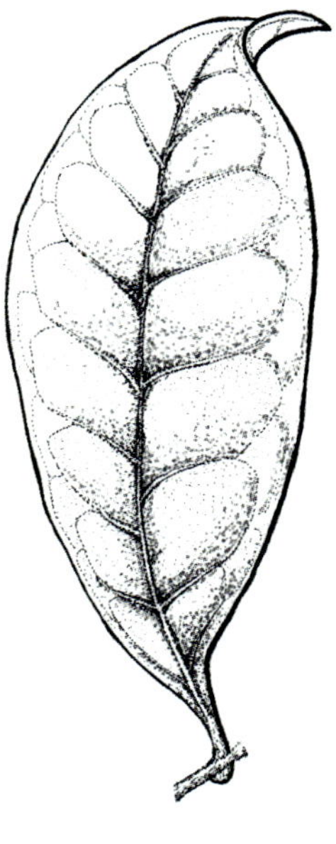

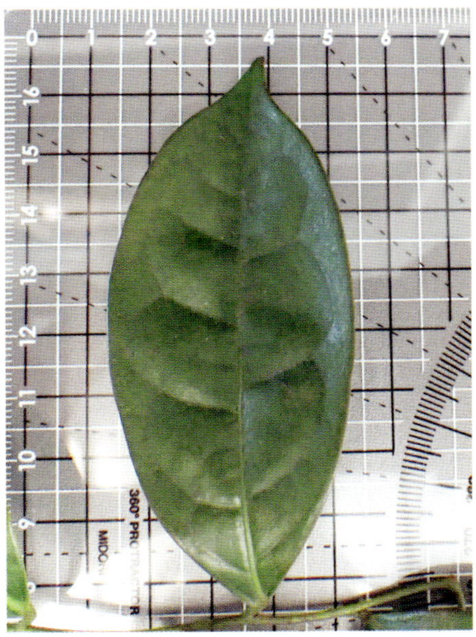

葉面 Leaf, adaxial

花 Flower

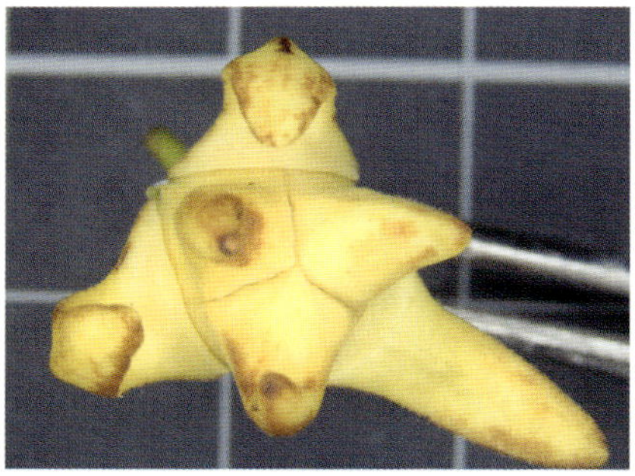

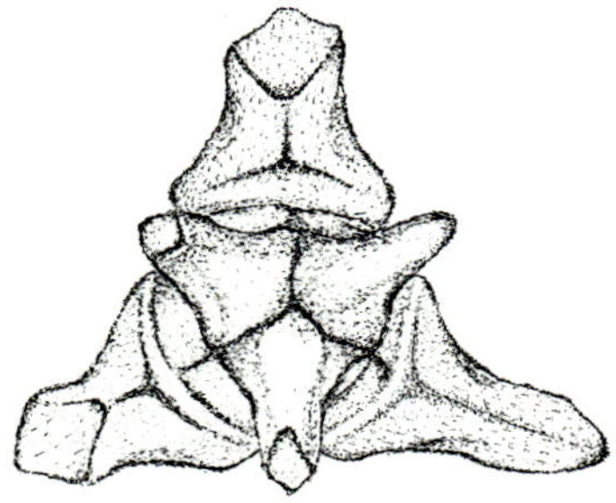

花，正面觀 Flower, frontal view

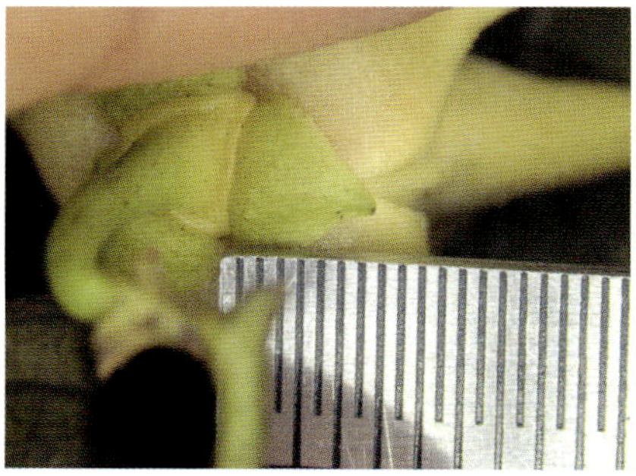

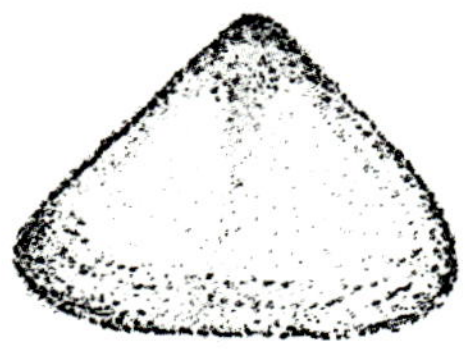

萼片 Sepal

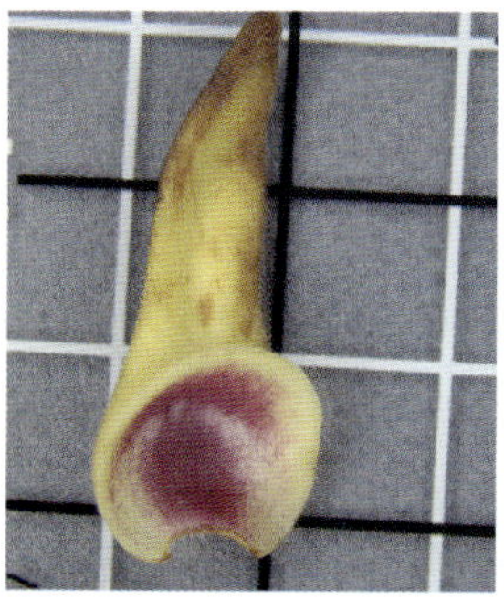

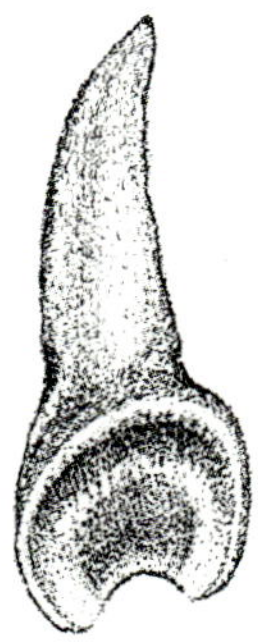

外輪花瓣 Outer petal

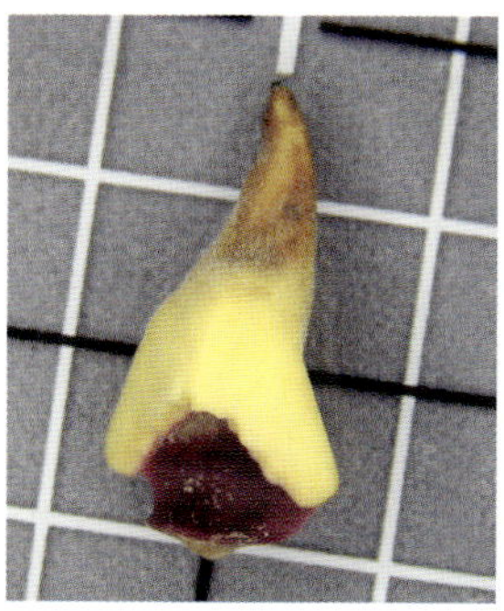

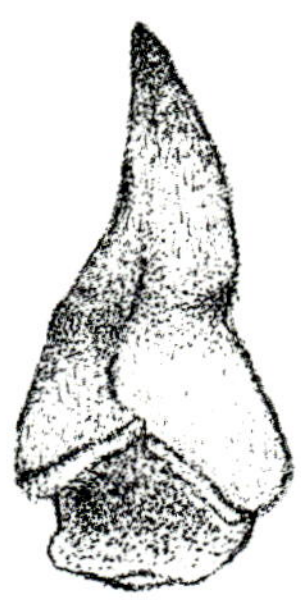

內輪花瓣 Inner petal

花 Flower

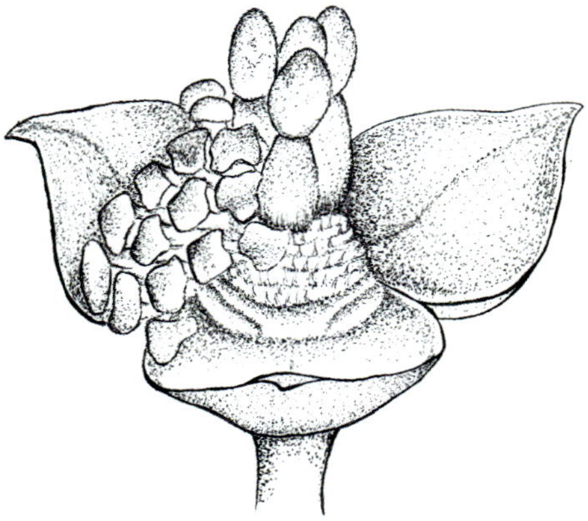
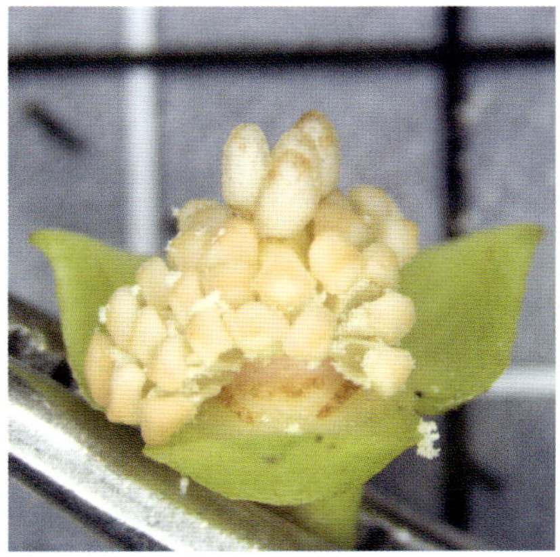

花解剖圖，展示數枚雄蕊及雌蕊
Dissected flower, showing some stamens and gynoecium

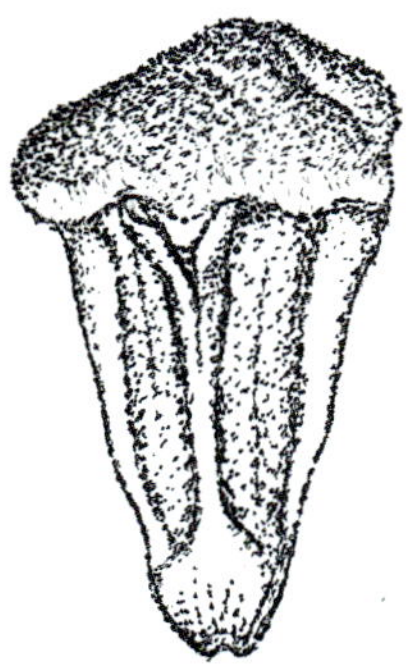
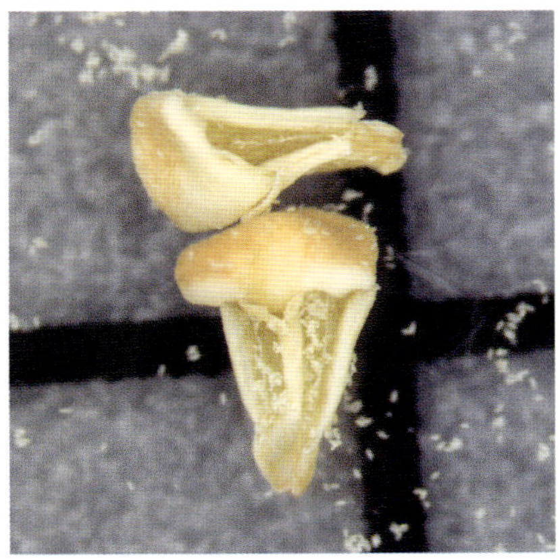

雄蕊 Stamens

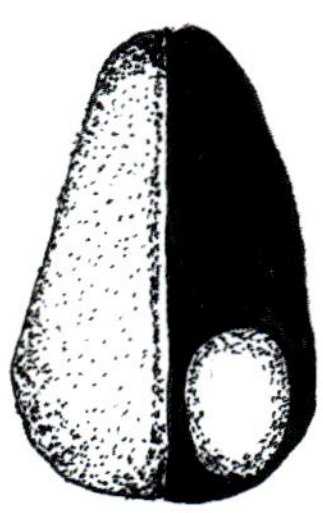
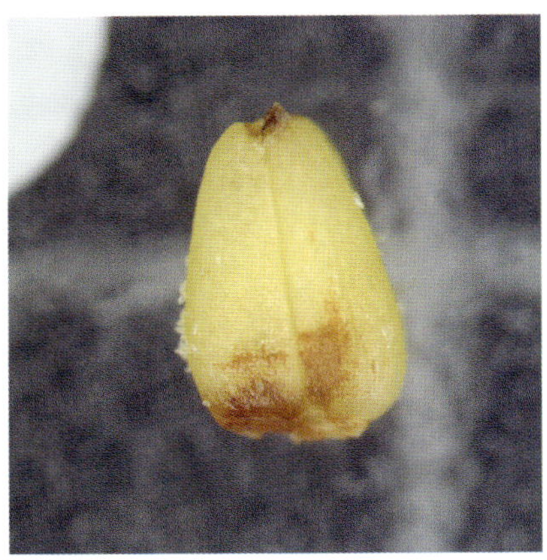

心皮 Carpel

Paris polyphylla var. *chinensis* (Franch.) H. Hara

華重樓

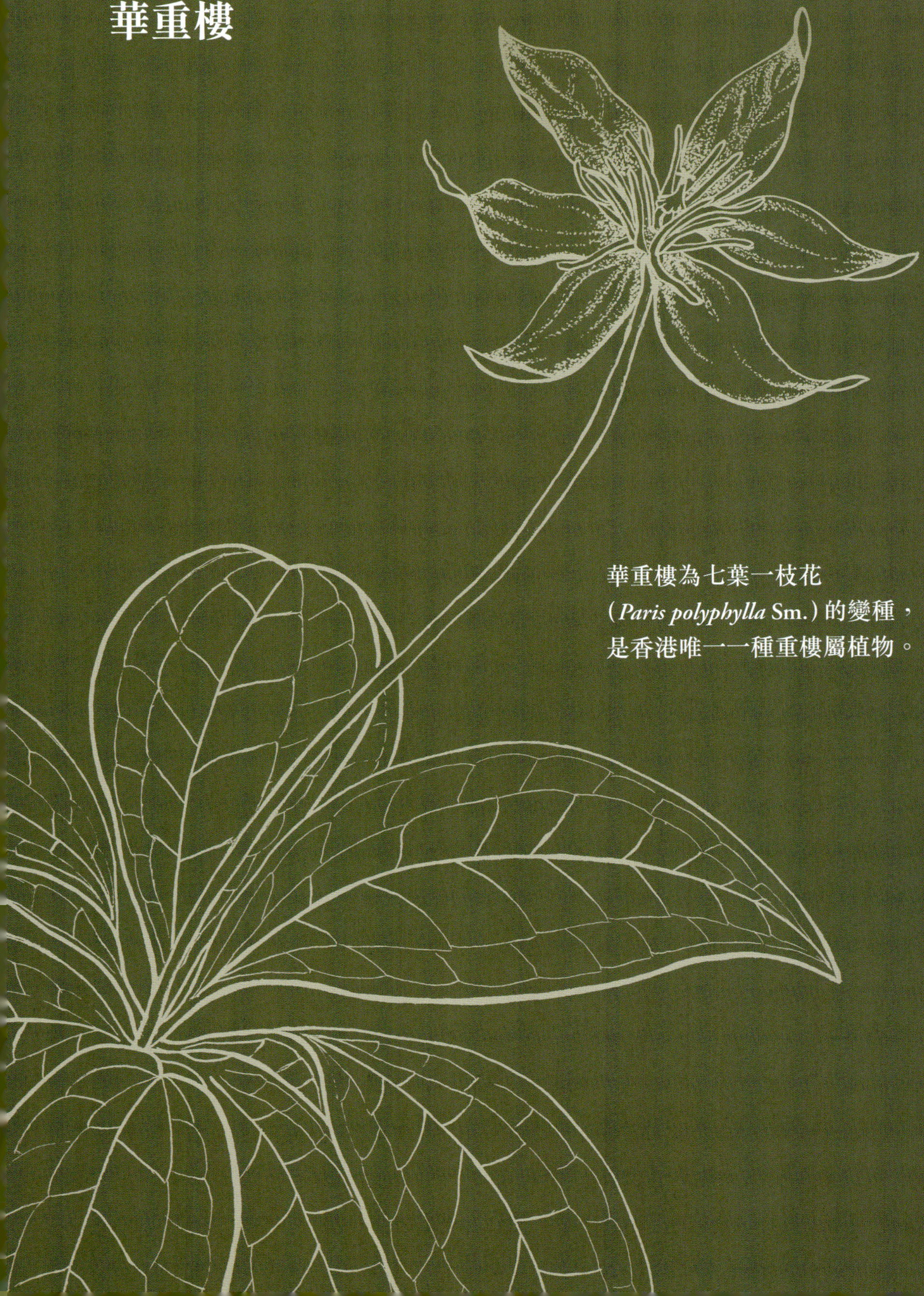

華重樓為七葉一枝花（*Paris polyphylla* Sm.）的變種，是香港唯一一種重樓屬植物。

採集者編號
Collector no.
Shiu Ying Hu 13102

採集年份
Year of collection
1973

0 1 2 3 4 5 6 7 8 9 10 cm
copyright reserved
胡秀英植物標本館
SHIU-YING HU HERBARIUM

CUHK06072

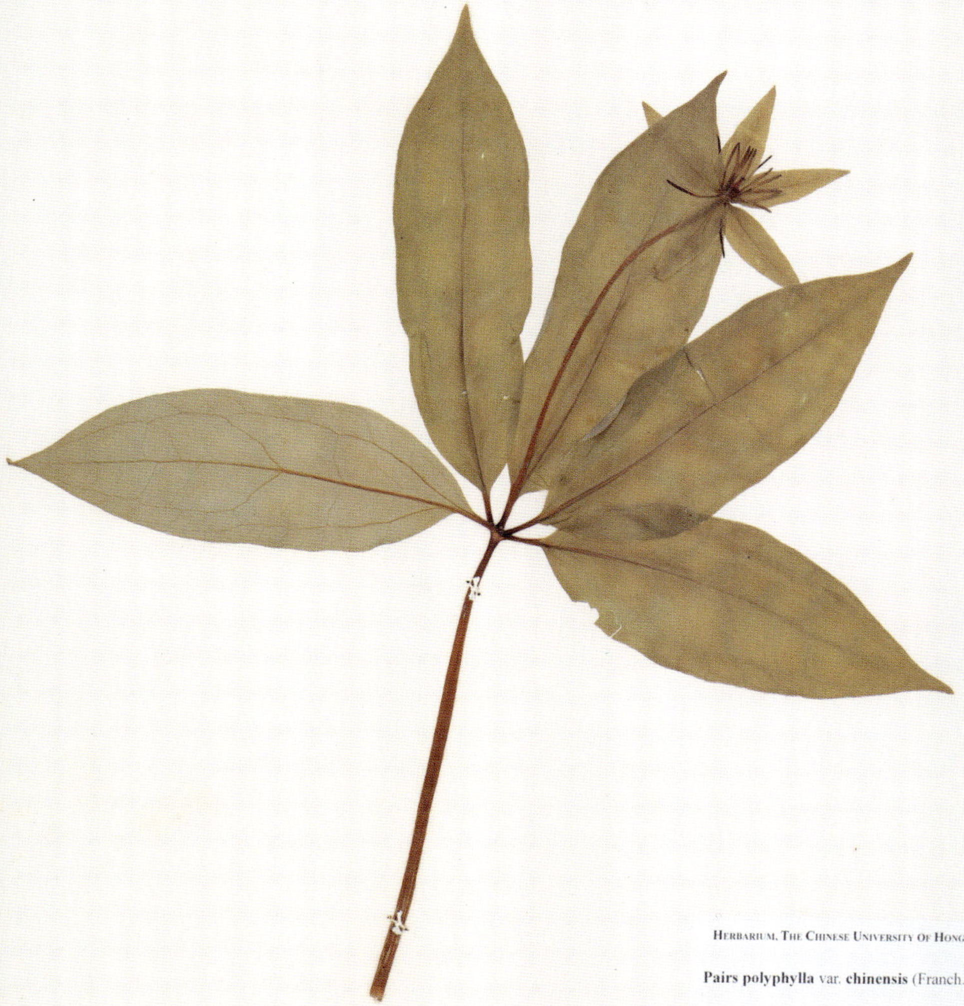

HERBARIUM, THE CHINESE UNIVERSITY OF HONG KONG

Pairs polyphylla var. **chinensis** (Franch.) Hara

(Wong Yan 13 Jun 2007)

Herbarium, The Chinese University of Hong Kong
Plants of Hong Kong

Paris chinensis Franch.

Perennial herb, growing between boulders in shade of tree, top of gorge;

Coll. Shiu Ying Hu 13102 March 17, 197X-3
Det. S. Y. Hu

華重樓

學名 *Paris polyphylla* var. *chinensis* (Franch.) H. Hara
中文名 華重樓
科名 黑藥花科
本地分布狀態 原生植物
繪圖編號 M. C. Li 002

花期	花色	果期	果色
5 至 7 月	黃綠色	8 至 11 月	未有記錄

華重樓為多年生的草本植物，主要分布於山林和峽谷，屬七葉一枝花（*Paris polyphylla* Sm.）的變種。根據香港標本室記載，華重樓是香港唯一一種重樓屬植物。

由於華重樓與其他同屬植物均具藥用價值，而被應用於傳統中醫學上，但人類的過度採摘，導致野生植物種群急遽下降。

華重樓已被列入《香港稀有及珍貴植物》中「易危」級別。

《中國植物誌》（英文版）的採用名稱
Paris polyphylla var. *chinensis* (Franchet) H. Hara（科名：百合科）

《香港植物誌》的採用名稱
Paris polyphylla Sm. var. *chinensis* (Franch.) H. Hara（科名：百合科）

參考文獻 References

1. Chen, X. Q., Liang, S. Y., Xu, J. M., & Tamura, M. N. (2000). Liliaceae. In Z. Y. Wu & P. H. Raven (Eds.), *Flora of China* (Vol. 24, pp. 73–263). Science Press & Missouri Botanical Garden Press.
2. Hong Kong Herbarium. (2022). *HK Herbarium-Paris polyphylla* var. *chinensis*. HK Plant Database. Retrieved March 17, 2022, from https://herbarium.gov.hk/en/hk-plant-database/plant-detail/index.html?pType=species&oID=7267
3. Missouri Botanical Garden. (2022a). *Name-Paris polyphylla* var. *chinensis*. Tropicos. Retrieved March 17, 2022, from https://www.tropicos.org/name/18405723

Paris polyphylla var. *chinensis* (Franch.) H. Hara

Scientific name	*Paris polyphylla* var. *chinensis* (Franch.) H. Hara
Chinese name	華重樓
Family	Melanthiaceae Batsch
Local distribution status	Native species
Illustration number	M. C. Li 002

FLOWERING PERIOD	COLOUR OF FLOWER	FRUITING PERIOD	COLOUR OF FRUIT
May to July	Yellowish green	August to November	Not recorded

Paris polyphylla var. *chinensis* (Franch.) H. Hara grows as a perennial herb, is mainly distributed in forests and ravines, and belongs to a variety of *Paris polyphylla* Sm. According to the records of the Hong Kong Herbarium, the only *Paris* species in Hong Kong is *Paris polyphylla* var. *chinensis* (Franch.) H. Hara.

Similar to the other *Paris* species, *Paris polyphylla* var. *chinensis* (Franch.) H. Hara has been applied in Chinese herbal medicine due to its medicinal properties. Its populations declined sharply due to overharvesting in the wild.

Paris polyphylla var. *chinensis* (Franch.) H. Hara is listed as 'Vulnerable' in *Rare and Precious Plants of Hong Kong.*

Name in *Flora of China*
Paris polyphylla var. *chinensis* (Franchet) H. Hara (Family: Liliaceae)

Name in *Flora of Hong Kong*
Paris polyphylla Sm. var. *chinensis* (Franch.) H. Hara (Family: Liliaceae)

4. Missouri Botanical Garden. (2022b). *Name-Paris polyphylla*. Tropicos. Retrieved March 17, 2022, from https://www.tropicos.org/name/18403525
5. Zheng, J. Y., Wang, H., Chen, X. X., Wang, P., Gao, P., Li, X. N., & Zhu, G. P. (2012). Microsatellite markers for assessing genetic diversity of the medicinal plant *Paris polyphylla* var. *chinensis* (Trilliaceae). *Genet. Mol. Res 11*(3), 1975–1980. http://dx.doi.org/10.4238/2012.August.6.1

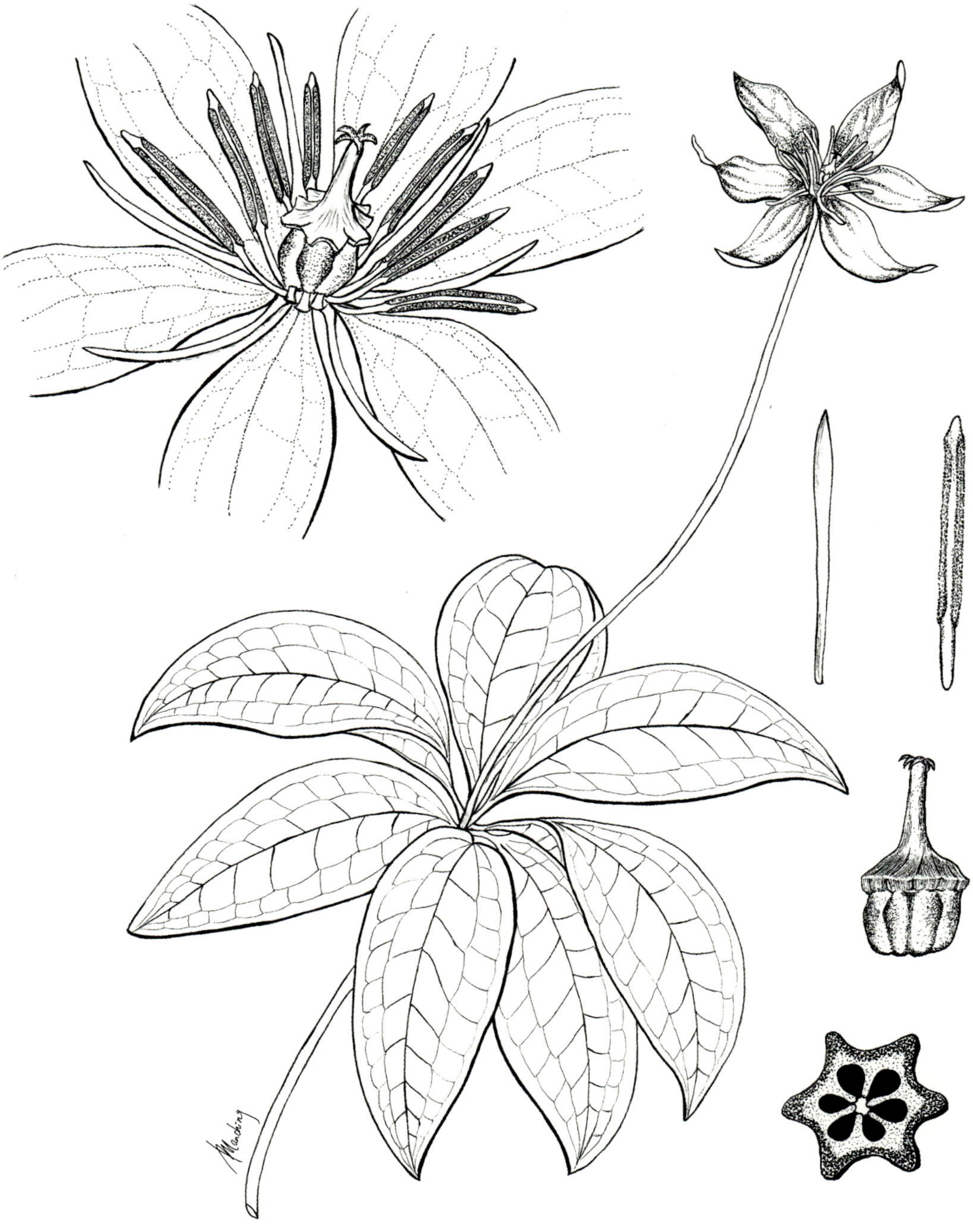

Paris polyphylla var. *chinensis* (Franch.) H. Hara 華重樓
Illustration 002 by M. C. Li
Authenticated by D. T. W. Lau

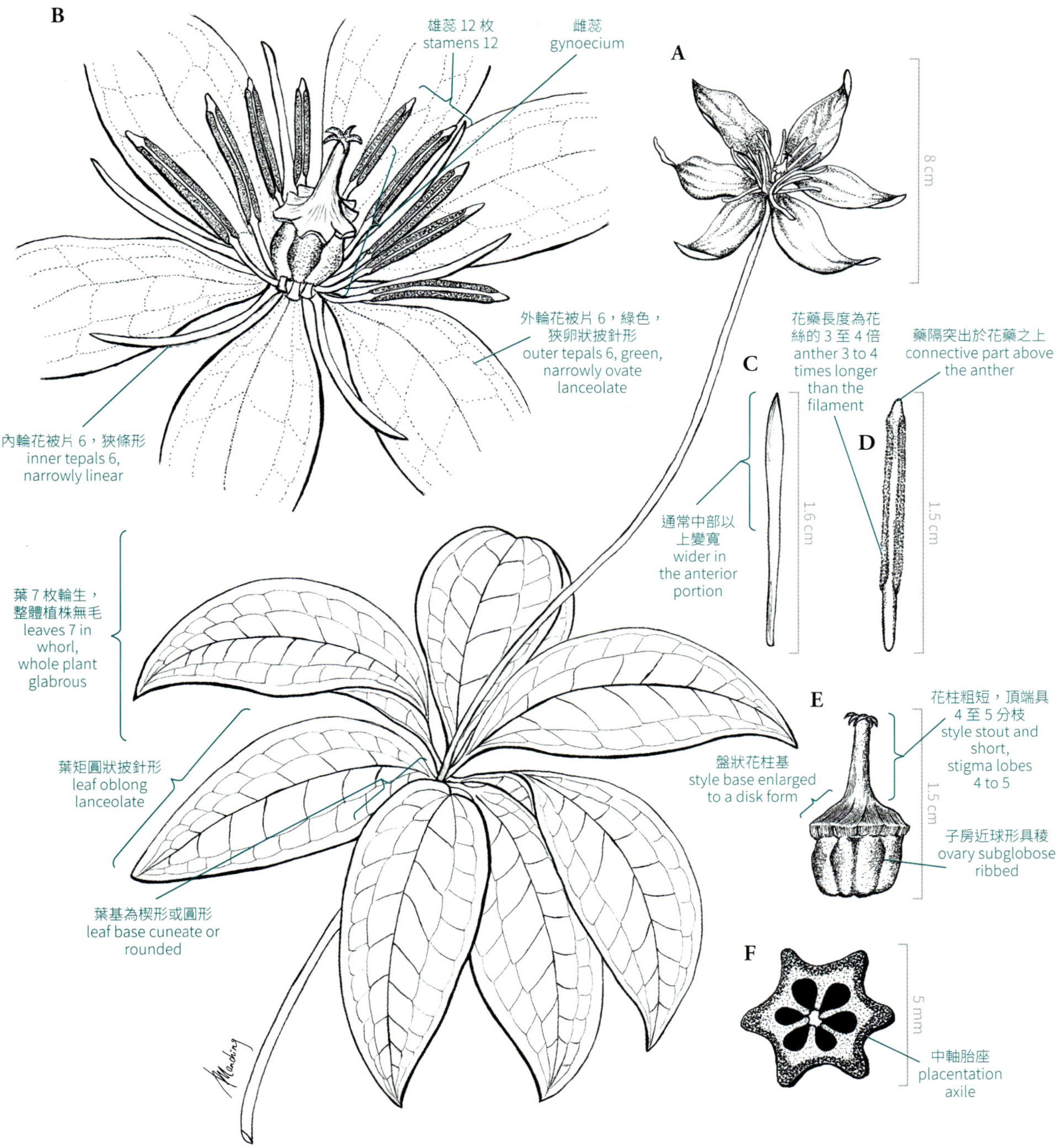

Paris polyphylla var. *chinensis* (Franch.) H. Hara 華重樓

Illustration 002 by M. C. Li
Authenticated by D. T. W. Lau

植物結構 PLANT STRUCTURES:
A. 花枝 Flowering branch **B.** 花，正面觀 Flower, frontal view **C.** 內輪花被片 Inner tepal **D.** 雄蕊 Stamen **E.** 雌蕊 Gynoecium **F.** 子房，橫切面 Ovary, in C. S.

參考標本 REFERENCE SPECIMENS: Rare and endangered plants (K. W. Lam, S. W. Shek, J. Y. Y. Lau, D. T. W. Lau) 3B; Shiu Ying Hu 13102; S. K. Tsang 072 (H1911); R. Blinkworth 3710 (K001119877)

參考文獻 REFERENCES:

Chen, X. Q., Liang, S. Y., Xu, J. M., & Tamura, M. N. (2000). Liliaceae. In Z. Y. Wu & P. H. Raven (Eds.), *Flora of China* (Vol. 24, pp. 73–263). Science Press & Missouri Botanical Garden Press.

Hu, Q. M. (2011). Liliaceae. In Hong Kong Herbarium, AFCD & South China Botanical Garden, CAS (Eds.), *Flora of Hong Kong* (Vol. 4, pp. 259–277). Agriculture, Fisheries and Conservation Department, HKSAR Government.

Stevens, P. F. (2022). Liliales. Angiosperm Phylogeny Website. Retrieved March 17, 2022, from http://www.mobot.org/MOBOT/research/APweb/welcome.html

Wang, F. T., & Tang, T. (1978). Liliaceae. In Delectis Florae Reipublicae Popularis Sinicae Agendae Academiae Sinicae Edita (Ed.), *Flora Reipublicae Popularis Sinicae* (Vol. 15(2), pp. 1–255). Science Press.

Zhang, S. Z. (2015). Liliaceae. In Fairy Lake Botanical Garden, Shenzhen & Chinese Academy of Sciences (Eds.), *Flora of Shenzhen* (Vol. 4, pp. 444–480). China Forestry Publishing House.

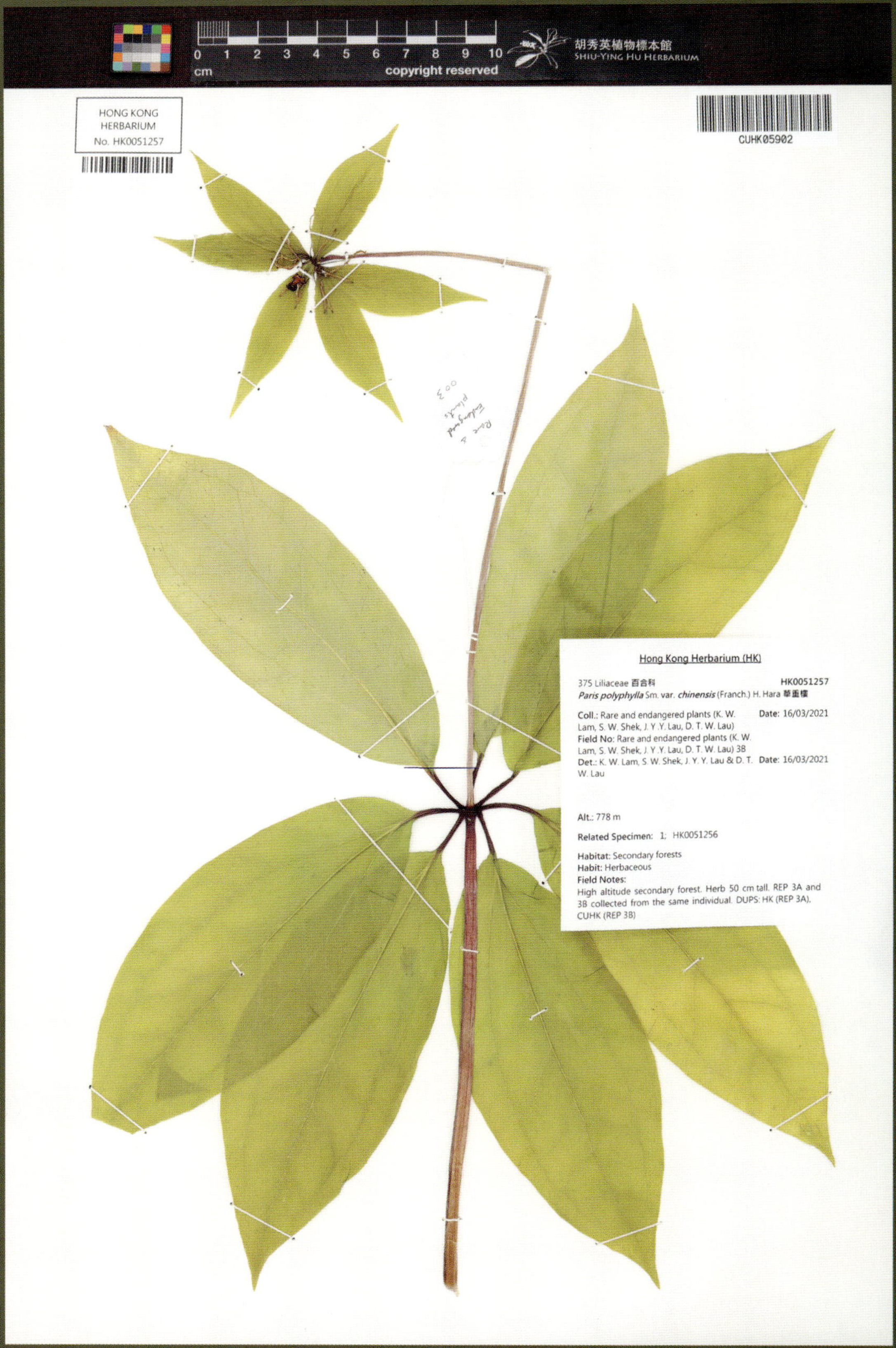

採集者編號 Collector no.
Rare and endangered plants (K. W. Lam, S. W. Shek, J. Y. Y. Lau, D. T. W. Lau) 3B

採集年份 Year of collection
2021

外觀 Overview

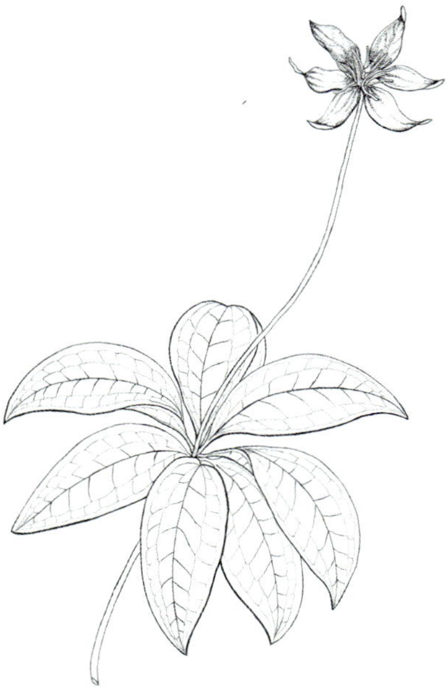

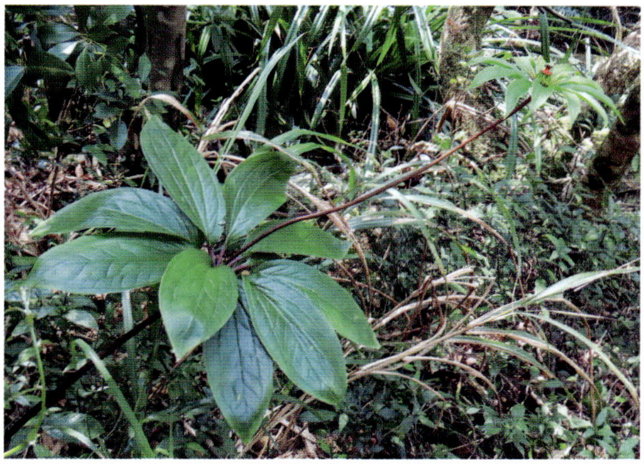

花枝 Flowering branch

葉 Leaf

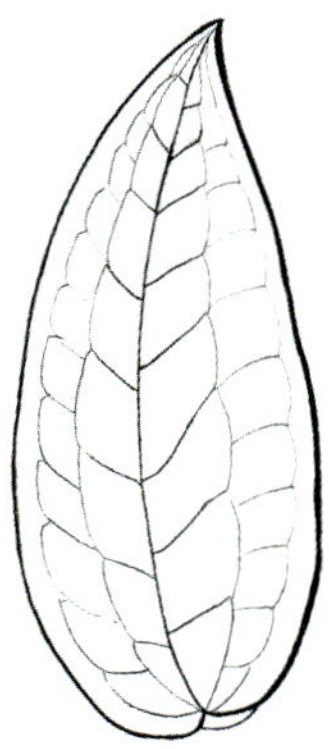

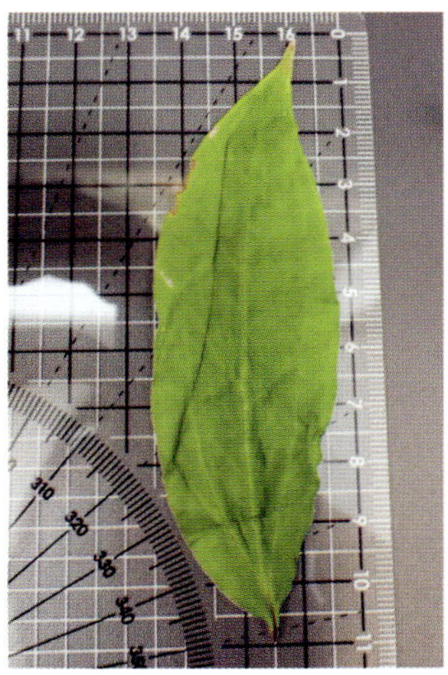

葉面 Leaf, adaxial

花 Flower

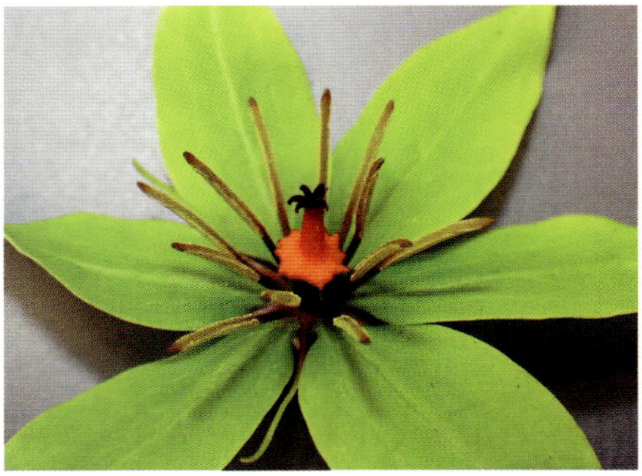
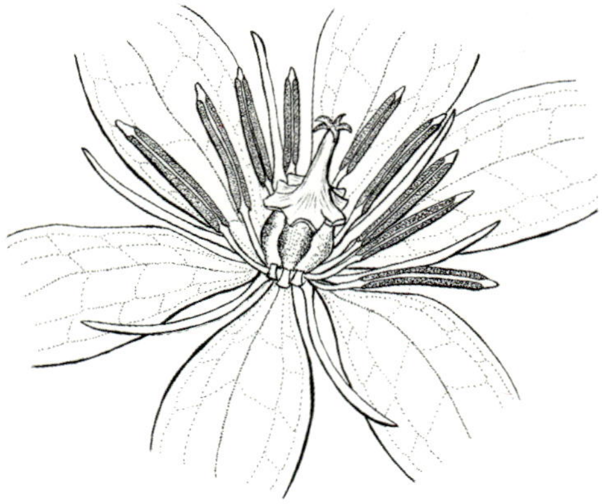

花，正面觀 Flower, frontal view

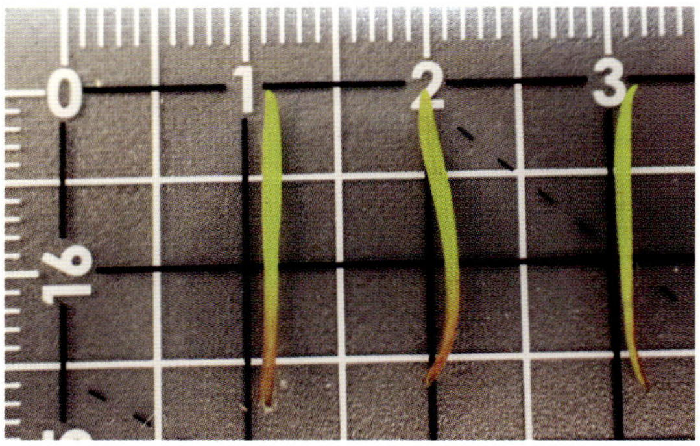

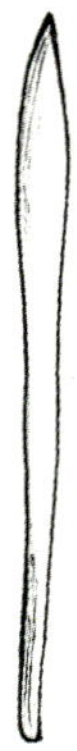

內輪花被片 Inner tepals

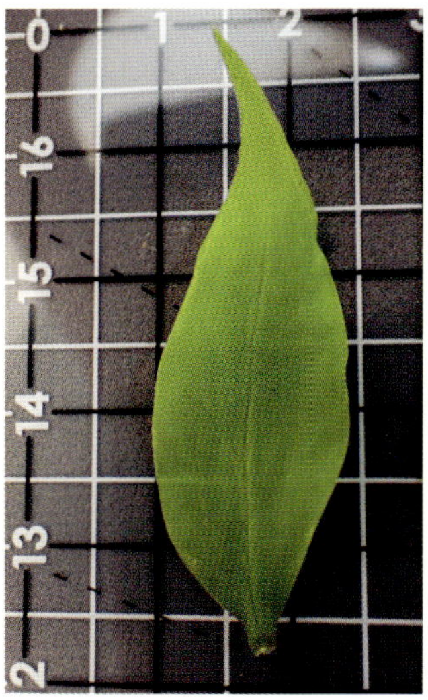

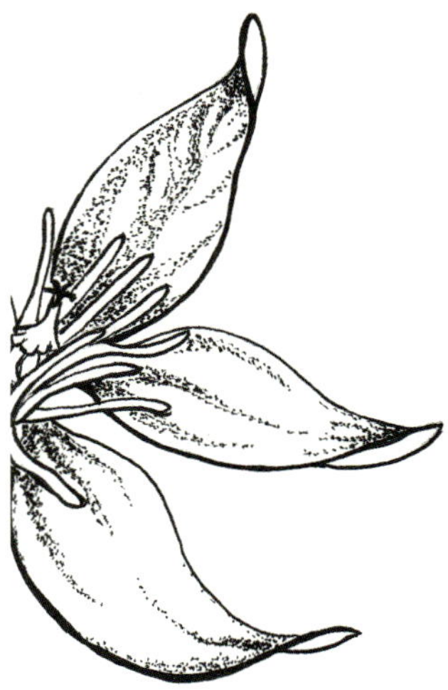

外輪花被片 Outer tepal

花 Flower

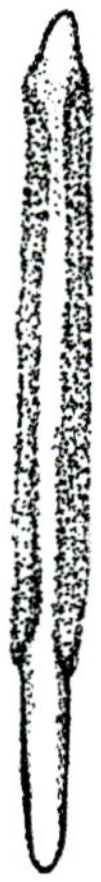

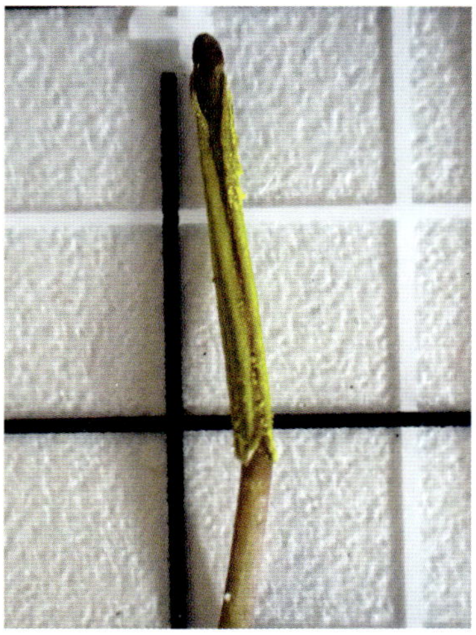

雄蕊 Stamen

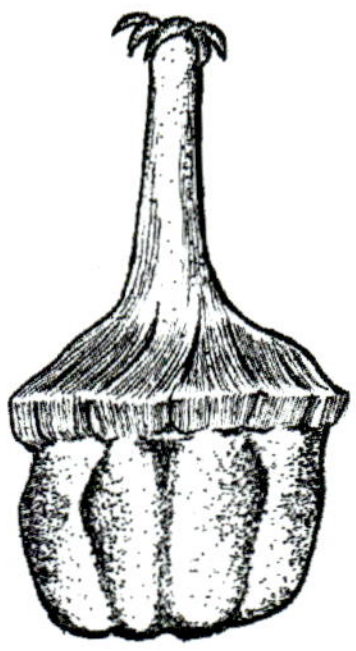

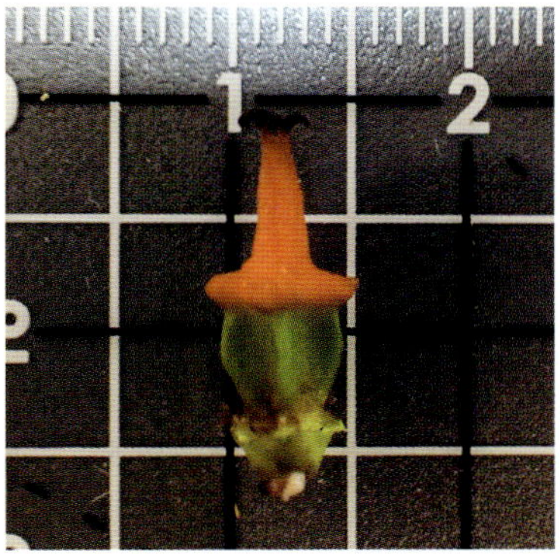

雌蕊 Gynoecium

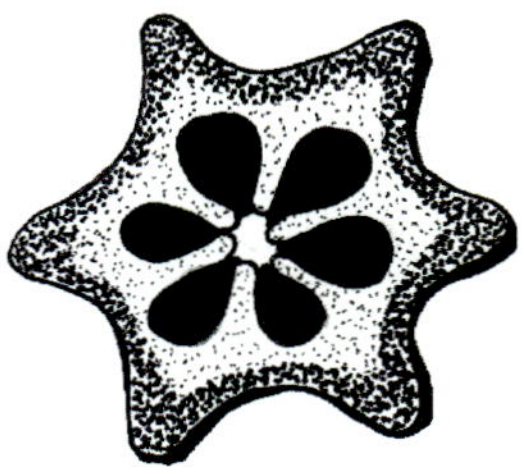

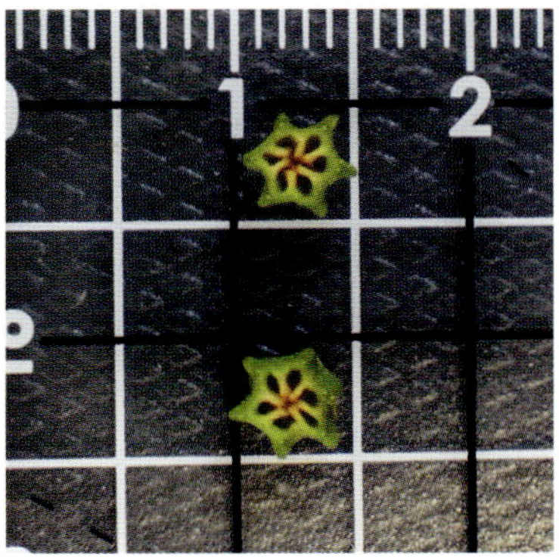

子房，橫切面 Ovary, in C. S.

Iris speculatrix Hance
小花鳶尾

小花鳶尾依靠東方蜜蜂
（*Apis cerana* Fabricius, 1793）
等昆蟲傳播花粉。

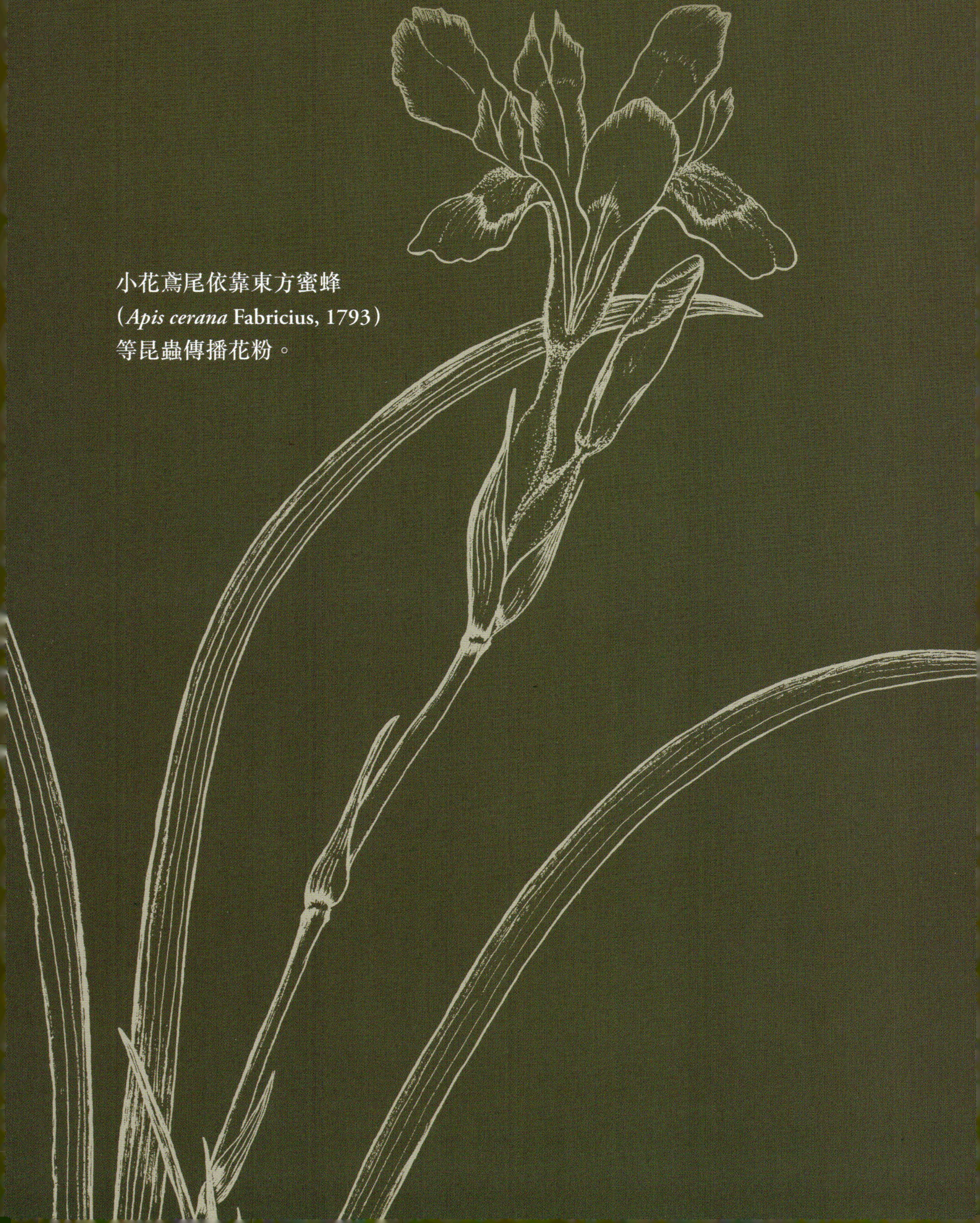

胡秀英教授採集的標本
Specimen Collected by Professor Shiu-Ying Hu

採集者編號
Collector no.
Hu & But 20422

採集年份
Year of collection
1992

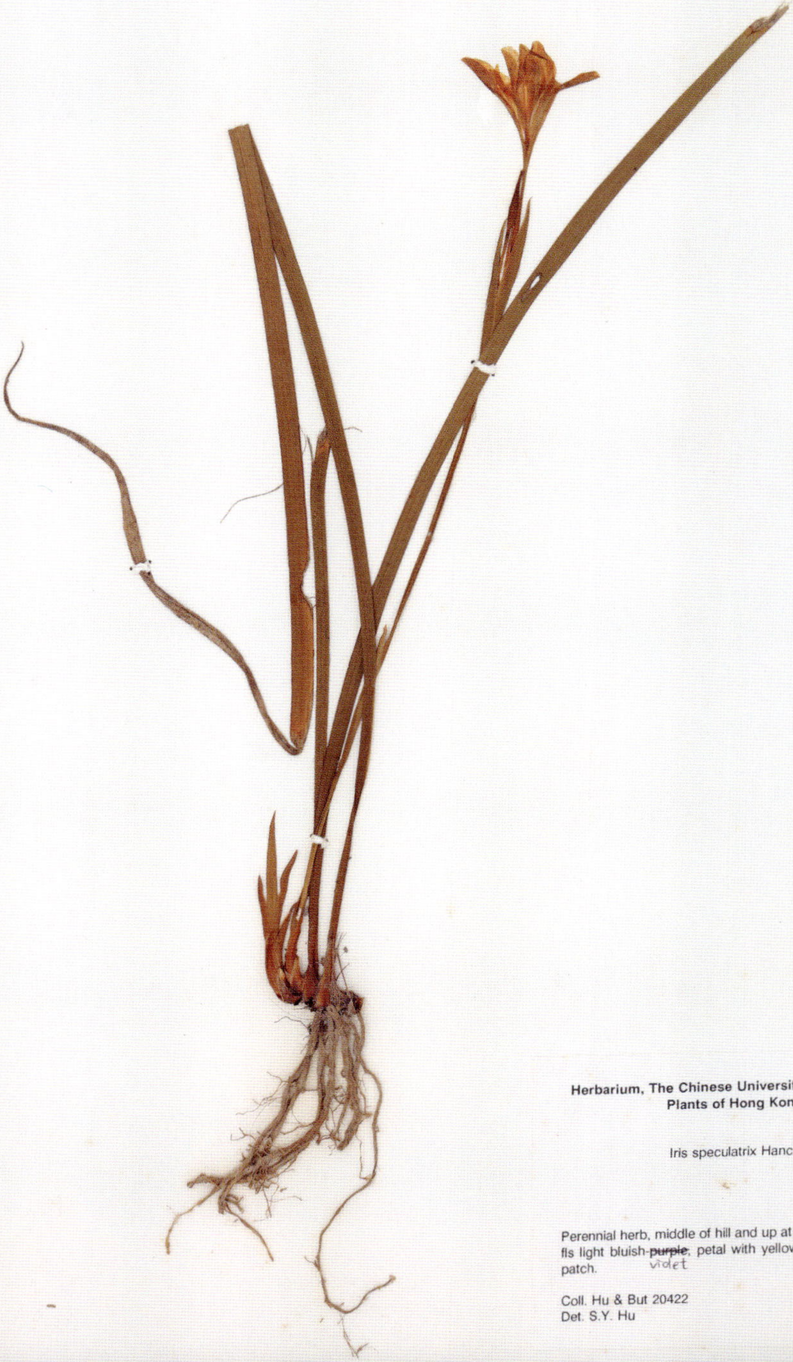

Herbarium, The Chinese University of Hong Kong
Plants of Hong Kong

Iris speculatrix Hance

Perennial herb, middle of hill and up at top, exposed area; fls light bluish-~~purple~~ violet, petal with yellow spot below white patch.

Coll. Hu & But 20422 Apr. 21, 1992
Det. S.Y. Hu

386

小花鳶尾

學名 *Iris speculatrix* Hance
中文名 小花鳶尾
科名 鳶尾科
本地分布狀態 原生植物
繪圖編號 M. C. Li 008

花期	花色	果期	果色
5月	紫色	7至8月	未有記錄

小花鳶尾屬多年生草本，擁有匍匐的根狀莖，生長於山林，其模式標本於1874年由 H. F. Hance 在香港島採集。

研究指出小花鳶尾主要依靠東方蜜蜂（*Apis cerana* Fabricius, 1793）等昆蟲傳播花粉。

小花鳶尾受香港法例第 96A章《林務規例》規管，於《香港稀有及珍貴植物》中屬「無危」級別。

《中國植物誌》（英文版）的採用名稱
Iris speculatrix Hance（科名：鳶尾科）

《香港植物誌》的採用名稱
Iris speculatrix Hance（科名：鳶尾科）

參考文獻 References

1. GBIF Secretariat. (2022). *Apis cerana* Fabricius, 1793. GBIF Backbone Taxonomy. Retrieved March 17, 2022, from https://www.gbif.org/species/1341979
2. Hong Kong Herbarium. (2022). *HK Herbarium-Iris speculatrix*. HK Plant Database. Retrieved March 17, 2022, from https://herbarium.gov.hk/en/hk-plant-database/plant-detail/index.html?pType=species&oID=3920
3. Missouri Botanical Garden. (2022). *Name-Iris speculatrix*. Tropicos. Retrieved March 17, 2022, from https://www.tropicos.org/name/16602435

Iris speculatrix Hance

Scientific name	*Iris speculatrix* Hance
Chinese name	小花鳶尾
Family	Iridaceae Juss.
Local distribution status	Native species
Illustration number	M. C. Li 008

FLOWERING PERIOD	COLOUR OF FLOWER	FRUITING PERIOD	COLOUR OF FRUIT
May	Violet	July to August	Not recorded

Iris speculatrix Hance is a perennial herb that consists of creeping rhizomes and grows in forests. In 1874, the type specimen was collected by H. F. Hance on Hong Kong Island.

Scientific research revealed that *Iris speculatrix* Hance mainly relies on insects like *Apis cerana* Fabricius, 1793 for pollination.

Iris speculatrix Hance has been protected under Forestry Regulations (Cap. 96A) and is classified as 'Least Concern' in *Rare and Precious Plants of Hong Kong*.

Name in *Flora of China*
Iris speculatrix Hance (Family: Iridaceae)

Name in *Flora of Hong Kong*
Iris speculatrix Hance (Family: Iridaceae)

4. Wang, G. Q., Li, D. Q., Zhang, J. P., & Xia, Y. P. (2014). Floral syndrome and breeding characters of *Iris speculatrix*. *Acta Agriculturae Zhejiangensis 26*(5), 1218–1222.
5. Zhao, Y. T., Noltie, H. J., & Mathew, B. (2000). Iridaceae. In Z. Y. Wu & P. H. Raven (Eds.), *Flora of China* (Vol. 24, pp. 297–313). Science Press & Missouri Botanical Garden Press.

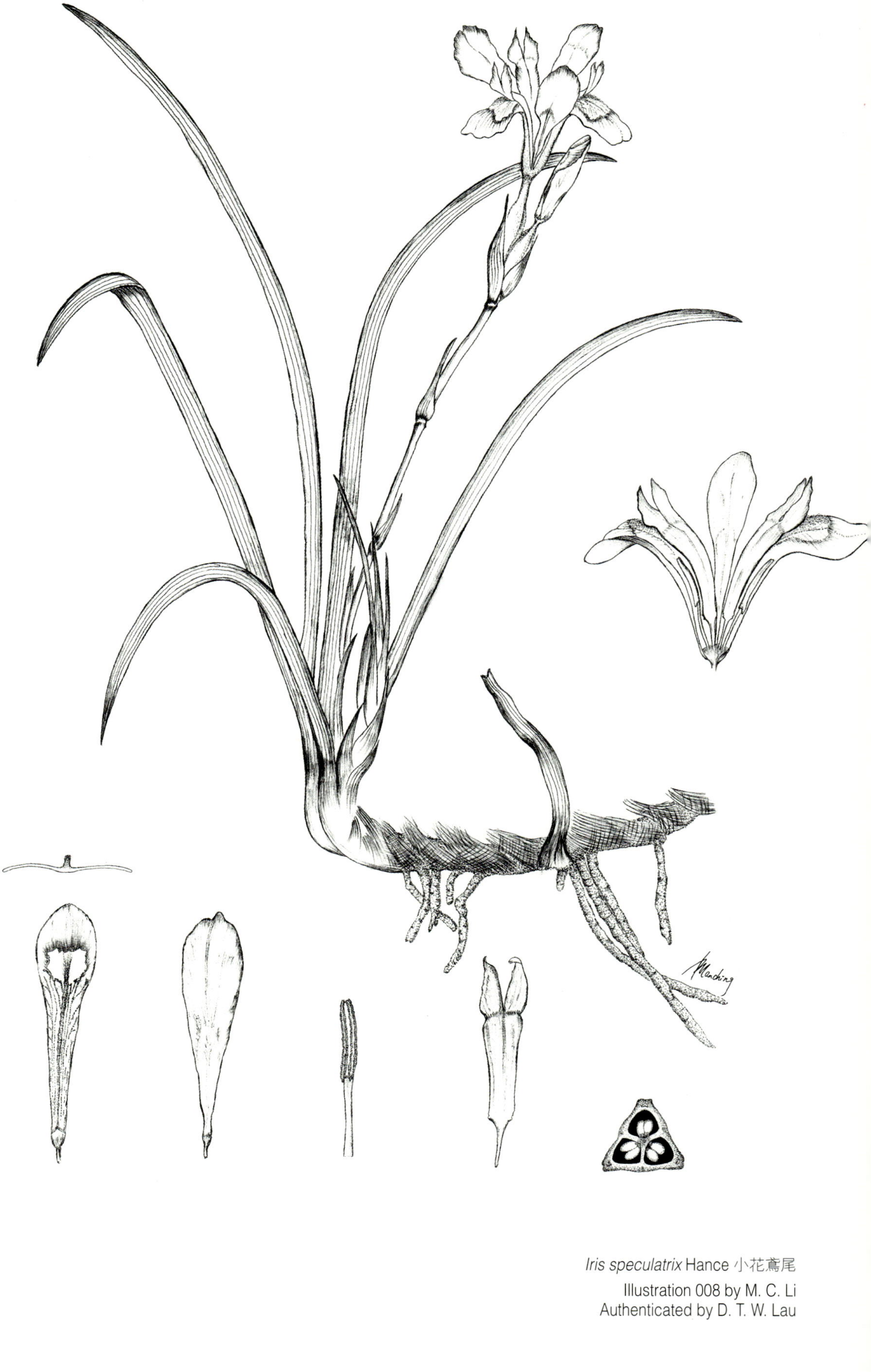

Iris speculatrix Hance 小花鳶尾
Illustration 008 by M. C. Li
Authenticated by D. T. W. Lau

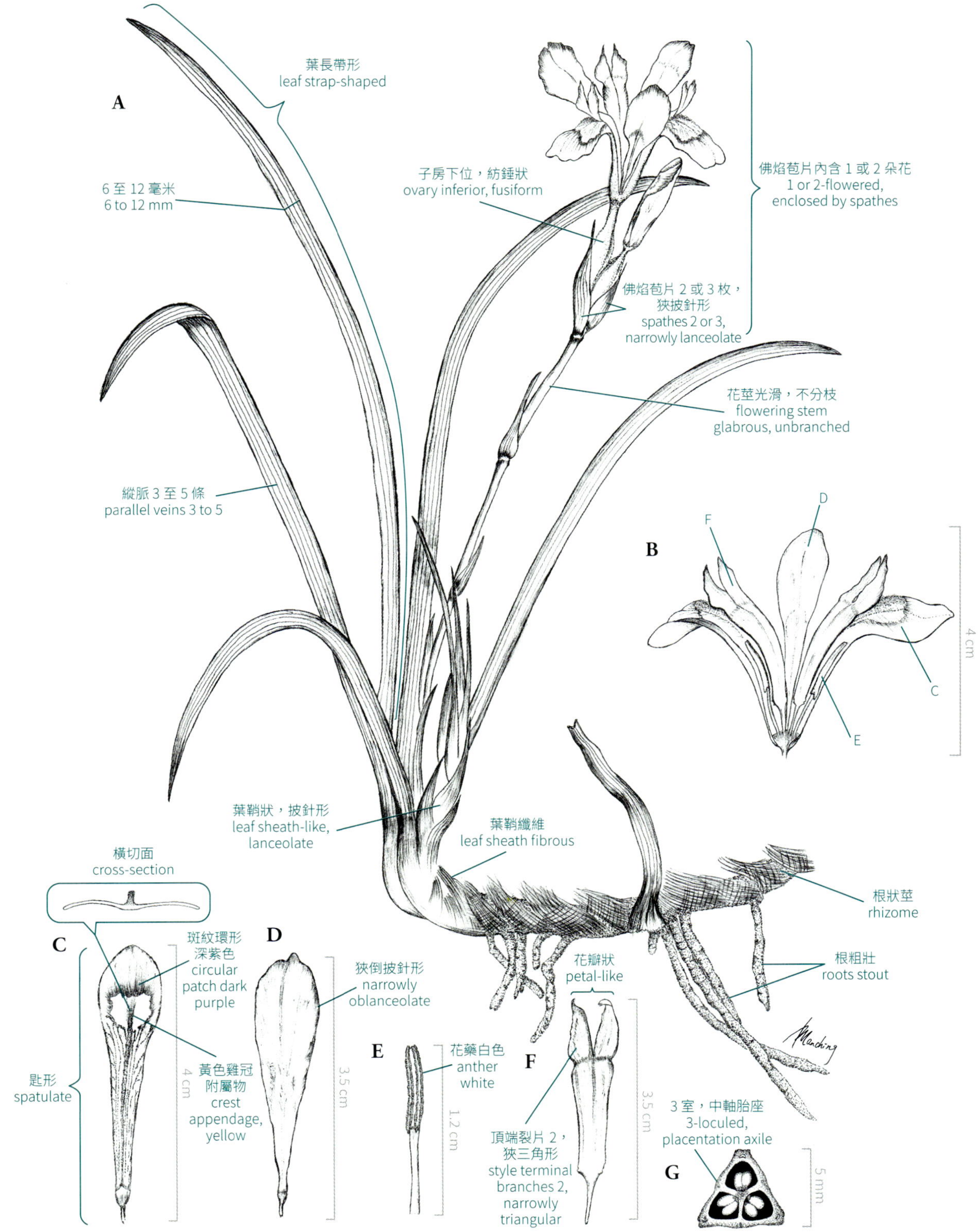

Iris speculatrix Hance 小花鳶尾

Illustration 008 by M. C. Li

Authenticated by D. T. W. Lau

植物結構 PLANT STRUCTURES:

A. 花枝 Flowering branch **B.** 花解剖圖，展示 2 片外花被裂片，1 片內花被裂片，2 枚雄蕊及 2 片花柱分枝 Dissected flower, showing 2 outer perianth segments, 1 inner perianth segment, 2 stamens and 2 style branches **C.** 外花被裂片 Outer perianth segment **D.** 內花被裂片 Inner perianth segment **E.** 雄蕊 Stamen **F.** 花柱分枝 Style branch **G.** 子房，橫切面 Ovary, in C. S.

參考標本 REFERENCE SPECIMENS: Rare and endangered plants (S. W. Shek, K. W, Lam, D. T. W. Lau, J. Y. Y. Lau) 9B; Hu & But 20422

參考文獻 REFERENCES:

Pei, C., & Ting, C. T. (1985). Iridaceae. In Delectis Florae Reipublicae Popularis Sinicae Agendae Academiae Sinicae Edita (Ed.), *Flora Reipublicae Popularis Sinicae* (Vol. 16(1), pp. 120–198). Science Press.

Stevens, P. F. (2022). Asparagales. Angiosperm Phylogeny Website. Retrieved March 17, 2022, from http://www.mobot.org/MOBOT/research/APweb/welcome.html

Wu, T. L. (2005). Iridaceae. In South China Botanical Garden, the Chinese Academy of Sciences (Ed.), *Flora of Guangdong* (Vol. 6, pp. 381–389). Guangdong Science and Technology Press.

Xia, N. H. (2011). Iridaceae. In Hong Kong Herbarium, AFCD & South China Botanical Garden, CAS (Eds.), *Flora of Hong Kong* (Vol. 4, pp. 277–279). Agriculture, Fisheries and Conservation Department, HKSAR Government.

Zhao, Y. T., Noltie, H. J., & Mathew, B. (2000). Iridaceae. In Z. Y. Wu & P. H. Raven (Eds.), *Flora of China* (Vol. 24, pp. 297–313). Science Press & Missouri Botanical Garden Press.

憑證標本
Voucher Specimen

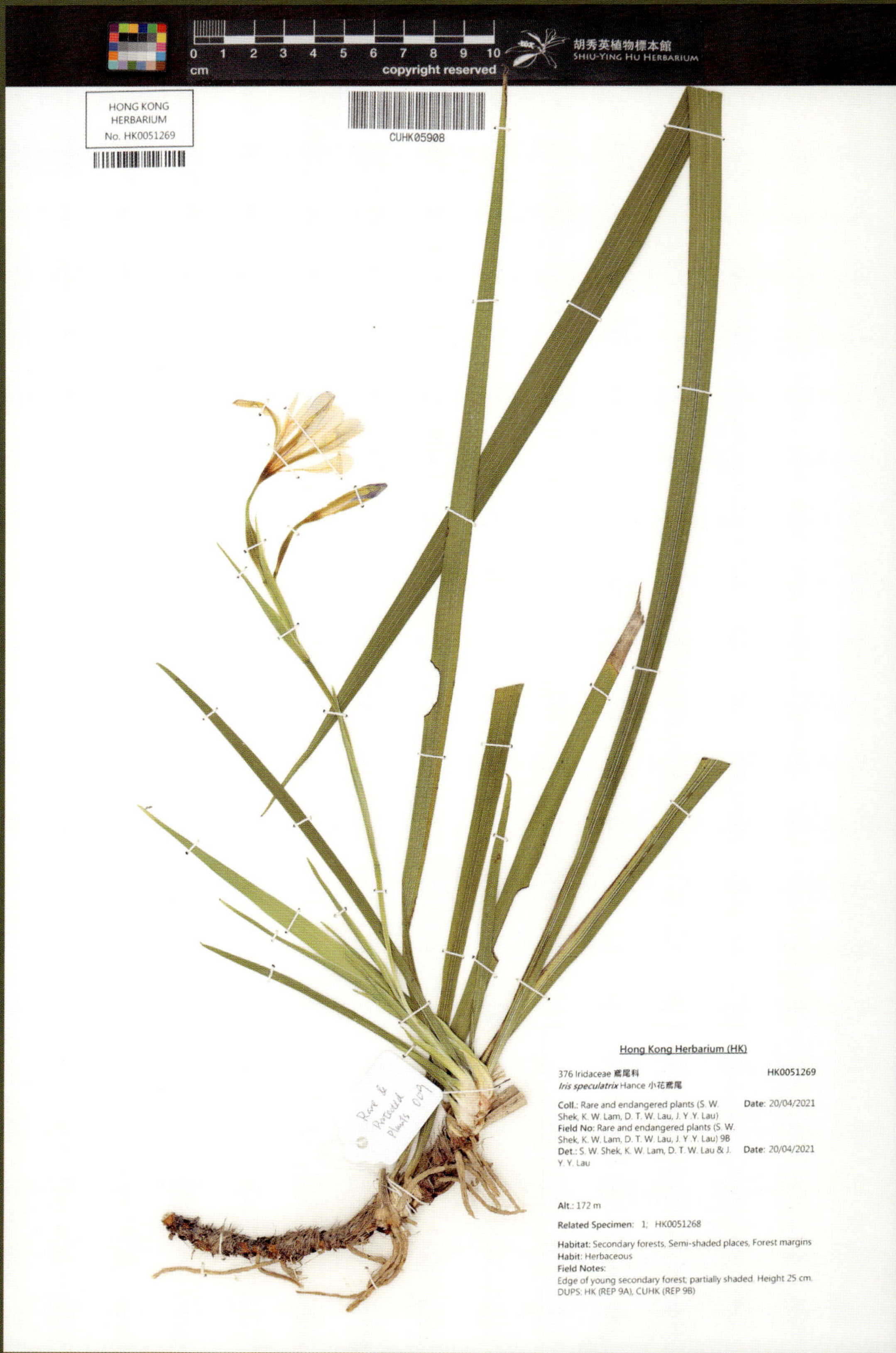

採集者編號 Collector no.
Rare and endangered plants (S. W. Shek, K. W, Lam, D. T. W. Lau, J. Y. Y. Lau) 9B

採集年份 Year of collection
2021

外觀 Overview

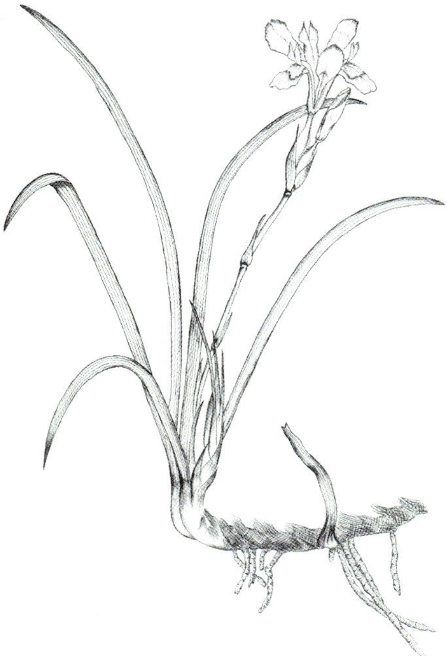

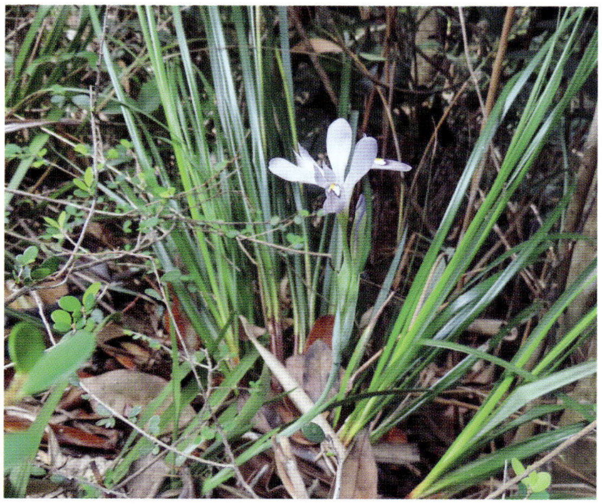

花枝 Flowering branch

葉 Leaf

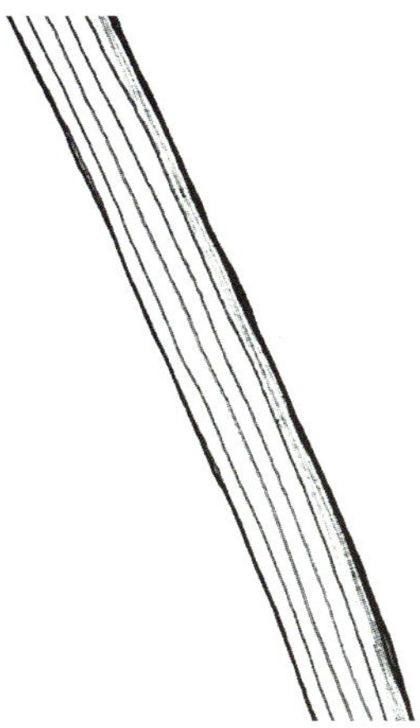

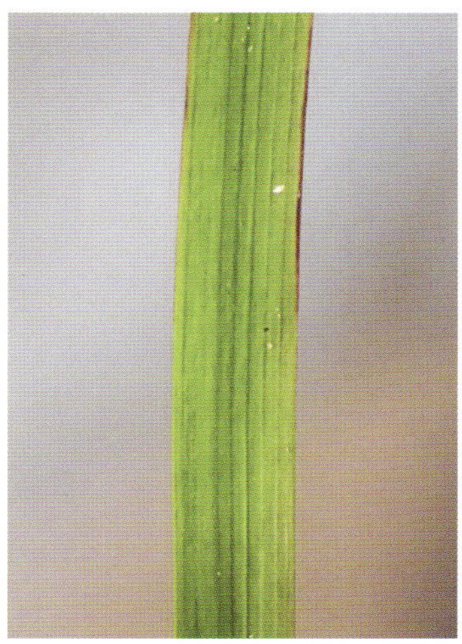

葉面 Leaf, adaxial

花 Flower

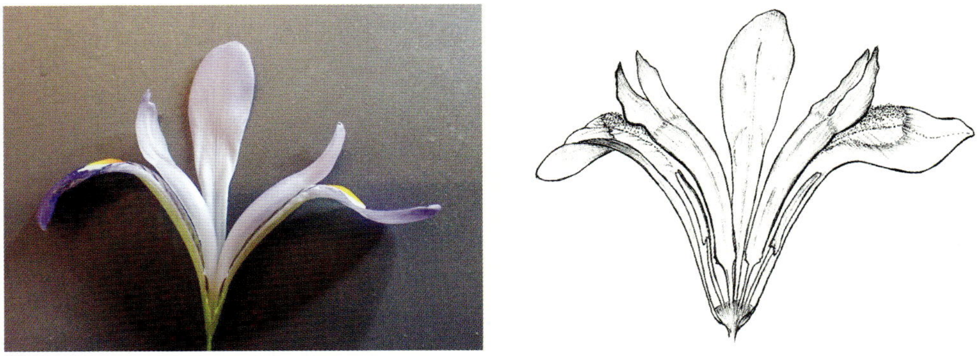

花解剖圖，展示 2 片外花被裂片，1 片內花被裂片，2 枚雄蕊及 2 片花柱分枝
Dissected flower, showing 2 outer perianth segments, 1 inner perianth segment, 2 stamens and 2 style branches

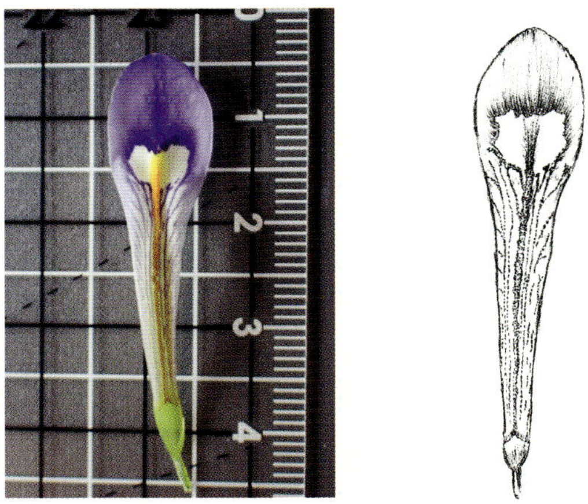

外花被裂片
Outer perianth segment

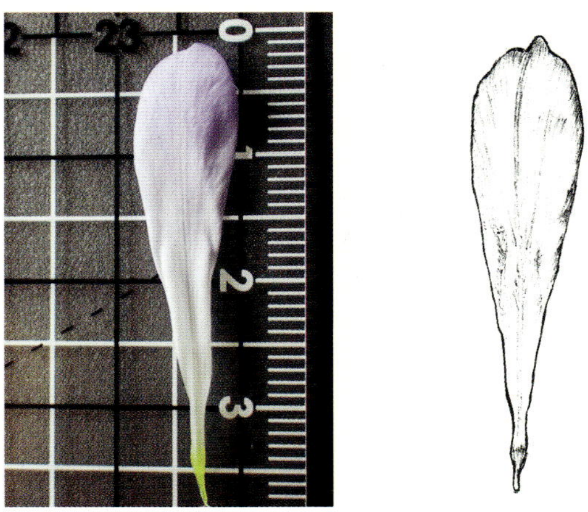

內花被裂片
Inner perianth segment

花 Flower

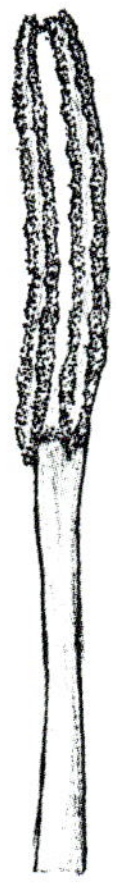
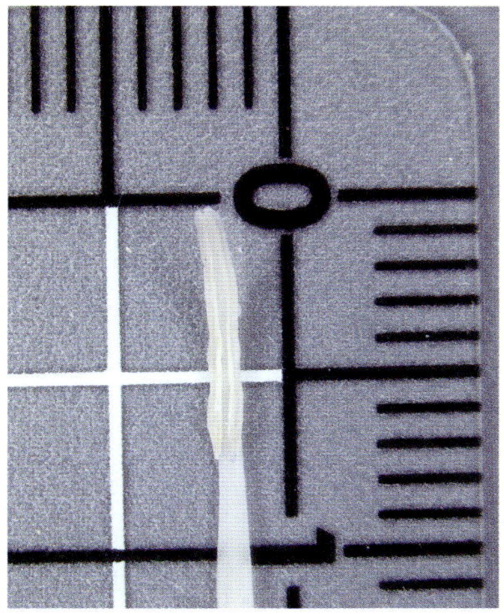

雄蕊 Stamen

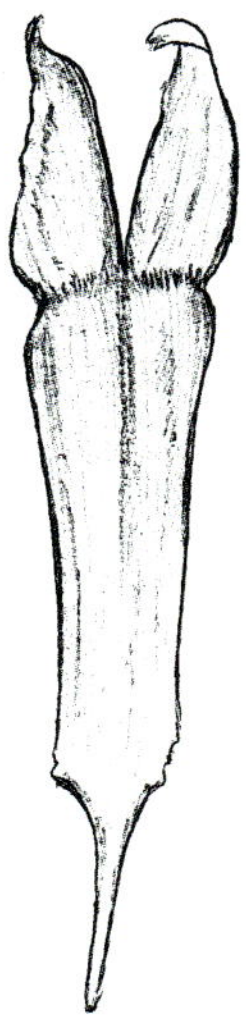
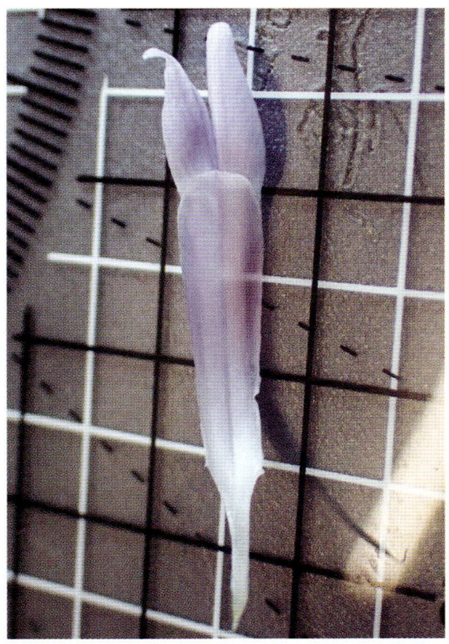

花柱分枝 Style branch

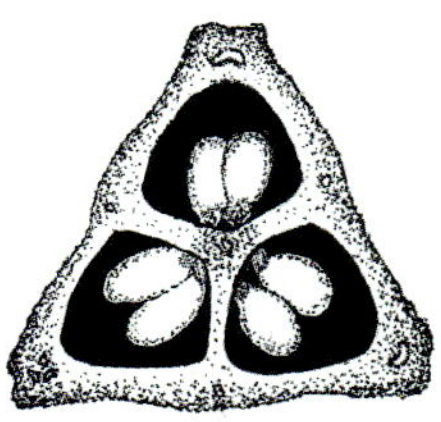

子房，橫切面 Ovary, in C. S.

Rhodoleia championii Hook.

紅花荷

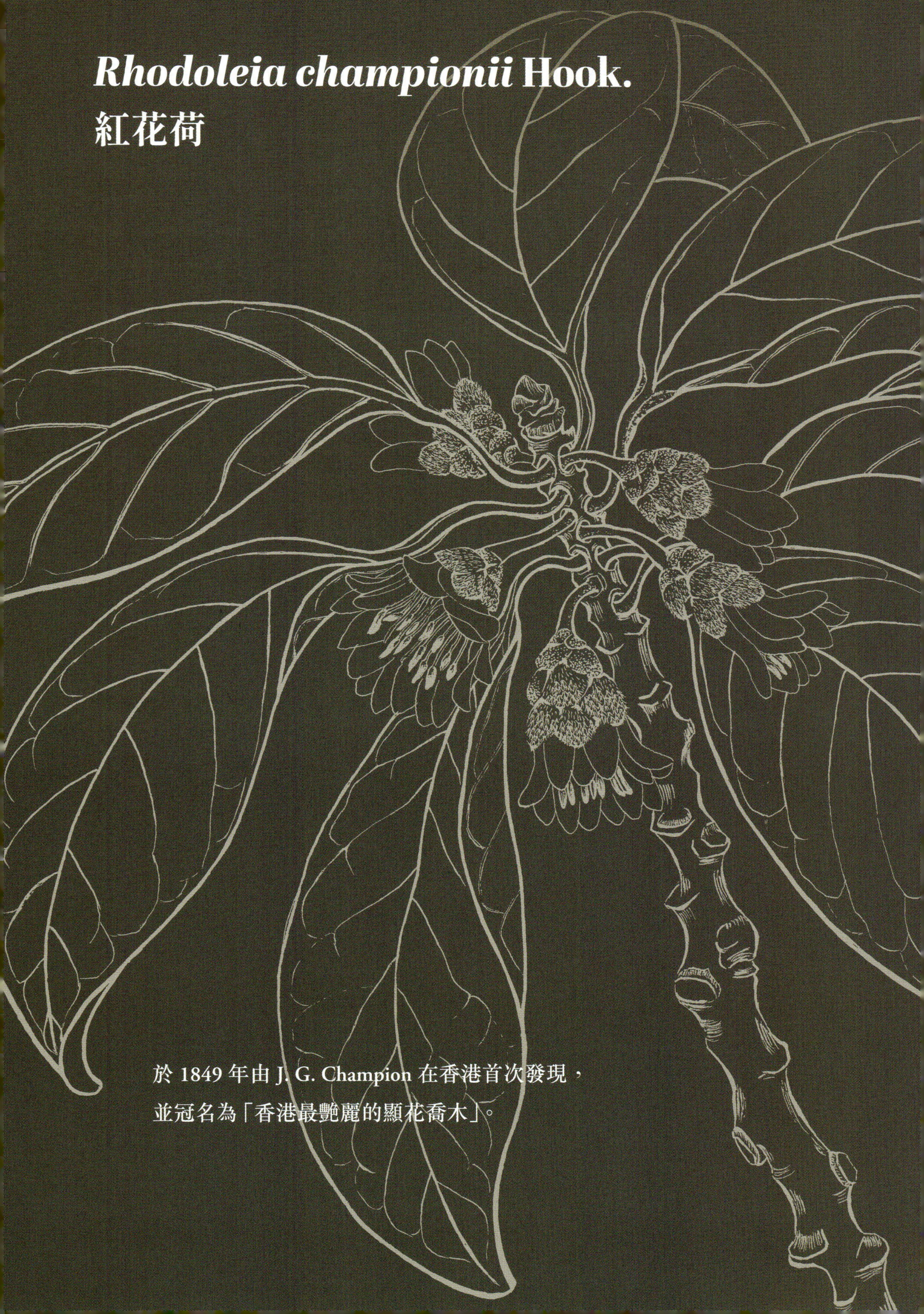

於 1849 年由 J. G. Champion 在香港首次發現，
並冠名為「香港最艷麗的顯花喬木」。

採集者編號
Collector no.
Shiu Ying Hu 12089

採集年份
Year of collection
1972

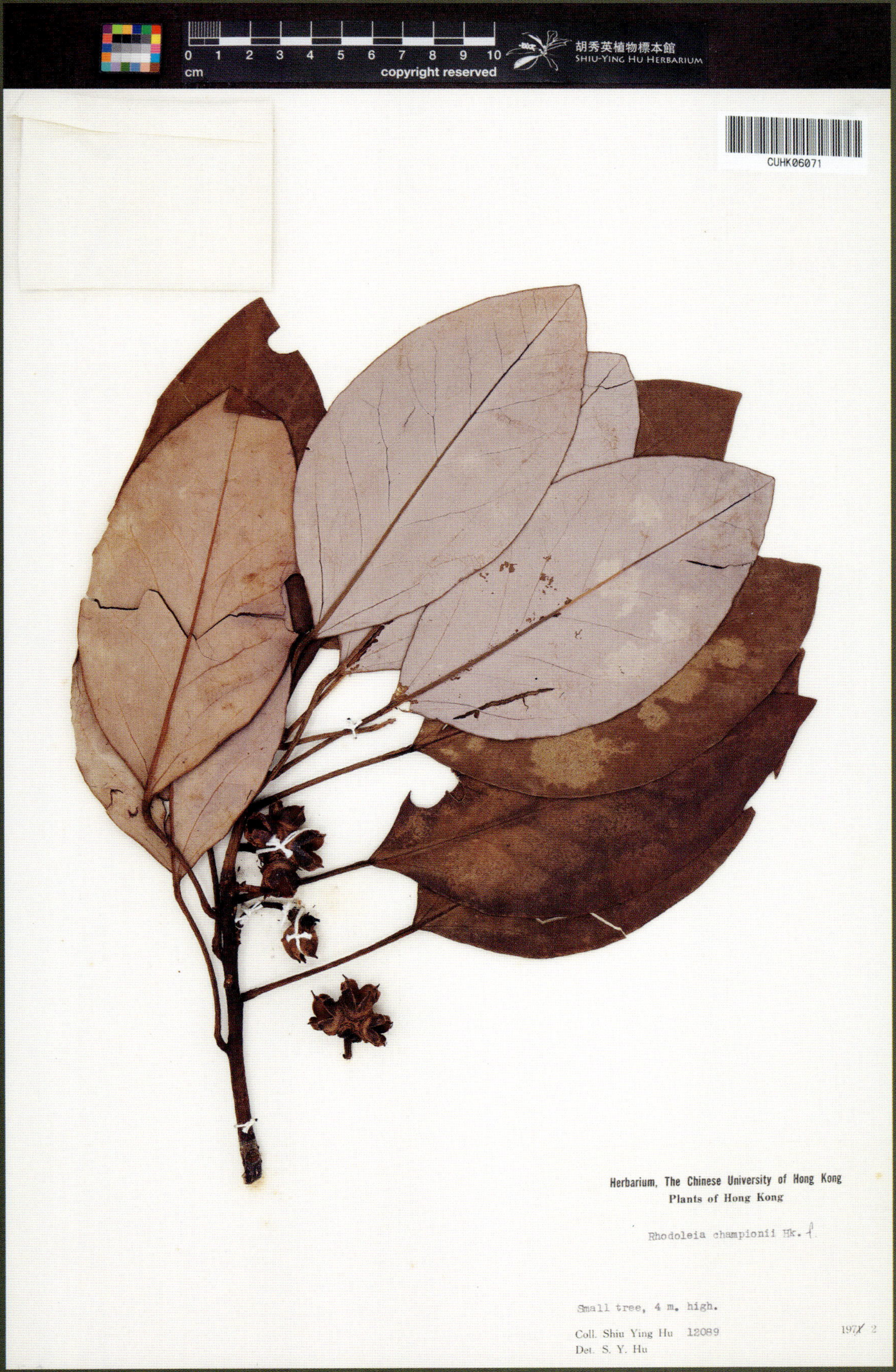
0 1 2 3 4 5 6 7 8 9 10
cm
copyright reserved
胡秀英植物標本館
SHIU-YING HU HERBARIUM
CUHK06071
Herbarium, The Chinese University of Hong Kong
Plants of Hong Kong
Rhodoleia championii Hk. f.
Small tree, 4 m. high.
Coll. Shiu Ying Hu 12089
197X 2
Det. S. Y. Hu

紅花荷

學名	*Rhodoleia championii* Hook.
中文名	紅花荷
科名	金縷梅科
本地分布狀態	原生植物
繪圖編號	M. C. Li 001

花期	花色	果期	果色
2 至 4 月	紅色	5 至 8 月	褐色

紅花荷生長於常綠闊葉林，其模式標本採自香港仔，於 1849 年由 J. G. Champion 在香港首次發現，並冠名為「香港最艷麗的顯花喬木」。

每逢 2 至 4 月的花季，鮮艷的紅花荷吸引叉尾太陽鳥（*Aethopyga christinae* Swinhoe, 1869）和暗綠繡眼鳥（*Zosterops japonicus* Temminck & Schlegel, 1845）等雀鳥採食花蜜並協助傳播花粉。研究發現紅花荷擁有堅挺的花托筒和總花梗，有效承托雀鳥啄食時所帶來的壓力。

紅花荷受香港法例第 96A章《林務規例》規管，並於《香港稀有及珍貴植物》中屬「易危」級別。

《中國植物誌》(英文版) 的採用名稱

Rhodoleia championii Hooker (科名：金縷梅科)

《香港植物誌》的採用名稱

Rhodoleia championii Hook. f. (科名：金縷梅科)

參考文獻 References

1. Chiu, B. T. (1960). Flowers of Hong Kong: Synopsis of a lecture delivered on November 2, 1960, based on Mr. F. A. Nixon's collection of colour transparencies. *Journal of the Hong Kong Branch of the Royal Asiatic Society 1*, 27–29.
2. GBIF Secretariat. (2021a). *Aethopyga christinae* Swinhoe, 1869. GBIF Backbone Taxonomy. Retrieved March 17, 2022, from https://www.gbif.org/species/2484678
3. GBIF Secretariat. (2021b). *Zosterops japonicus* Temminck & Schlegel, 1845. GBIF Backbone Taxonomy. Retrieved March 17, 2022, from https://www.gbif.org/species/9300456
4. Gu, L., Luo, Z., Zhang, D., & Renner, S. S. (2010). Passerine pollination of *Rhodoleia championii* (Hamamelidaceae) in subtropical China. *Biotropica 42*(3), 336–341. https://doi.org/10.1111/j.1744-7429.2009.00585.x

Rhodoleia championii Hook.

Scientific name	*Rhodoleia championii* Hook.
Chinese name	紅花荷
Family	Hamamelidaceae R. Br.
Local distribution status	Native species
Illustration number	M. C. Li 001

FLOWERING PERIOD	COLOUR OF FLOWER	FRUITING PERIOD	COLOUR OF FRUIT
February to April	Red	May to August	Brown

The evergreen broad-leaved forest nurtures the growth of *Rhodoleia championii* Hook., which is regarded as 'the handsomest of Hong Kong's flowering trees'. It was first found in Hong Kong in 1849 by J. G. Champion. The type specimen was collected in Aberdeen.

Throughout every flowering period from February to April, the blossoms of *Rhodoleia championii* Hook. attract birds like *Aethopyga christinae* Swinhoe, 1869 and *Zosterops japonicus* Temminck & Schlegel, 1845 for foraging and pollination. A scientific study found that the robust hypanthium and peduncle of *Rhodoleia championii* Hook. can withstand the pressure of the pecking of birds.

Rhodoleia championii Hook. is regulated under Forestry Regulations (Cap. 96A) and is classified as 'Vulnerable' in *Rare and Precious Plants of Hong Kong*.

Name in *Flora of China*
Rhodoleia championii Hooker (Family: Hamamelidaceae)

Name in *Flora of Hong Kong*
Rhodoleia championii Hook. f. (Family: Hamamelidaceae)

5. Hong Kong Herbarium. (2022). *HK Herbarium-Rhodoleia championii*. HK Plant Database. Retrieved March 17, 2022, from https://herbarium.gov.hk/en/hk-plant-database/plant-detail/index.html?pType=species&oID=5131
6. Missouri Botanical Garden. (2022). *Name-Rhodoleia championii*. Tropicos. Retrieved March 17, 2022, from https://www.tropicos.org/name/15100096
7. Zhang, Z. Y., Zhang, H. D., & Endress, P. K. (2003). Hamamelidaceae. In Z. Y. Wu, P. H. Raven & D. Y. Hong (Eds.), *Flora of China* (Vol. 9, pp. 18–42). Science Press & Missouri Botanical Garden Press.

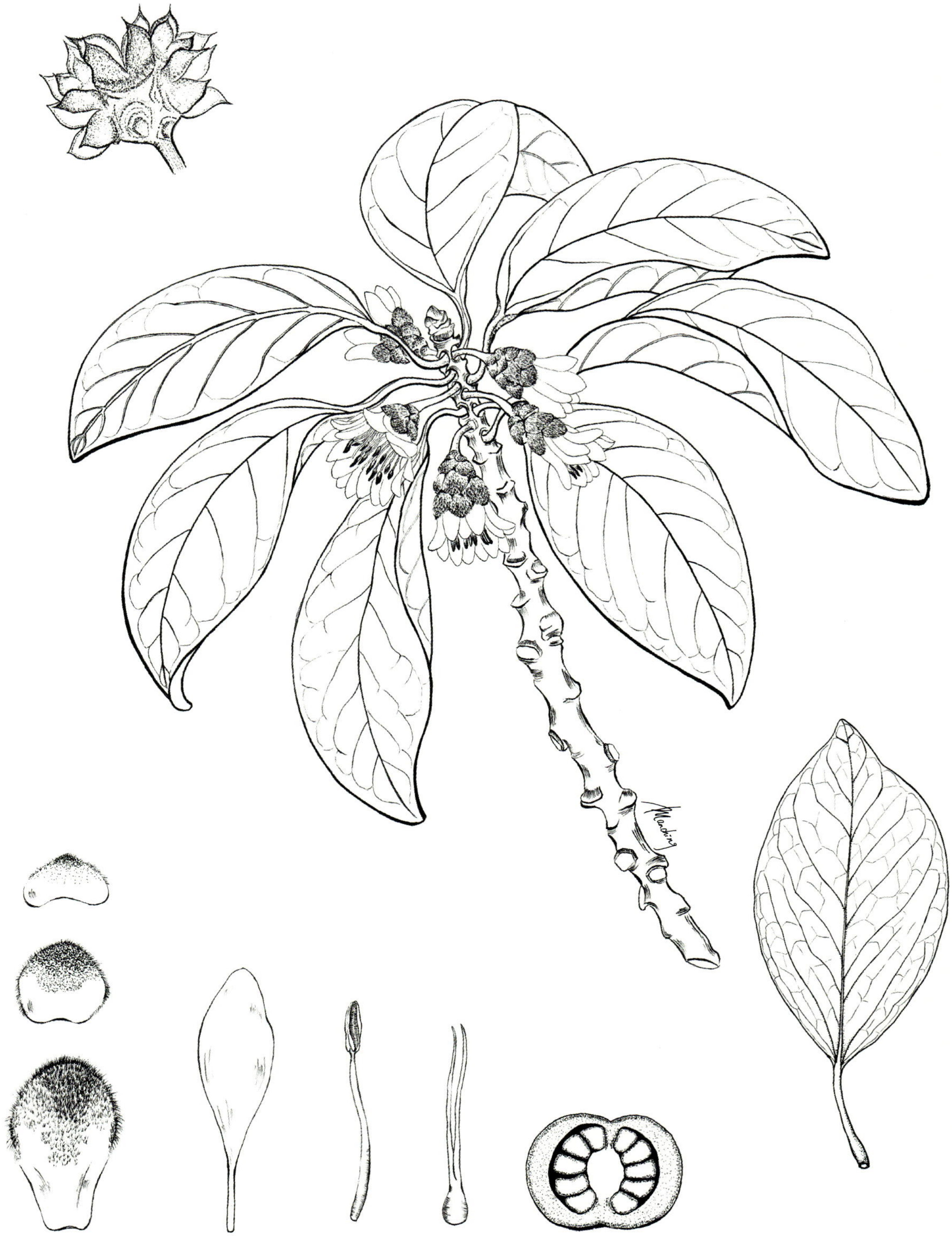

Rhodoleia championii Hook. 紅花荷

Illustration 001 by M. C. Li
Authenticated by D. T. W. Lau

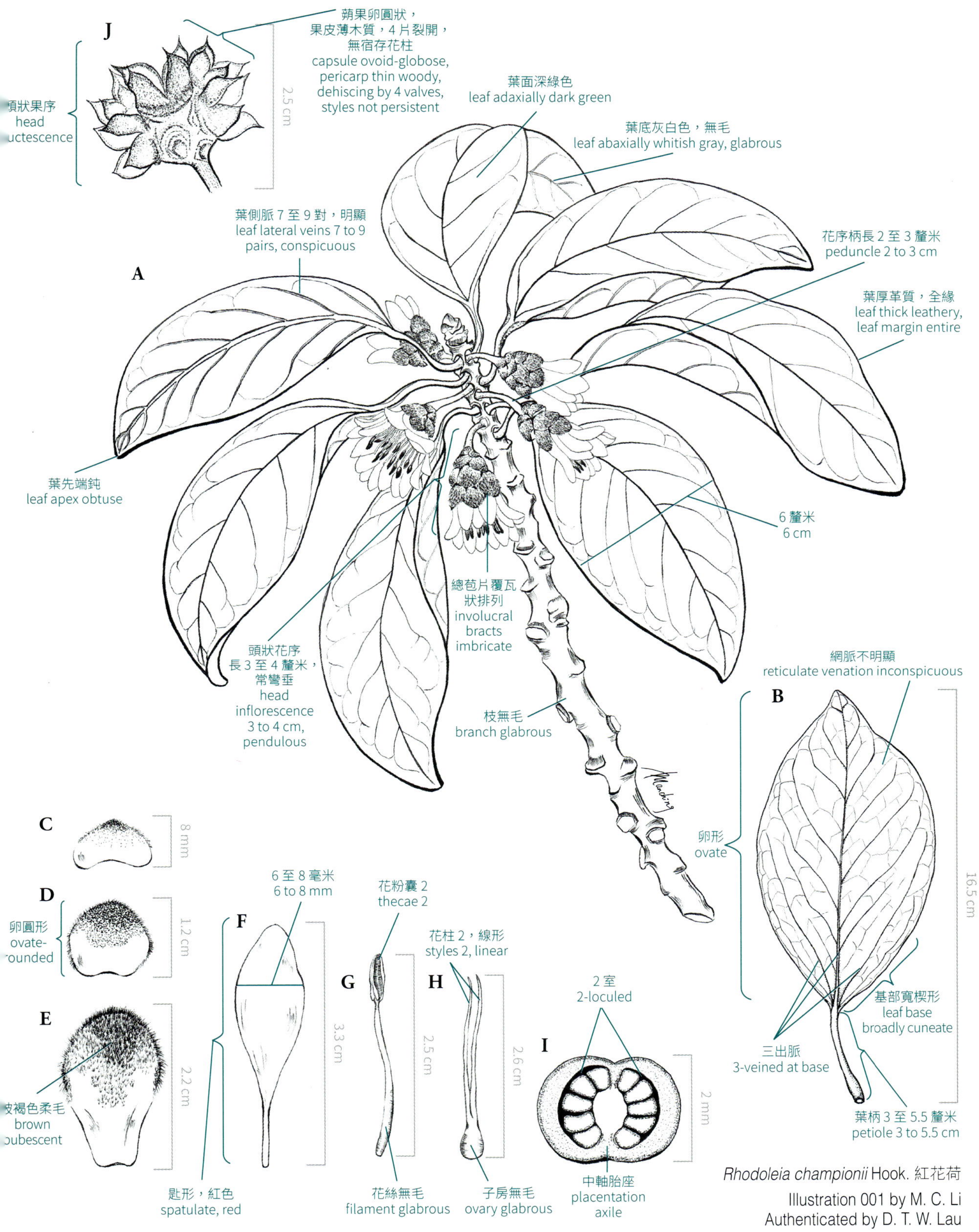

Rhodoleia championii Hook. 紅花荷
Illustration 001 by M. C. Li
Authenticated by D. T. W. Lau

植物結構 PLANT STRUCTURES:
A. 花枝 Flowering branch **B.** 葉底 Leaf abaxial **C–E.** 苞片 General shape of bracts **F.** 花瓣 Petal **G.** 雄蕊 Stamen **H.** 雌蕊 Gynoecium **I.** 子房，橫切面 Ovary, in C. S. **J.** 果序，5 個蒴果 Infructescence, with 5 capsules

參考標本 REFERENCE SPECIMENS: M. C. Li 020 (CUSLSH 2931); Shiu Ying Hu 12089; Champion, #s.n. (P00749074)

參考文獻 REFERENCES:

Chang, H. T. (1979). Hamamelidaceae. In Delectis Florae Reipublicae Popularis Sinicae Agendae Academiae Sinicae Edita (Ed.), *Flora Reipublicae Popularis Sinicae* (Vol. 35(2), pp. 36–116). Science Press.
Chen, F. H. (1987). Hamamelidaceae. In South China Botanical Garden, the Chinese Academy of Sciences (Ed.), *Flora of Guangdong* (Vol. 1, pp. 151–167). Guangdong Science and Technology Press.
Stevens, P. F. (2022). Saxifragales. Angiosperm Phylogeny Website. Retrieved March 17, 2022, from http://www.mobot.org/MOBOT/research/APweb/welcome.html
Xia, N. H. (2007). Hamamelidaceae. In Hong Kong Herbarium, AFCD & South China Botanical Garden, CAS (Eds.), *Flora of Hong Kong* (Vol. 1, pp. 88–95). Agriculture, Fisheries and Conservation Department, HKSAR Government.
Zhang, Z. Y., Zhang, H. D., & Endress, P. K. (2003). Hamamelidaceae. In Z. Y. Wu, P. H. Raven & D. Y. Hong (Eds.), *Flora of China* (Vol. 9, pp. 18–42). Science Press & Missouri Botanical Garden Press.
Zhang, Z. Y. (2017). Hamamelidaceae. In Fairy Lake Botanical Garden, Shenzhen & Chinese Academy of Sciences (Eds.), *Flora of Shenzhen* (Vol. 1, pp. 397–407). China Forestry Publishing House.

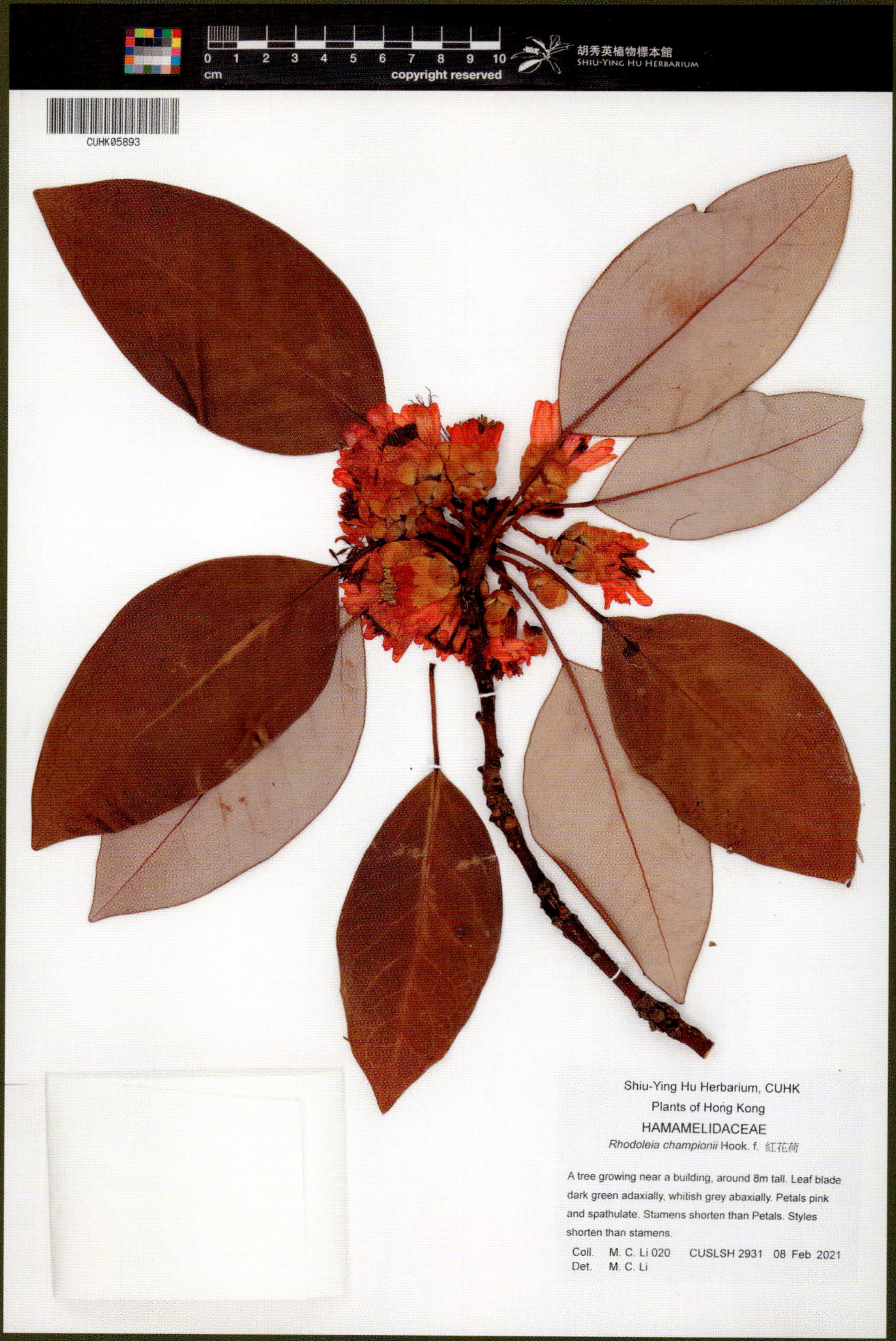

採集者編號 Collector no.
M. C. Li 020

採集年份 Year of collection
2021

外觀 Overview

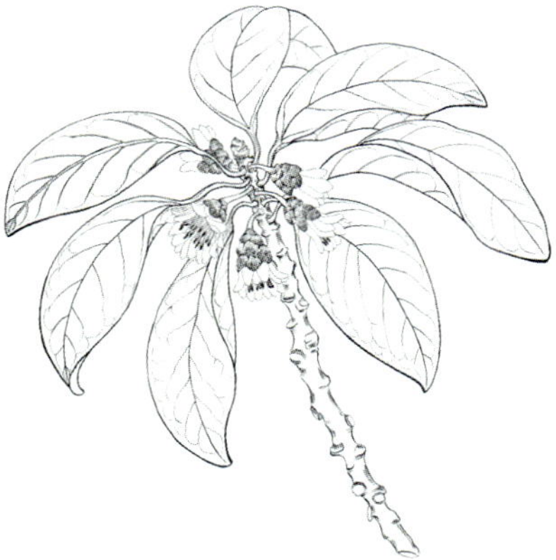

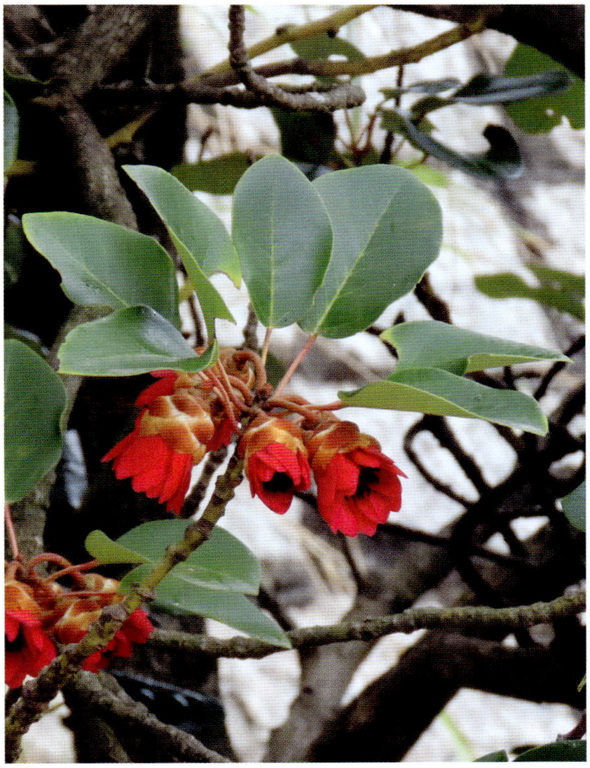

花枝 Flowering branch

葉 Leaf

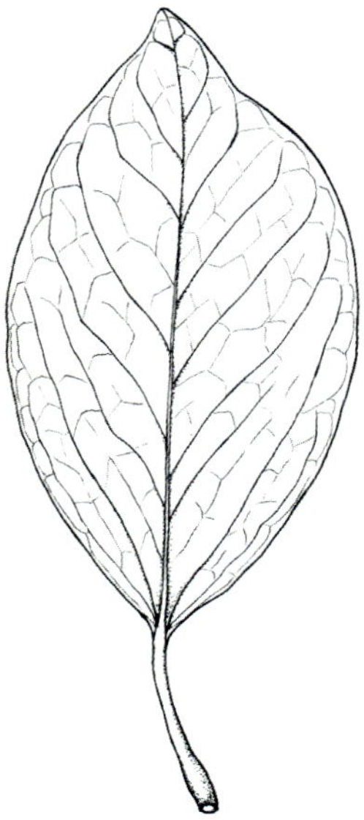

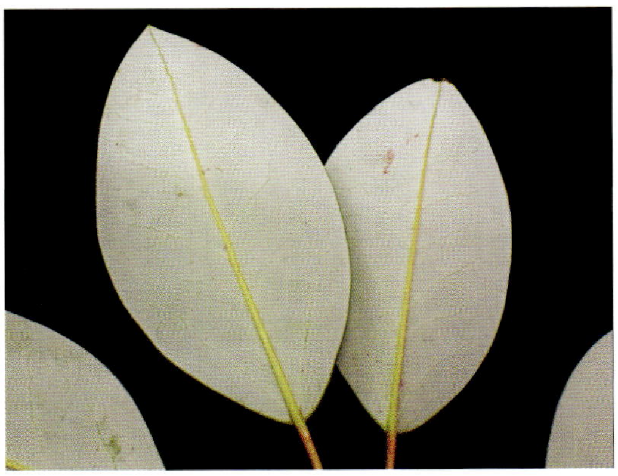

葉底 Leaf, abaxial

花 Flower

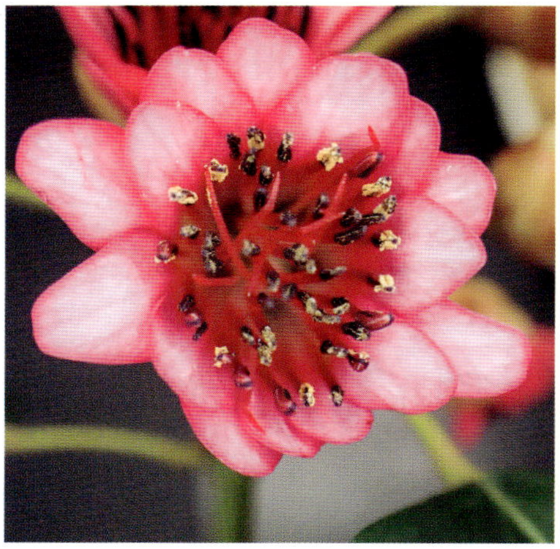

花，正面觀 Flower, frontal view

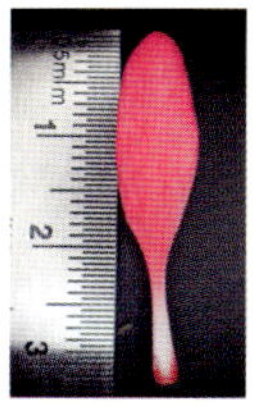

花瓣
Petal

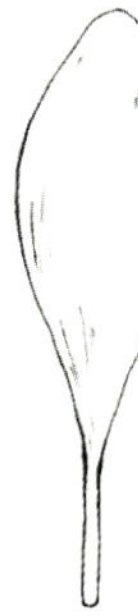

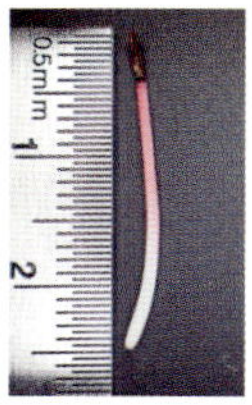

雄蕊
Stamen

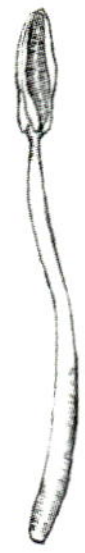

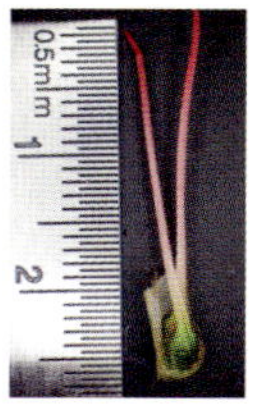

雌蕊
Gynoecium

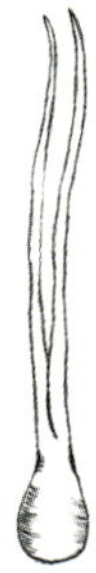

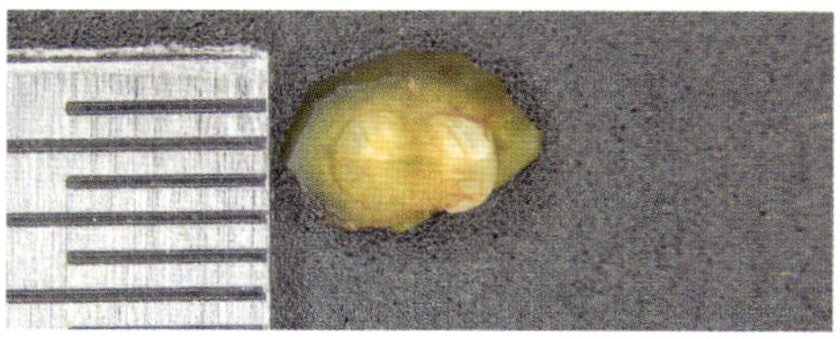

子房，橫切面 Ovary, in C. S.

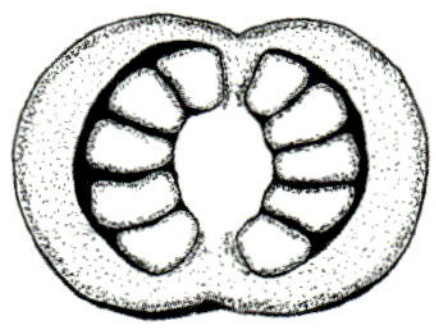

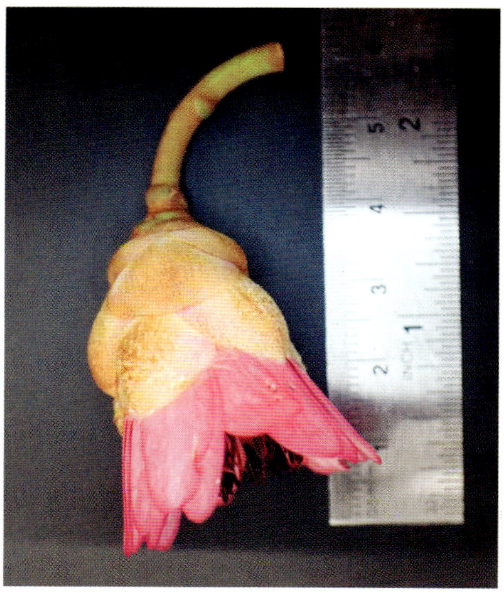

花，側面觀 Flower, lateral view

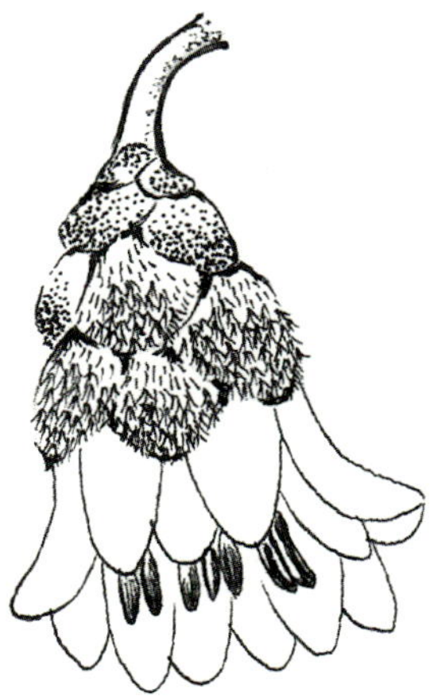

花 Flower

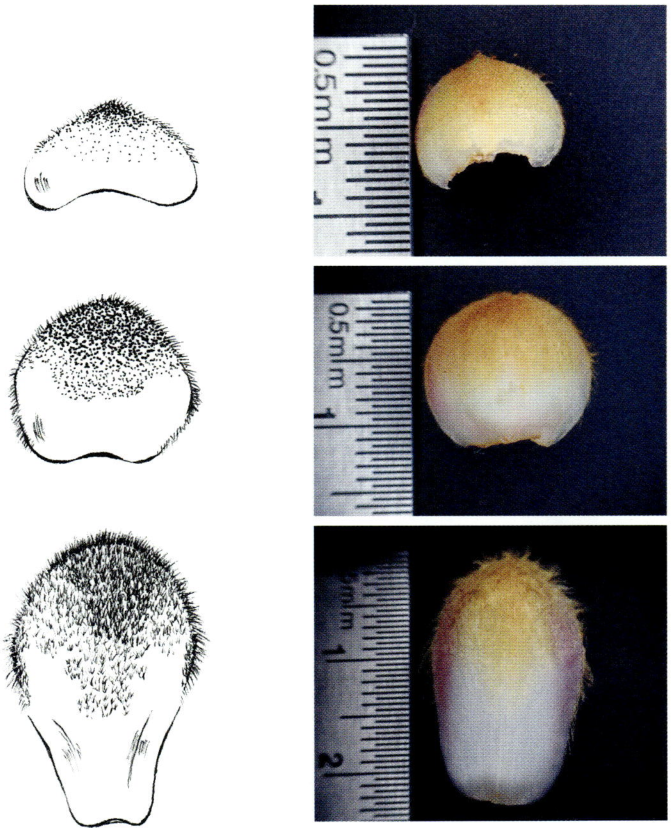

苞片 General shape of bracts

果 Fruit

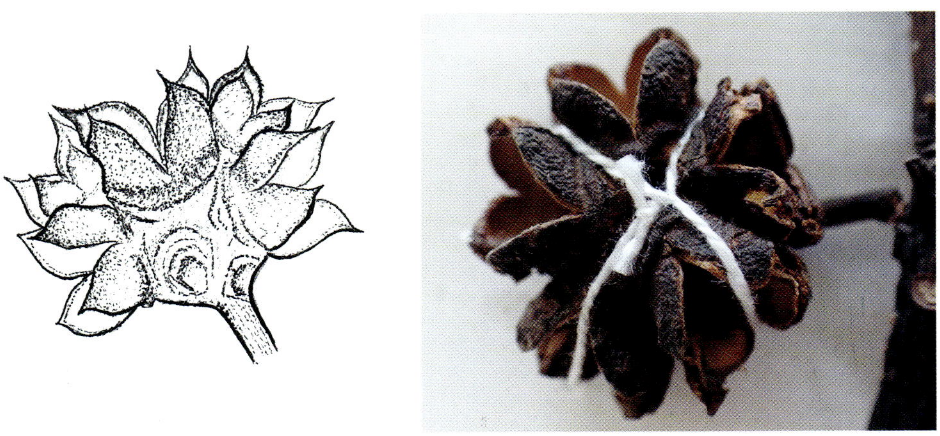

蒴果 Capsules

Dalbergia assamica Benth.
南嶺黃檀

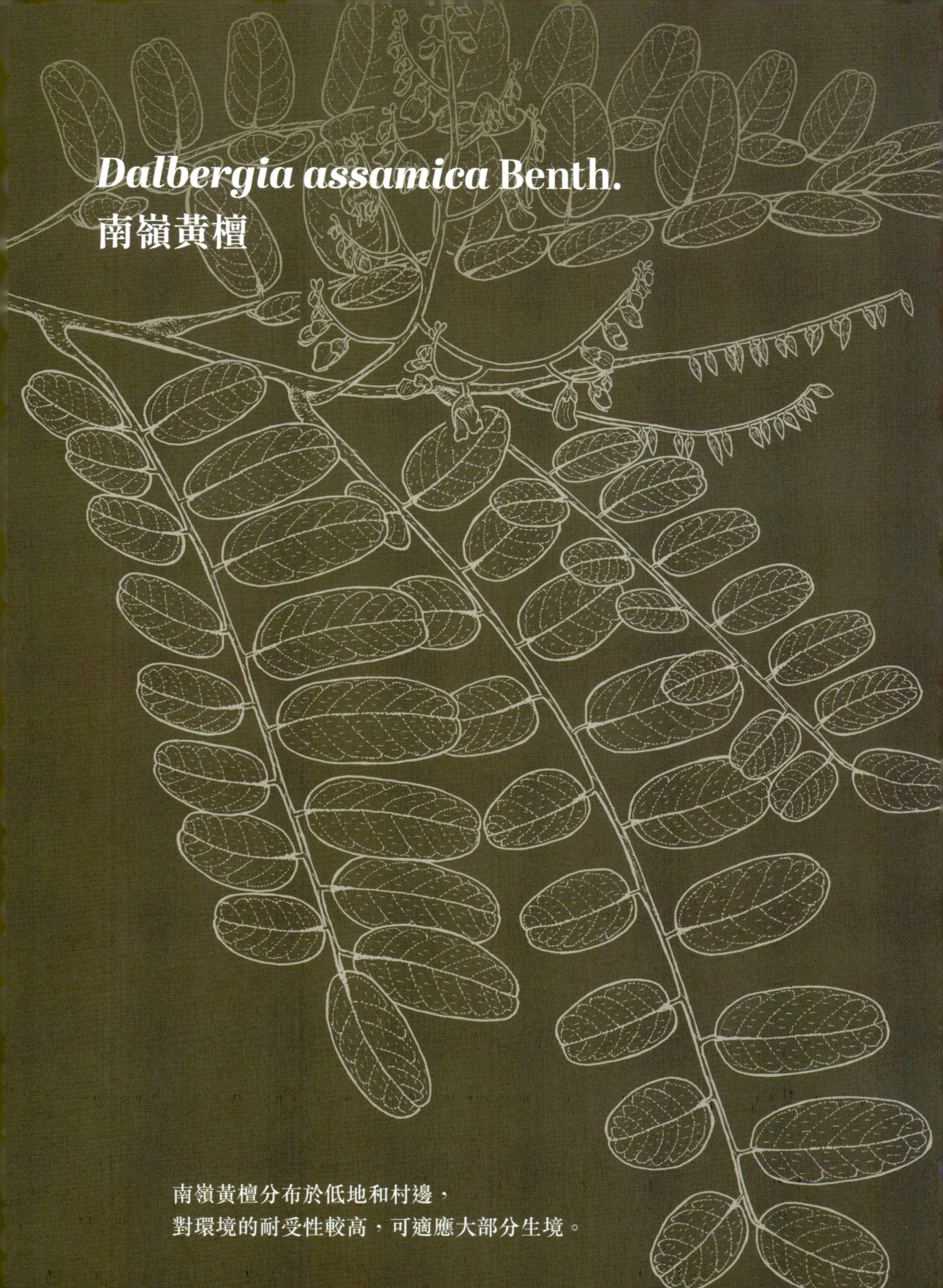

南嶺黃檀分布於低地和村邊，
對環境的耐受性較高，可適應大部分生境。

南嶺黃檀

學名	*Dalbergia assamica* Benth.
中文名	南嶺黃檀
科名	豆科
本地分布狀態	原生植物
繪圖編號	M. C. Li 010

花期	花色	果期	果色
5 至 10 月	白色	9 至 12 月	綠色

南嶺黃檀分布於低地和村邊，對環境的耐受性較高，可適應大部分生境。在阿薩姆邦，人們會種植南嶺黃檀以用作遮蔭和防風。

南嶺黃檀受香港法例第 586 章《保護瀕危動植物物種條例》規管，於國際自然保護聯盟瀕危物種紅色名錄中被列為「無危」物種。

《中國植物誌》(英文版) 的採用名稱
Dalbergia assamica Bentham (科名：豆科)

《香港植物誌》的採用名稱
Dalbergia assamica Benth. (科名：蝶形花科)

參考文獻 References

1. Hong Kong Herbarium. (2022). *HK Herbarium-Dalbergia assamica*. HK Plant Database. Retrieved March 17, 2022, from https://herbarium.gov.hk/en/hk-plant-database/plant-detail/index.html?pType=species&oID=3672
2. IUCN. (2012). *Dalbergia assamica*. The IUCN Red List of Threatened Species. Retrieved March 17, 2022, from https://www.iucnredlist.org/species/19892099/20054530
3. Missouri Botanical Garden. (2022). *Name-Dalbergia assamica* Benth. Tropicos. Retrieved March 17, 2022, from https://www.tropicos.org/name/13017948

Dalbergia assamica Benth.

Scientific name	*Dalbergia assamica* Benth.
Chinese name	南嶺黃檀
Family	Fabaceae Lindl.
Local distribution status	Native species
Illustration number	M. C. Li 010

FLOWERING PERIOD	COLOUR OF FLOWER	FRUITING PERIOD	COLOUR OF FRUIT
May to October	White	September to December	Green

Dalbergia assamica Benth. is distributed in lowlands and around villages. Its tolerance to environmental stress is relatively high and it can adapt to most habitats. In Assam, people grow *Dalbergia assamica* Benth. to provide shade and shelter.

Dalbergia assamica Benth. is regulated under the Protection of Endangered Species of Animals and Plants Ordinance (Cap. 586) and is listed as 'Least Concern' in the IUCN Red List of Threatened Species.

Name in *Flora of China*
Dalbergia assamica Bentham (Family: Fabaceae)

Name in *Flora of Hong Kong*
Dalbergia assamica Benth. (Family: Fabaceae)

4. Xu, L. R., Chen, D. Z., Zhu, X. Y., Huang, P. H., Wei, Z., Sa, R., Zhang, D. X., Bao, B. J., Wu, D. L., Sun, H., Gao, X. F., Liu, Y. X., Chang, Z. Y., Li, J. Q., Zhang, M. L., Podlech, D., Ohashi, H., Larsen, K., Welsh, S. L., Vincent, M. A., Gilbert, M. G., Pedley, L., Schrire, B. D., Yakovlev, G. P., Thulin, M., Nielsen, I. C., Choi, B. H., Turland, N. J., Polhill, R. M., Larsen, S. S., Ding, H., Iokawa, Y., Melanie, W. D., Kenicer, G., Nemoto, T., Lock, M., Salinas, A. D., Kramina, T. E., Brach, A. R., Bartholomew, B., & Sokoloff, D. D. (2010). Fabaceae. In Z. Y. Wu, P. H. Raven & D. Y. Hong (Eds.), *Flora of China* (Vol. 10, pp. 1–577). Science Press & Missouri Botanical Garden Press.

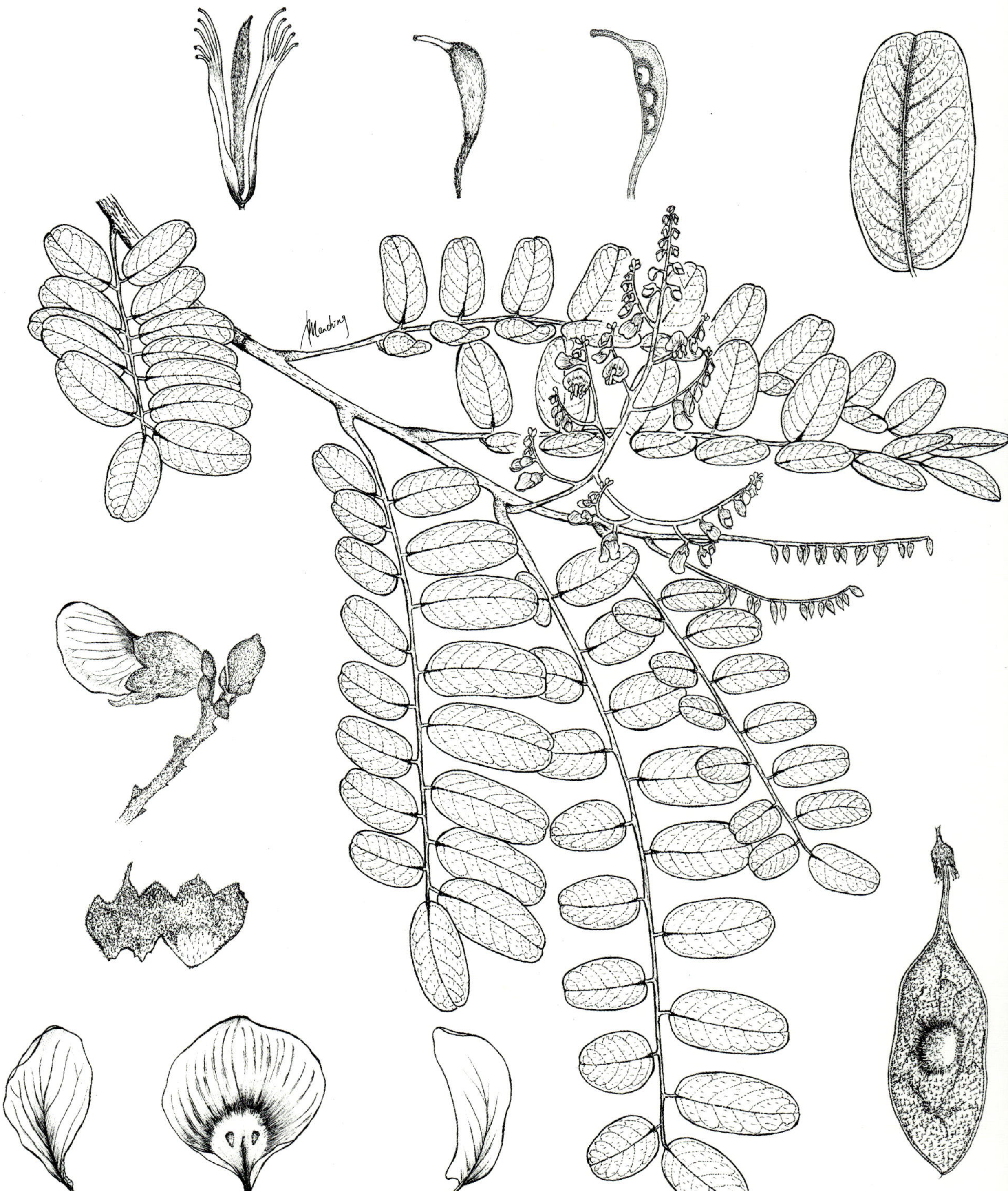

Dalbergia assamica Benth. 南嶺黃檀
Illustration 010 by M. C. Li
Authenticated by D. T. W. Lau

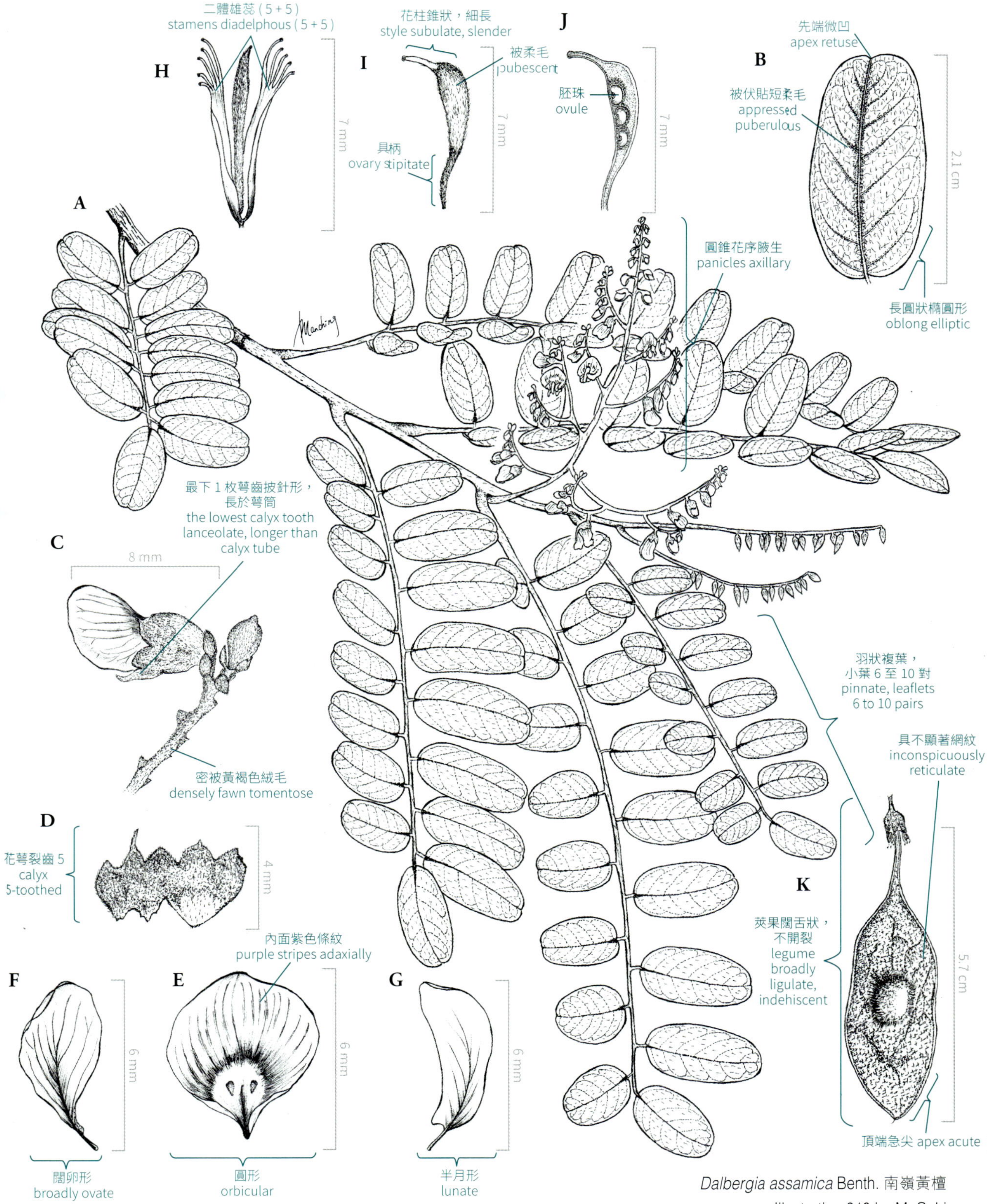

Dalbergia assamica Benth. 南嶺黃檀

Illustration 010 by M. C. Li

Authenticated by D. T. W. Lau

植物結構 PLANT STRUCTURES:

A. 花枝 Flowering branch **B.** 葉底 Leaf, abaxial **C.** 花，側面觀 Flower, lateral view **D.** 花萼 Calyx **E.** 旗瓣 Standard **F.** 翼瓣 Wing **G.** 龍骨瓣 Keel **H.** 雄蕊及雌蕊 Androecium with gynoecium **I.** 雌蕊 Gynoecium **J.** 雌蕊，縱切面 Gynoecium, in L. S. **K.** 果實 Fruit

參考標本 REFERENCE SPECIMENS: Rare and endangered plants (D. T. W. Lau, J. Y. Y. Lau, S. W. Shek, K. W. Lam) 20B; D. T. W. Lau 242 (CUSLSH2171)

參考文獻 REFERENCES:

Li, P. Q. (2010). Fabaceae. In Fairy Lake Botanical Garden, Shenzhen & Chinese Academy of Sciences (Eds.), *Flora of Shenzhen* (Vol. 2, pp. 278–379). China Forestry Publishing House.

Stevens, P. F. (2022). Fabales. Angiosperm Phylogeny Website. Retrieved March 17, 2022, from http://www.mobot.org/MOBOT/research/APweb/welcome.html

Wei, Z. (1994). Leguminosae. In Delectis Florae Reipublicae Popularis Sinicae Agendae Academiae Sinicae Edita (Ed.), *Flora Reipublicae Popularis Sinicae* (Vol. 40, pp. 1–327). Science Press.

Wu, D. L. (2008). Fabaceae. In Hong Kong Herbarium, AFCD & South China Botanical Garden, CAS (Eds.), *Flora of Hong Kong* (Vol. 2, pp. 59–119). Agriculture, Fisheries and Conservation Department, HKSAR Government.

Xu, L. R., Chen, D. Z., Zhu, X. Y., Huang, P. H., Wei, Z., Sa, R., Zhang, D. X., Bao, B. J., Wu, D. L., Sun, H., Gao, X. F., Liu, Y. X., Chang, Z. Y., Li, J. Q., Zhang, M. L., Podlech, D., Ohashi, H., Larsen, K., Welsh, S. L., Vincent, M. A., Gilbert, M. G., Pedley, L., Schrire, B. D., Yakovlev, G. P., Thulin, M., Nielsen, I. C., Choi, B. H., Turland, N. J., Polhill, R. M., Larsen, S. S., Ding, H., Iokawa, Y., Melanie, W. D., Kenicer, G., Nemoto, T., Lock, M., Salinas, A. D., Kramina, T. E., Brach, A. R., Bartholomew, B., & Sokoloff, D. D. (2010). Fabaceae. In Z. Y. Wu, P. H. Raven & D. Y. Hong (Eds.), *Flora of China* (Vol. 10, pp. 1–577). Science Press & Missouri Botanical Garden Press.

憑證標本
Voucher Specimen

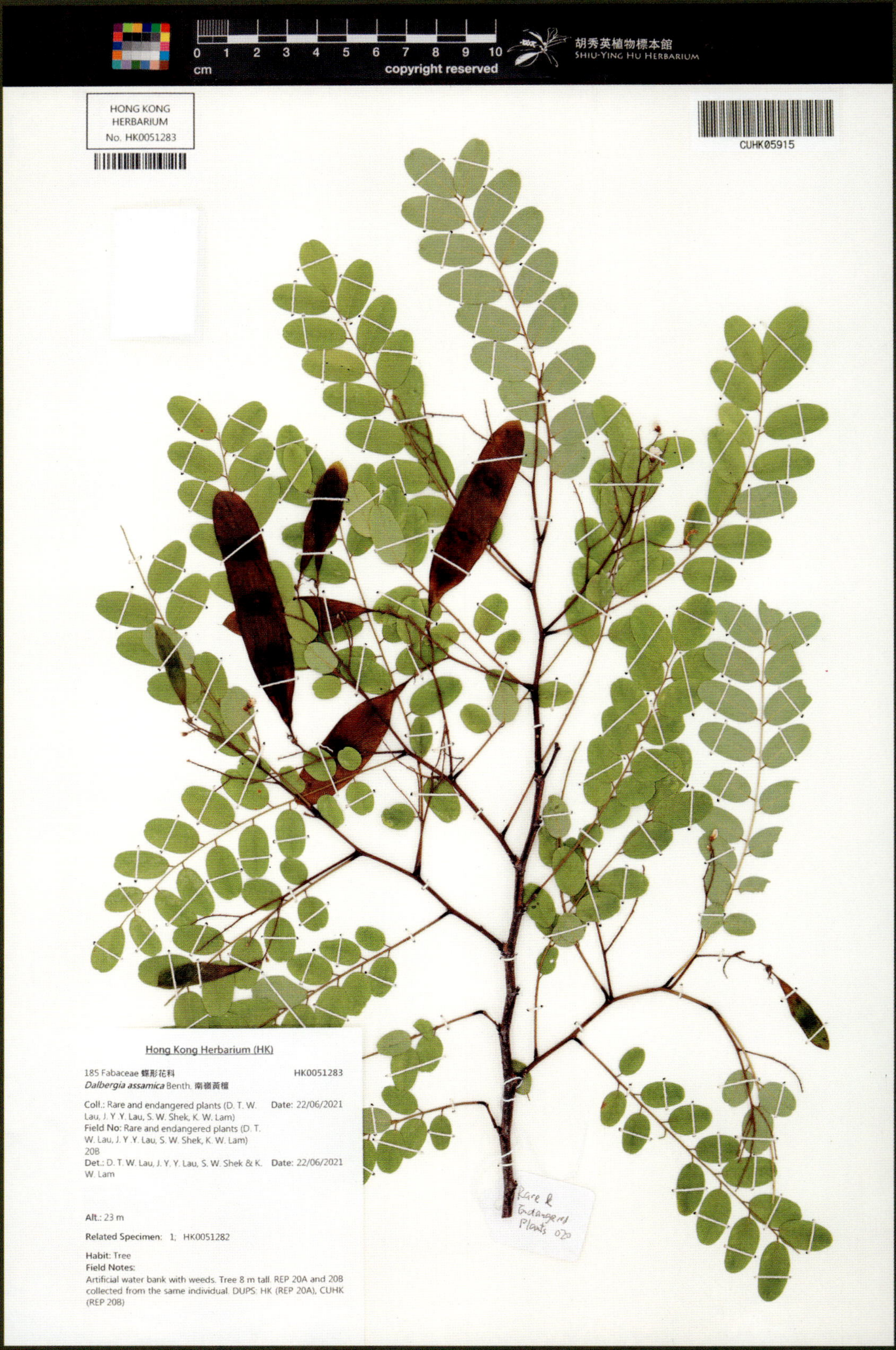

採集者編號 Collector no.
Rare and endangered plants (D. T. W. Lau, J. Y. Y. Lau, S. W. Shek, K. W. Lam) 20B

採集年份 Year of collection
2021

外觀 Overview

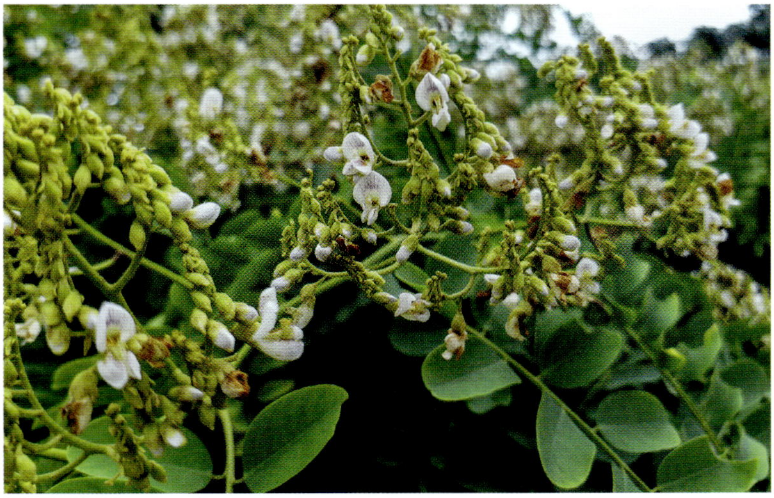

花枝 Flowering branch

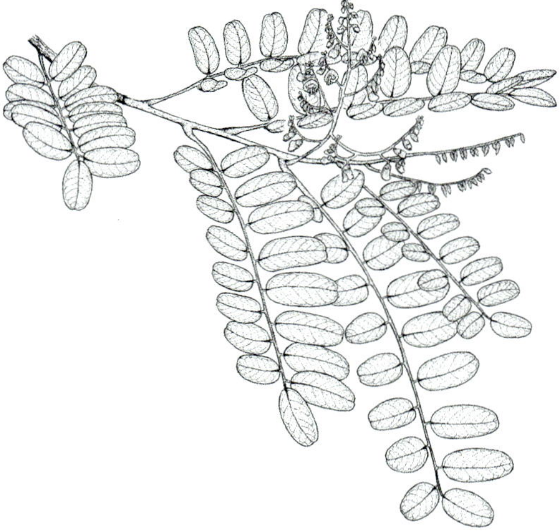

葉 Leaf

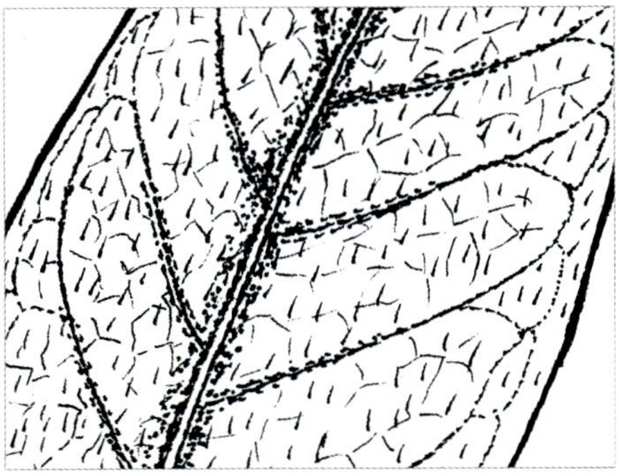

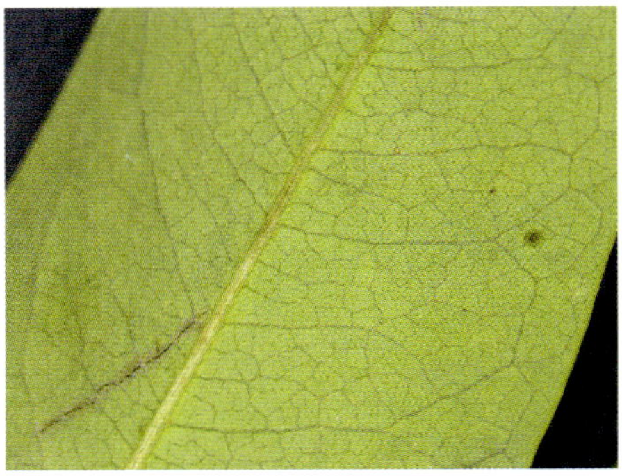

小葉葉底 Leaflet, abaxial

花 Flower

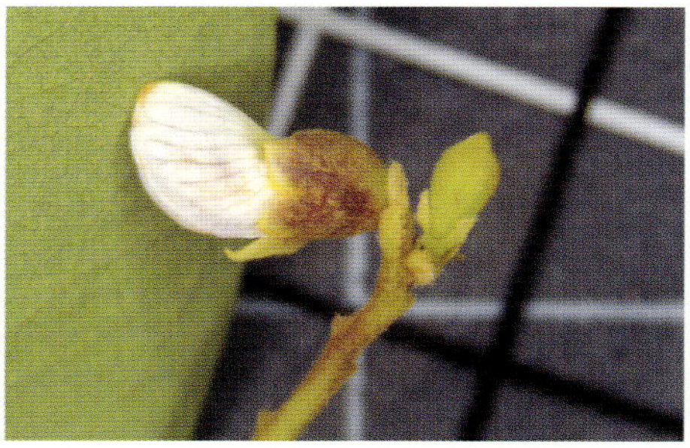

花，側面觀 Flower, lateral view

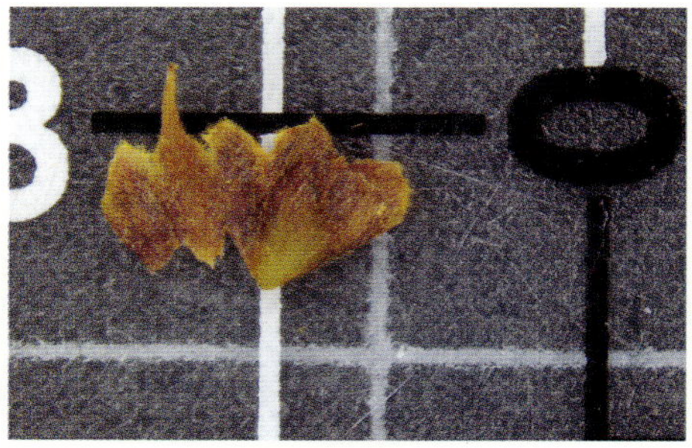

花萼 Calyx

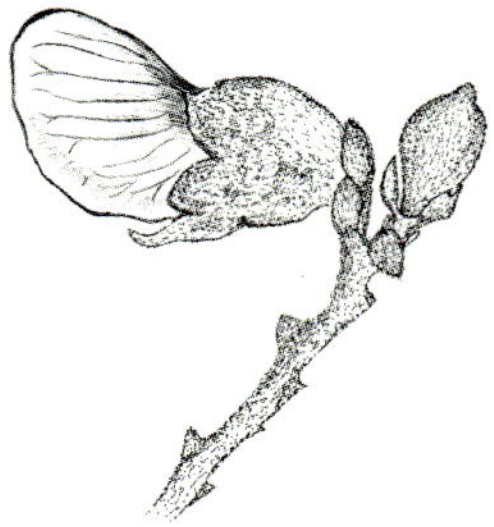

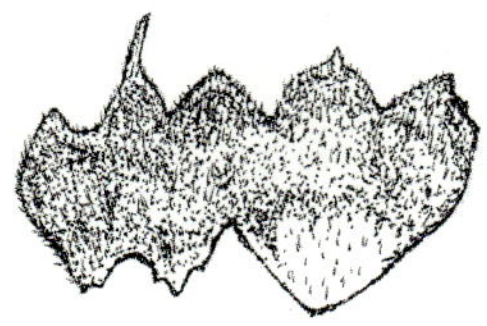

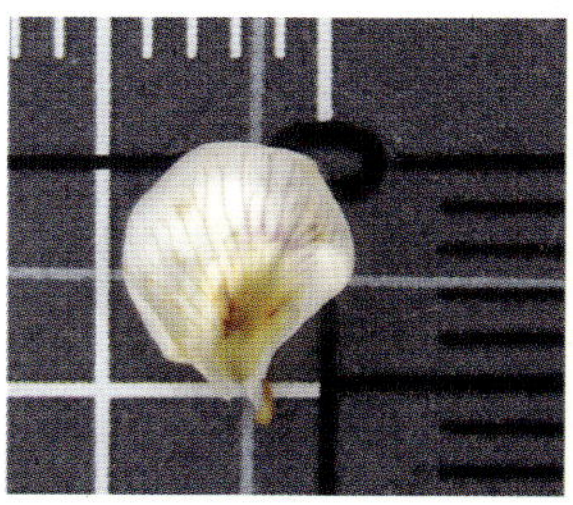

旗瓣 Standard

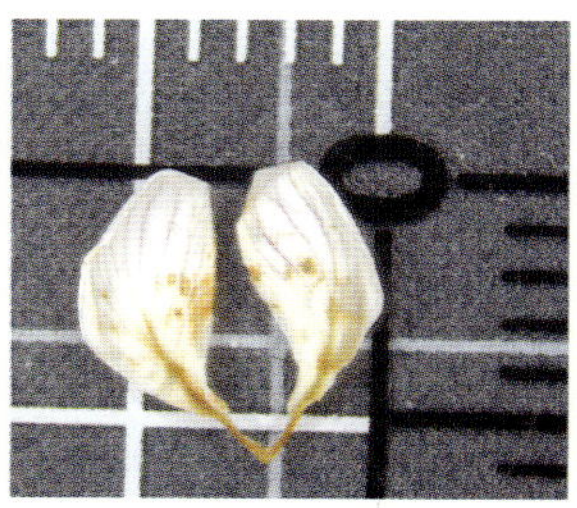

翼瓣 Wings

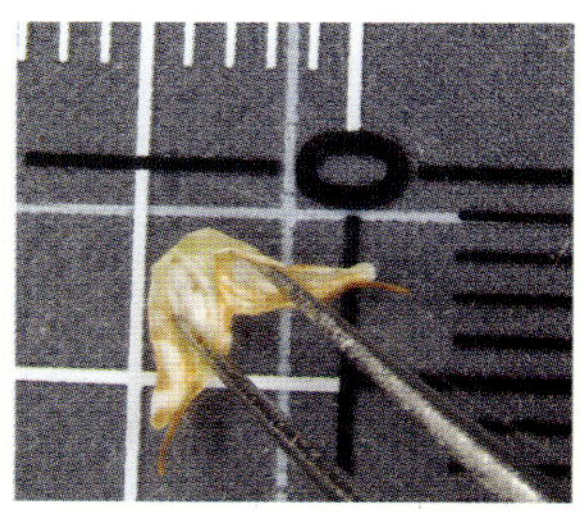

龍骨瓣 Keels

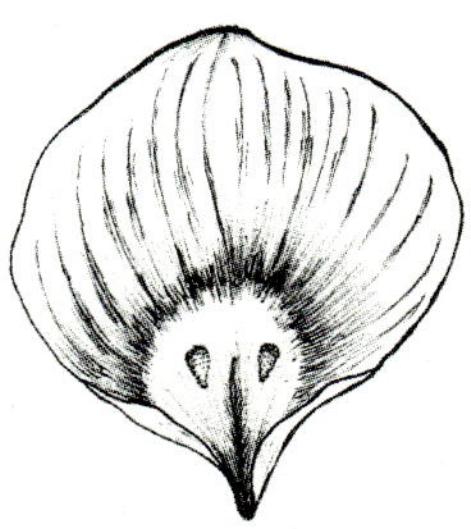

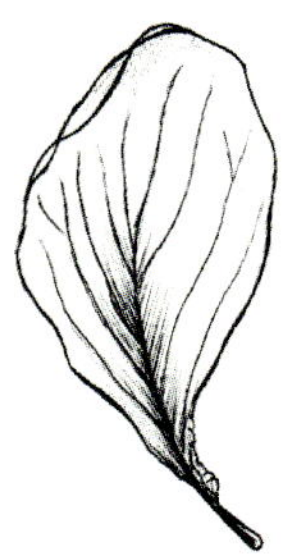

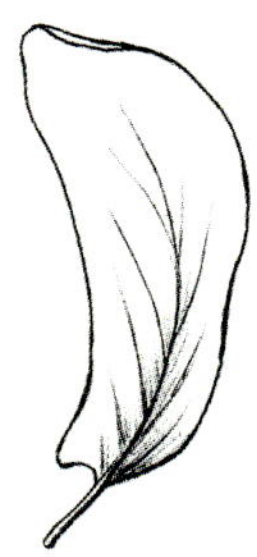

花 Flower

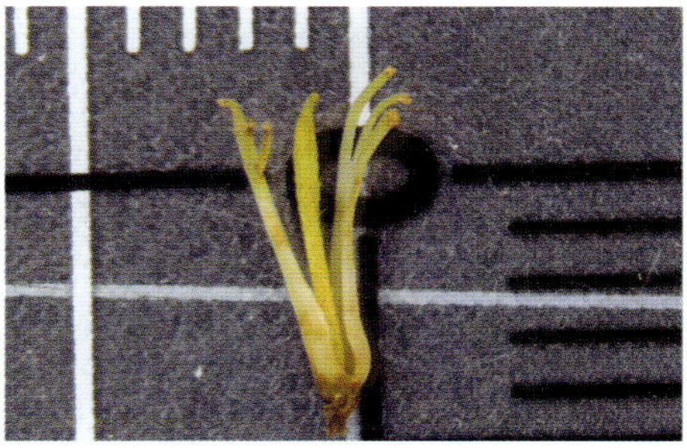

雄蕊及雌蕊
Androecium with gynoecium

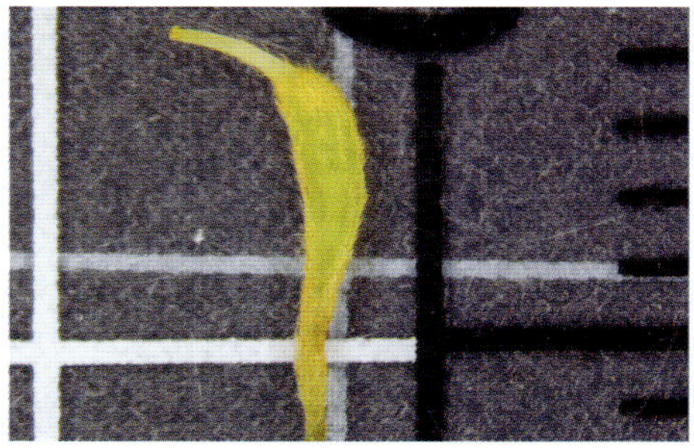

雌蕊 Gynoecium

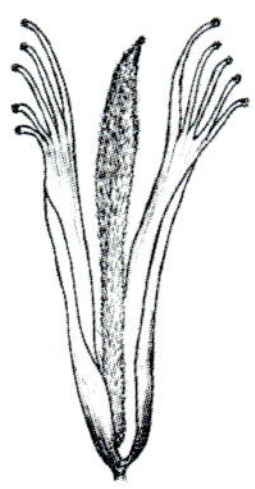

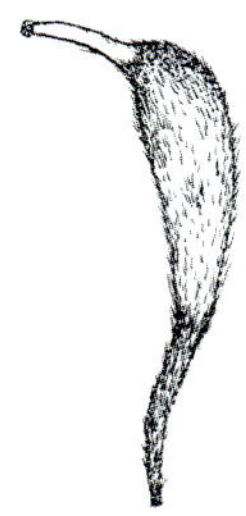

果 Fruit

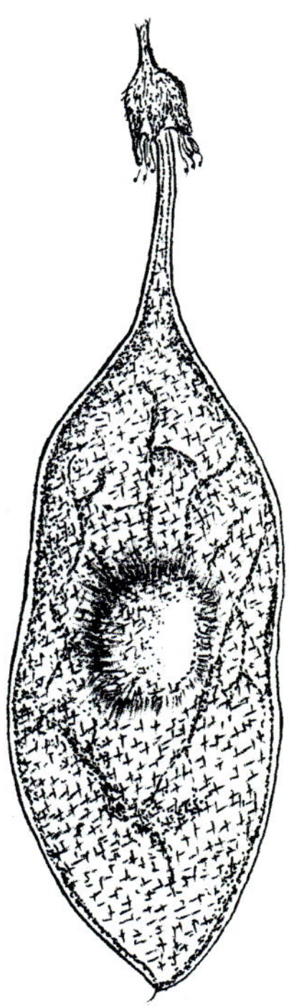

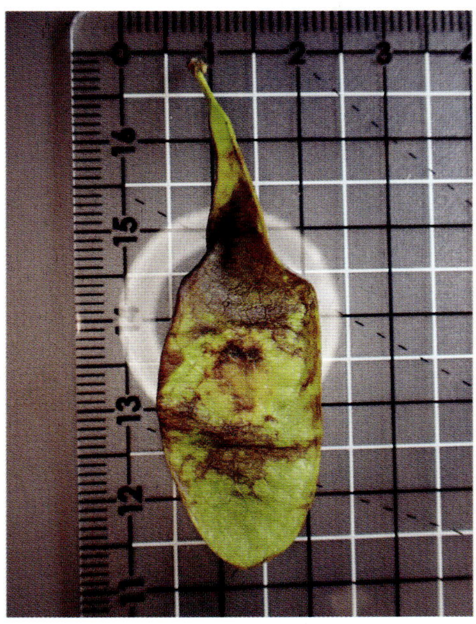

莢果 Legume

Dalbergia candenatensis (Dennst.) Prain
彎枝黃檀

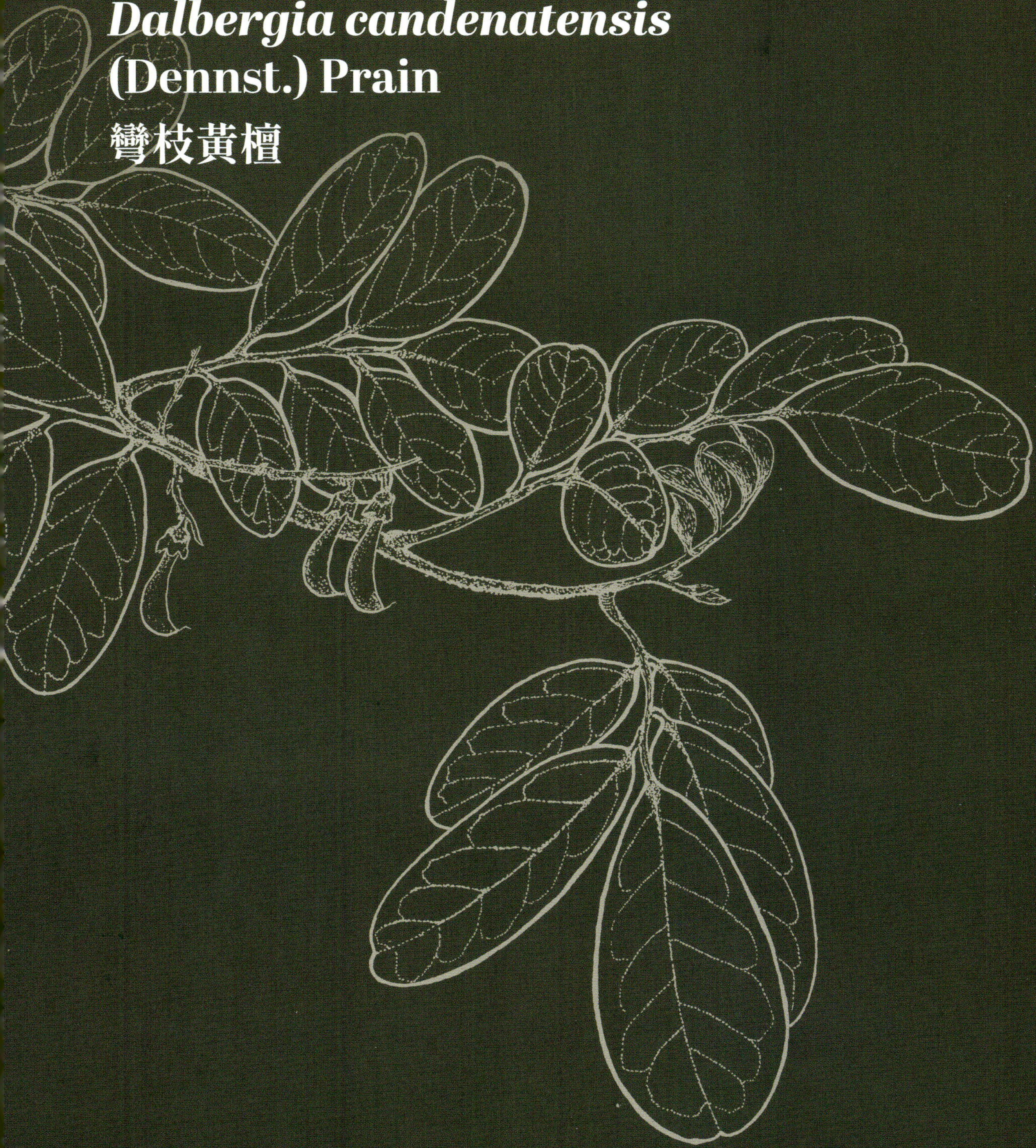

彎枝黃檀為木質攀援灌木，
已列入《瀕危野生動植物種國際貿易公約》附錄II。

彎枝黃檀

學名	*Dalbergia candenatensis* (Dennst.) Prain
中文名	彎枝黃檀
科名	豆科
本地分布狀態	原生植物
繪圖編號	M. C. Li 009

花期	花色	果期	果色
6月	白色	7至8月	綠色

彎枝黃檀為木質攀援灌木，生長於山林。

黃檀木一向被視作珍貴木材，過度的市場需求導致不少同屬的黃檀品種被非法砍伐。

黃檀屬的植物包括彎枝黃檀已被列入《瀕危野生動植物種國際貿易公約》附錄II，並受香港法例第586章《保護瀕危動植物物種條例》規管。

《中國植物誌》(英文版)的採用名稱

Dalbergia candenatensis (Dennstedt) Prain（科名：豆科）

《香港植物誌》的採用名稱

Dalbergia candenatensis (Dennst.) Prain（科名：蝶形花科）

參考文獻 References

1. Hong Kong Herbarium. (2022). *HK Herbarium-Dalbergia candenatensis*. HK Plant Database. Retrieved March 17, 2022, from https://herbarium.gov.hk/en/hk-plant-database/plant-detail/index.html?pType=species&oID=3671
2. Li, Q. W., Wu, J. H., Wang, Y. S., Lian, X. M., Wu, F. L., Zhou, L., Huang, Z. B., & Zhu, S. (2017). The phylogenetic analysis of *Dalbergia* (Fabaceae: Papilionaceae) based on different DNA barcodes. *Holzforschung 71*(12), 939–949. https://doi.org/10.1515/hf-2017-0052
3. Missouri Botanical Garden. (2022). *Name-Dalbergia candenatensis*. Tropicos. Retrieved March 17, 2022, from https://www.tropicos.org/name/13022636

Dalbergia candenatensis (Dennst.) Prain

Scientific name	*Dalbergia candenatensis* (Dennst.) Prain
Chinese name	彎枝黃檀
Family	Fabaceae Lindl.
Local distribution status	Native species
Illustration number	M. C. Li 009

FLOWERING PERIOD	COLOUR OF FLOWER	FRUITING PERIOD	COLOUR OF FRUIT
June	White	July to August	Green

Dalbergia candenatensis (Dennst.) Prain is a woody scandent shrub, growing in forests. The rosewood species is regarded as a luxury timber, illegal collection and tree felling were driven by strong market demand.

Dalbergia species, including *Dalbergia candenatensis* (Dennst.) Prain, have been included in CITES Appendix II, and are regulated under the Protection of Endangered Species of Animals and Plants Ordinance (Cap. 586).

Name in *Flora of China*
Dalbergia candenatensis (Dennstedt) Prain (Family: Fabaceae)

Name in *Flora of Hong Kong*
Dalbergia candenatensis (Dennst.) Prain (Family: Fabaceae)

4. The CITES Secretariat. (n.d.). *Dalbergia candenatensis.* Convention on International Trade in Endangered Species of Wild Fauna and Flora (CITES). Retrieved March 17, 2022, from https://cites.org/eng/taxonomy/term/11362

5. Xu, L. R., Chen, D. Z., Zhu, X. Y., Huang, P. H., Wei, Z., Sa, R., Zhang, D. X., Bao, B. J., Wu, D. L., Sun, H., Gao, X. F., Liu, Y. X., Chang, Z. Y., Li, J. Q., Zhang, M. L., Podlech, D., Ohashi, H., Larsen, K., Welsh, S. L., Vincent, M. A., Gilbert, M. G., Pedley, L., Schrire, B. D., Yakovlev, G. P., Thulin, M., Nielsen, I. C., Choi, B. H., Turland, N. J., Polhill, R. M., Larsen, S. S., Ding, H., Iokawa, Y., Melanie, W. D., Kenicer, G., Nemoto, T., Lock, M., Salinas, A. D., Kramina, T. E., Brach, A. R., Bartholomew, B., & Sokoloff, D. D. (2010). Fabaceae. In Z. Y. Wu, P. H. Raven & D. Y. Hong (Eds.), *Flora of China* (Vol. 10, pp. 1–577). Science Press & Missouri Botanical Garden Press.

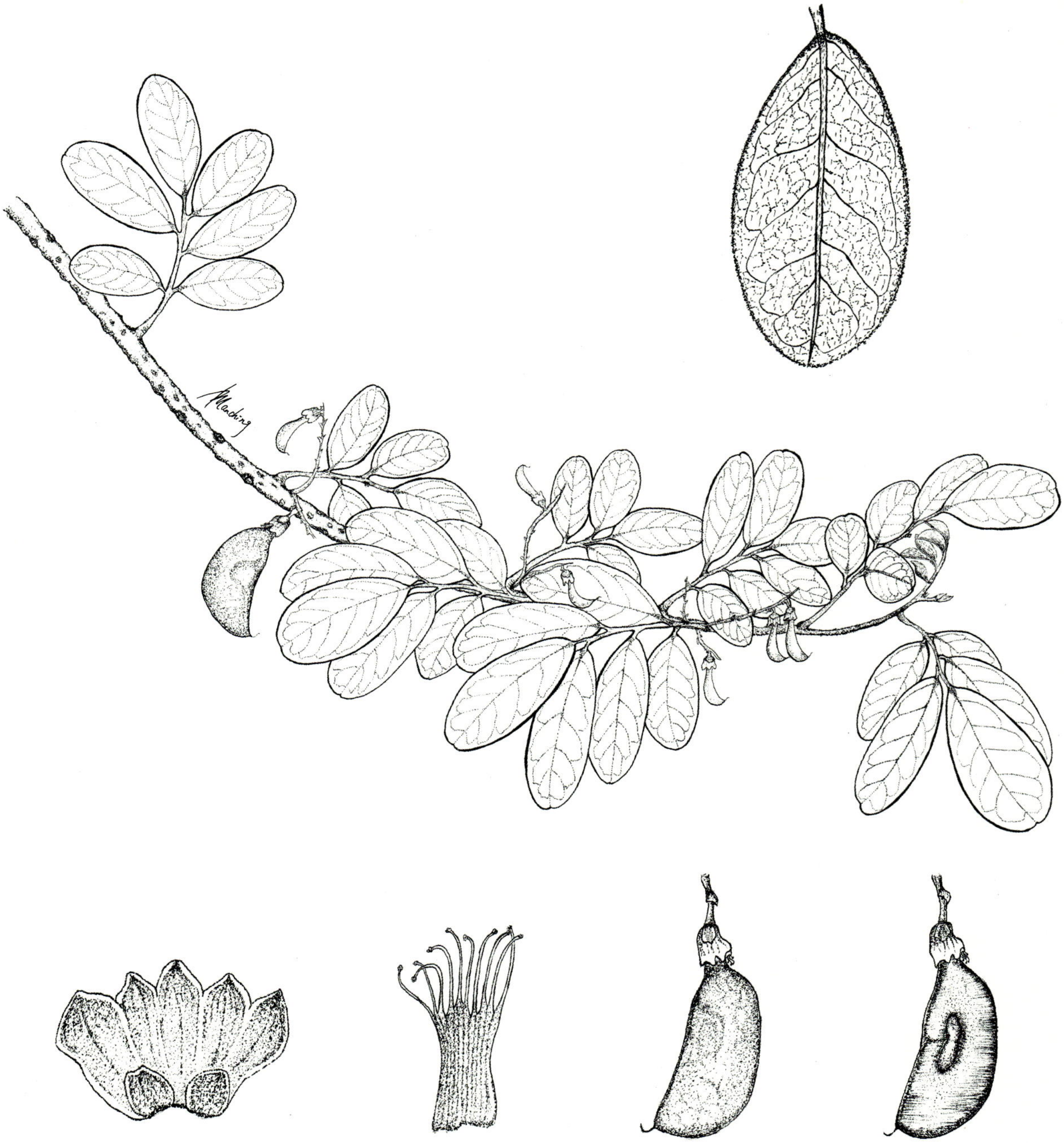

Dalbergia candenatensis (Dennst.) Prain 彎枝黃檀

Illustration 009 by M. C. Li
Authenticated by D. T. W. Lau

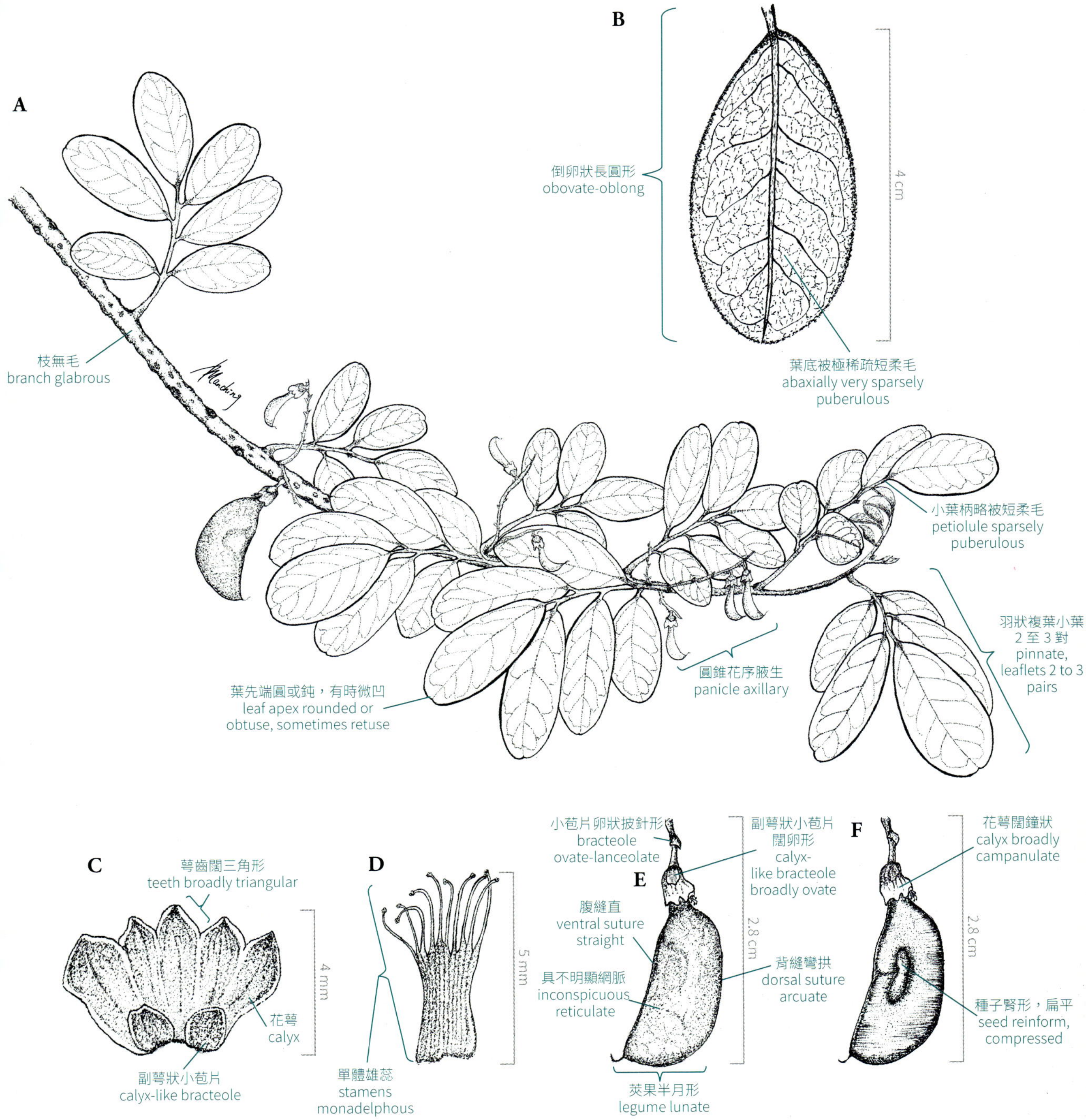

Dalbergia candenatensis (Dennst.) Prain 彎枝黃檀

Illustration 009 by M. C. Li
Authenticated by D. T. W. Lau

植物結構 PLANT STRUCTURES:
A. 果枝 Fruiting branch **B.** 葉底 Leaf, abaxial **C.** 花萼及小苞片 Calyx with bracteoles **D.** 雄蕊 Androecium **E.** 莢果 Legume **F.** 莢果，縱切面 Legume, in L. S.

參考標本 REFERENCE SPECIMENS: Rare and endangered plants (D. T. W. Lau, J. Y. Y. Lau, S. W. Shek, K. W. Lam) 16B

參考文獻 REFERENCES:

Li, P. Q. (2010). Fabaceae. In Fairy Lake Botanical Garden, Shenzhen & Chinese Academy of Sciences (Eds.), *Flora of Shenzhen* (Vol. 2, pp. 278–379). China Forestry Publishing House.

Stevens, P. F. (2022). Fabales. Angiosperm Phylogeny Website. Retrieved March 17, 2022, from http://www.mobot.org/MOBOT/research/APweb/welcome.html

Wei, Z. (1994). Leguminosae. In Delectis Florae Reipublicae Popularis Sinicae Agendae Academiae Sinicae Edita (Ed.), *Flora Reipublicae Popularis Sinicae* (Vol. 40, pp. 1–327). Science Press.

Wu, D. L. (2008). Fabaceae. In Hong Kong Herbarium, AFCD & South China Botanical Garden, CAS (Eds.), *Flora of Hong Kong* (Vol. 2, pp. 59–119). Agriculture, Fisheries and Conservation Department, HKSAR Government.

Xu, L. R., Chen, D. Z., Zhu, X. Y., Huang, P. H., Wei, Z., Sa, R., Zhang, D. X., Bao, B. J., Wu, D. L., Sun, H., Gao, X. F., Liu, Y. X., Chang, Z. Y., Li, J. Q., Zhang, M. L., Podlech, D., Ohashi, H., Larsen, K., Welsh, S. L., Vincent, M. A., Gilbert, M. G., Pedley, L., Schrire, B. D., Yakovlev, G. P., Thulin, M., Nielsen, I. C., Choi, B. H., Turland, N. J., Polhill, R. M., Larsen, S. S., Ding, H., Iokawa, Y., Melanie, W. D., Kenicer, G., Nemoto, T., Lock, M., Salinas, A. D., Kramina, T. E., Brach, A. R., Bartholomew, B., & Sokoloff, D. D. (2010). Fabaceae. In Z. Y. Wu, P. H. Raven & D. Y. Hong (Eds.), *Flora of China* (Vol. 10, pp. 1–577). Science Press & Missouri Botanical Garden Press.

憑證標本
Voucher Specimen

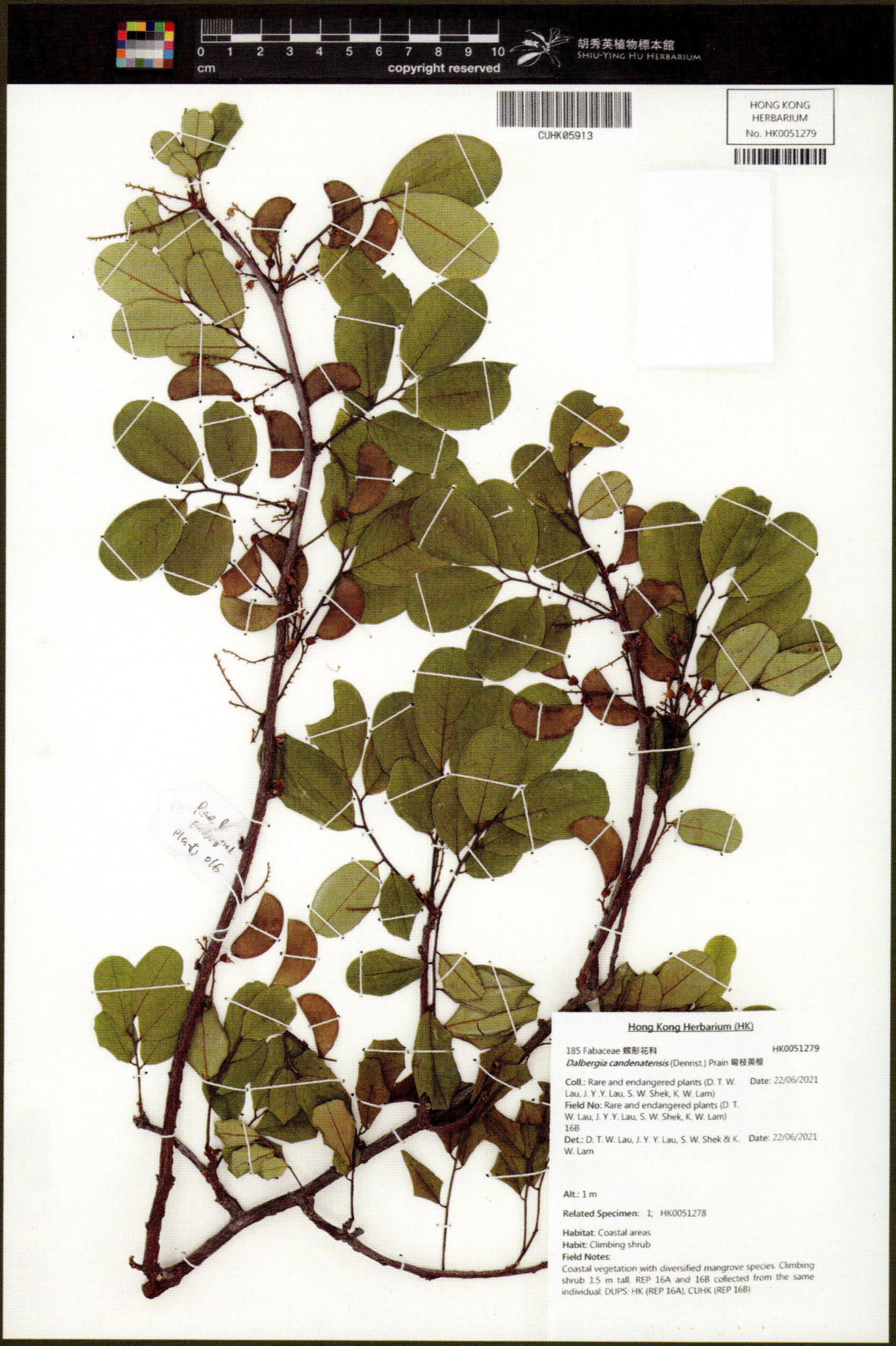

採集者編號 Collector no.
Rare and endangered plants (D. T. W. Lau, J. Y. Y. Lau, S. W. Shek, K. W. Lam) 16B

採集年份 Year of collection
2021

外觀 Overview

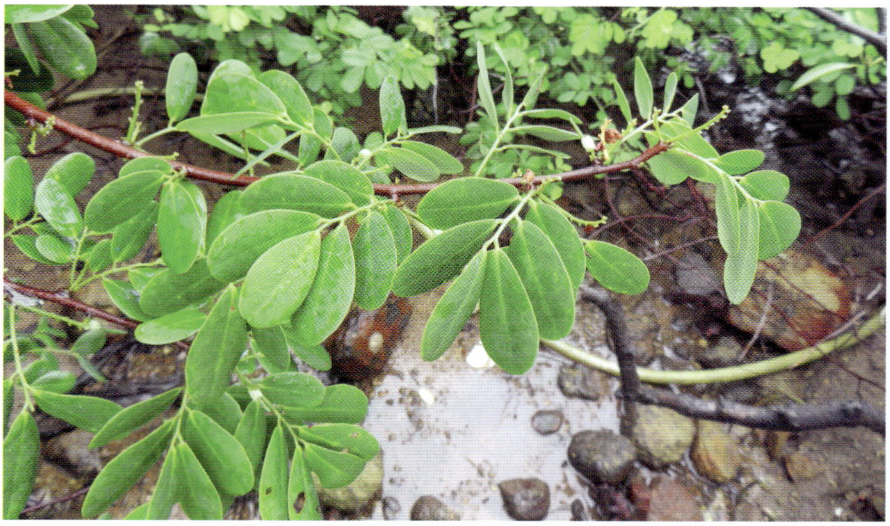

果枝 Fruiting branch

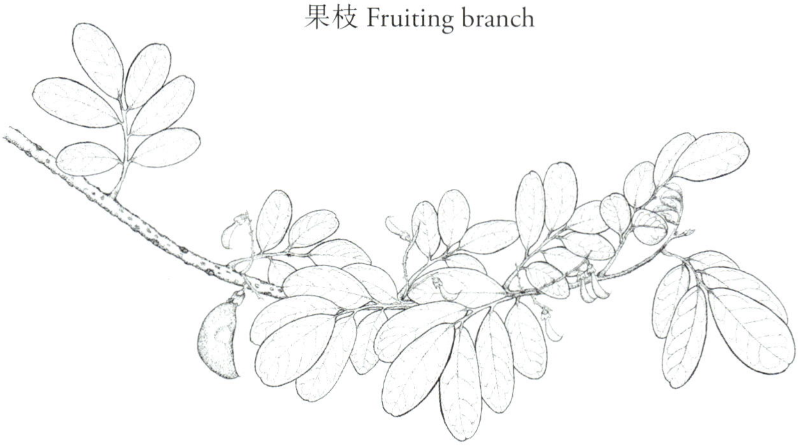

葉 Leaf

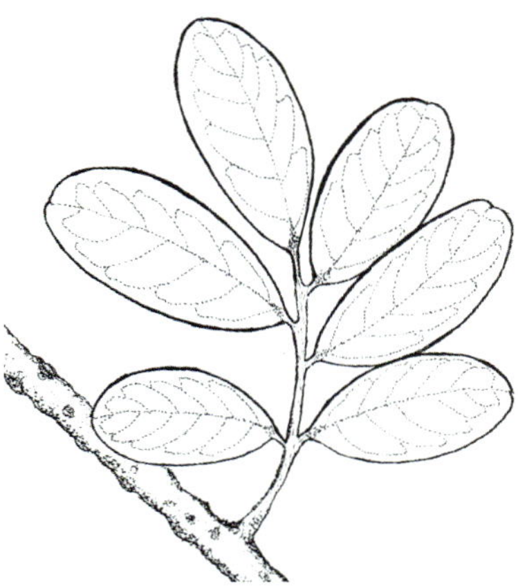

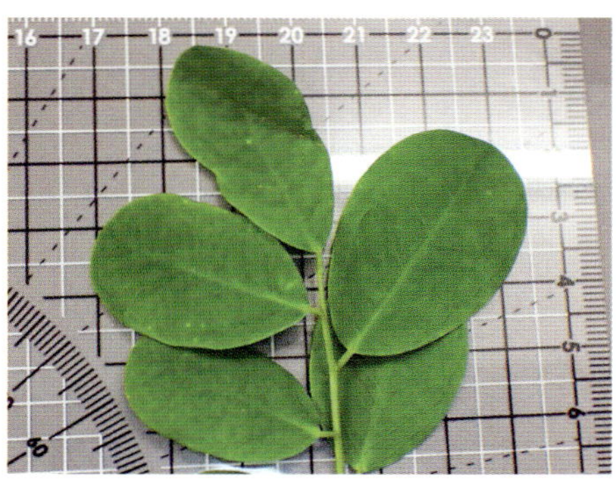

小葉葉面 Leaflets, adaxial

花 Flower

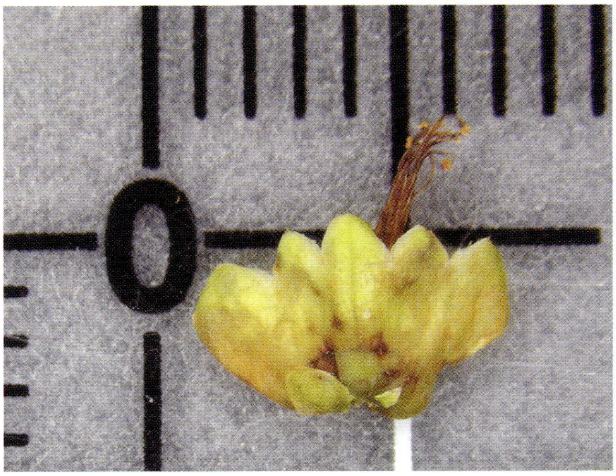
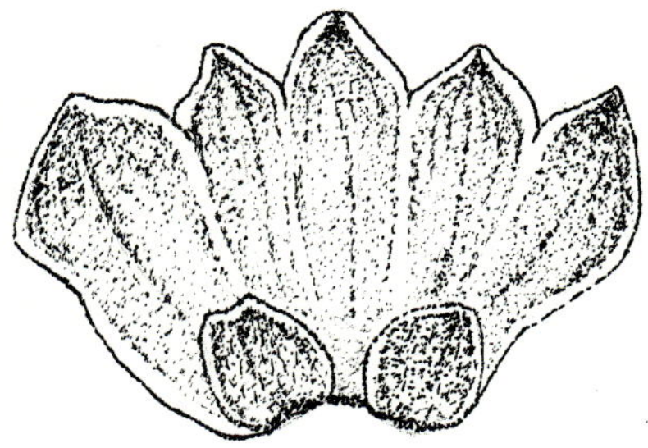

花萼及小苞片 Calyx with bracteoles

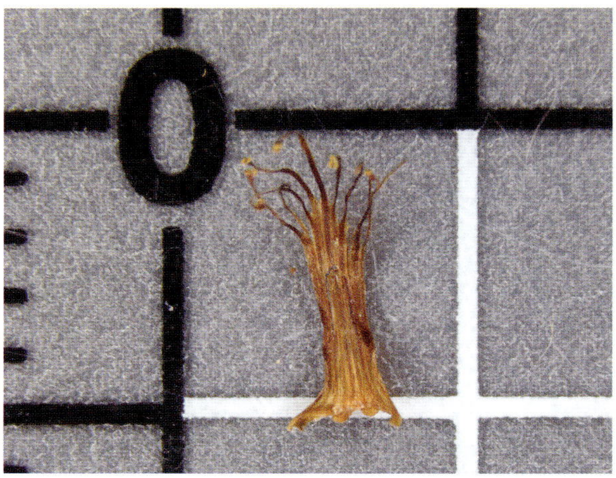
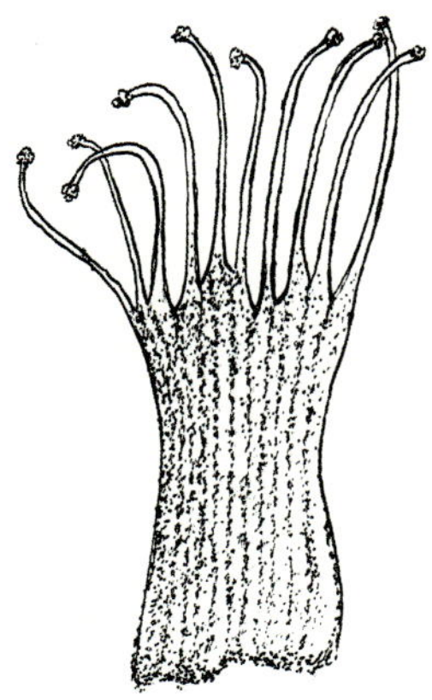

雄蕊 Androecium

果 Fruit

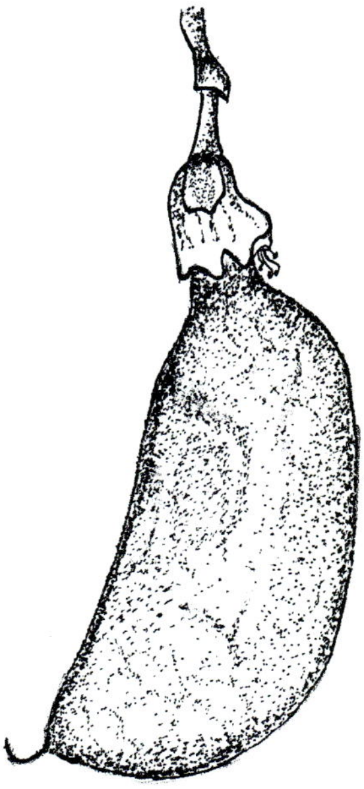

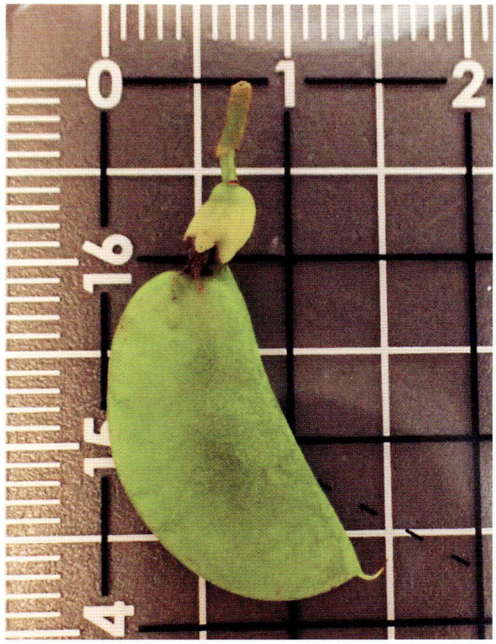

莢果 Legume

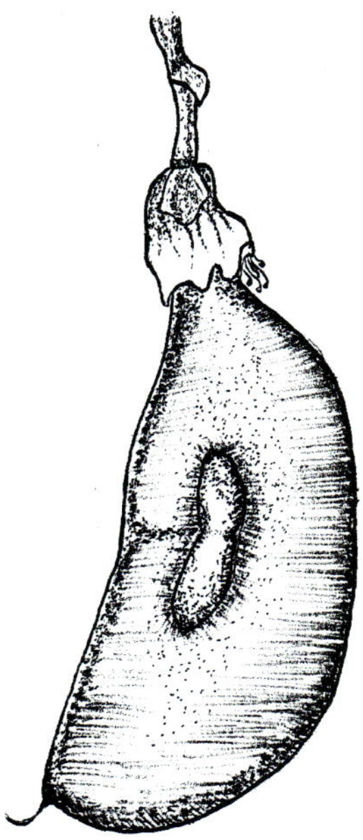

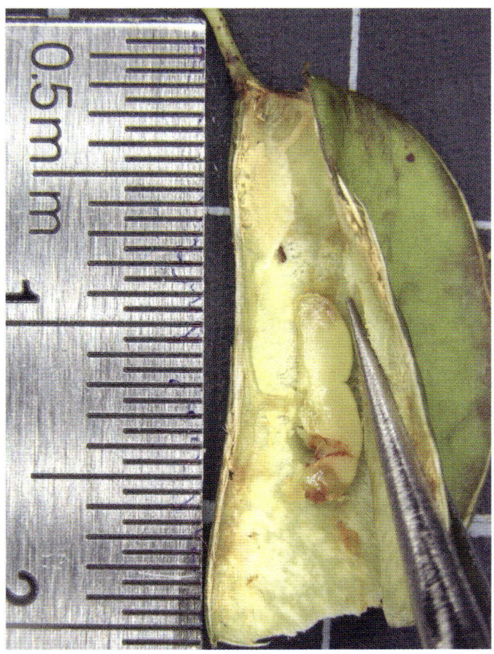

莢果，縱切面，展示種子
Legume, in L. S., showing the seed

Polygala hongkongensis Hemsl.
香港遠志

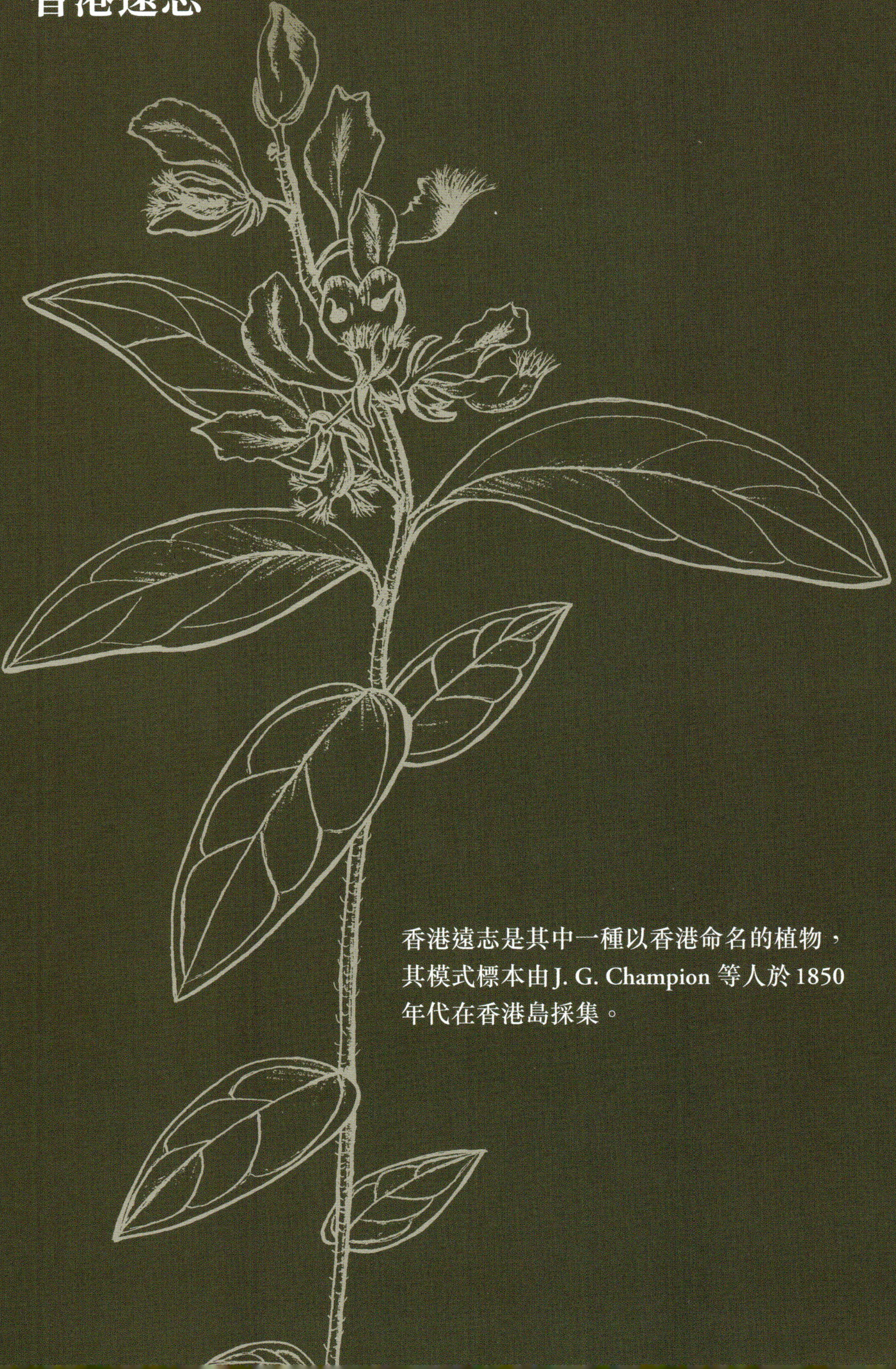

香港遠志是其中一種以香港命名的植物，
其模式標本由J. G. Champion 等人於 1850
年代在香港島採集。

胡秀英教授採集的標本
Specimen Collected by Professor Shiu-Ying Hu

採集者編號
Collector no.
Hu & But 20444

採集年份
Year of collection
1992

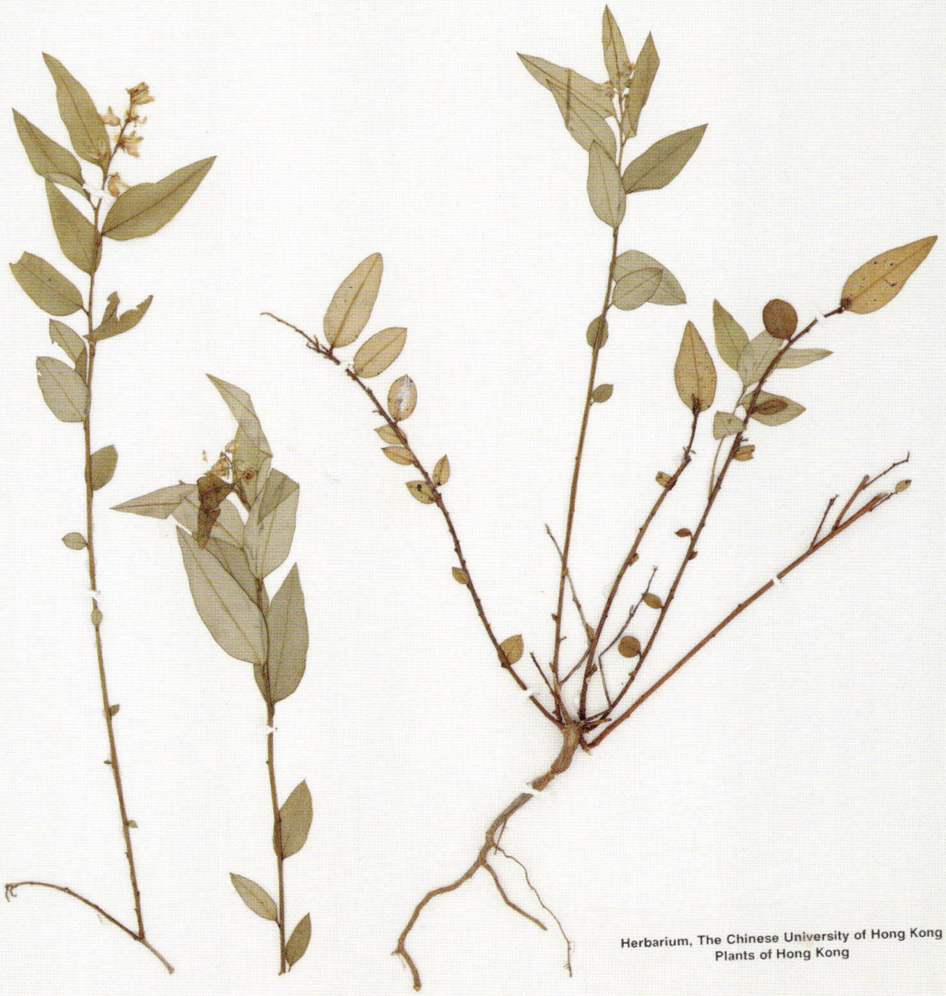

Herbarium, The Chinese University of Hong Kong
Plants of Hong Kong

Polygala hongkongensis Hemsl.

Violet Hill, H.K.

Perennial herbs, aerial stems 10-45 cm long, emerged from caudex; fls white with blue tint.

Coll. Hu & But 20444 Apr. 21, 1992
Det. S.Y. Hu

香港遠志

學名	*Polygala hongkongensis* Hemsl.
中文名	香港遠志
科名	遠志科
本地分布狀態	原生植物
繪圖編號	M. C. Li 003

花期	花色	果期	果色
5至6月	紫色	6至7月	未有記錄

香港遠志是其中一種以香港命名的多年生草本植物，其模式標本由 J. G. Champion 等人於 1850 年代在香港島採集。

其近親遠志（*Polygala tenuifolia* Willd.）是《中國藥典》中的常用中藥，早於 16 世紀《本草綱目》中已有記載其益智強志的功效，研究人員近年亦正開發香港遠志於相關方面的應用。

《中國植物誌》(英文版) 的採用名稱
Polygala hongkongensis Hemsley (科名: 遠志科)

《香港植物誌》的採用名稱
Polygala hongkongensis Hemsl. (科名: 遠志科)

參考文獻 References

1. Chen, S. K., Ma, H. Y., & Parnell, J. A. N. (2008). Polygalaceae. In Z. Y. Wu, P. H. Raven & D. Y. Hong (Eds.), *Flora of China* (Vol. 11, pp. 139–159). Science Press & Missouri Botanical Garden Press.
2. Hong Kong Herbarium. (2022). *HK Herbarium-Polygala hongkongensis.* HK Plant Database. Retrieved March 17, 2022, from https://herbarium.gov.hk/en/hk-plant-database/plant-detail/index.html?pType=species&oID=7571
3. Missouri Botanical Garden. (2022a). *Name-Polygala hongkongensis.* Tropicos. Retrieved March 17, 2022, from https://www.tropicos.org/name/50056178

Polygala hongkongensis Hemsl.

Scientific name	*Polygala hongkongensis* Hemsl.
Chinese name	香港遠志
Family	Polygalaceae Hoffmanns. & Link
Local distribution status	Native species
Illustration number	M. C. Li 003

FLOWERING PERIOD	COLOUR OF FLOWER	FRUITING PERIOD	COLOUR OF FRUIT
May to June	Violet	June to July	Not recorded

Polygala hongkongensis Hemsl. is one of the perennial herbs named after Hong Kong. During the 1850s, the type specimens were collected by J. G. Champion and others on Hong Kong Island.

Within the same genus, *Polygala tenuifolia* Willd. is officially listed in the *Pharmacopoeia of the People's Republic of China* and has been widely adopted as traditional Chinese medicine. Its memory-enhancing properties were recorded in *The Compendium of Materia Medica*, published during the 16th century. In recent years, researchers have also put effort into developing the medicinal use of *Polygala hongkongensis* Hemsl.

Name in *Flora of China*
Polygala hongkongensis Hemsley (Family: Polygalaceae)

Name in *Flora of Hong Kong*
Polygala hongkongensis Hemsl. (Family: Polygalaceae)

4. Missouri Botanical Garden. (2022b). *Name-Polygala tenuifolia*. Tropicos. Retrieved March 17, 2022, from https://www.tropicos.org/name/25900699

5. Zhan, H. T., Li, H. Y., Zeng, X. X., & Wu, J. F. (2012). Effects of mechanism of *Polygala hongkongensis* Hemsl extracts on learning and memory in Alzheimer's disease model rats. *Journal of Apoplexy and Nervous Diseases 29*(3), 243–245.

6. Zhang, M., Gao, X. H., Sun, X. M., Xu, J. T., Li, X. Y., Shi, Y. J., & Fan, X. T. (2007). Medical value and development of *Polygala tenuifolia*. *Journal of Beihua University (Natural Science) 8*(4), 322–327.

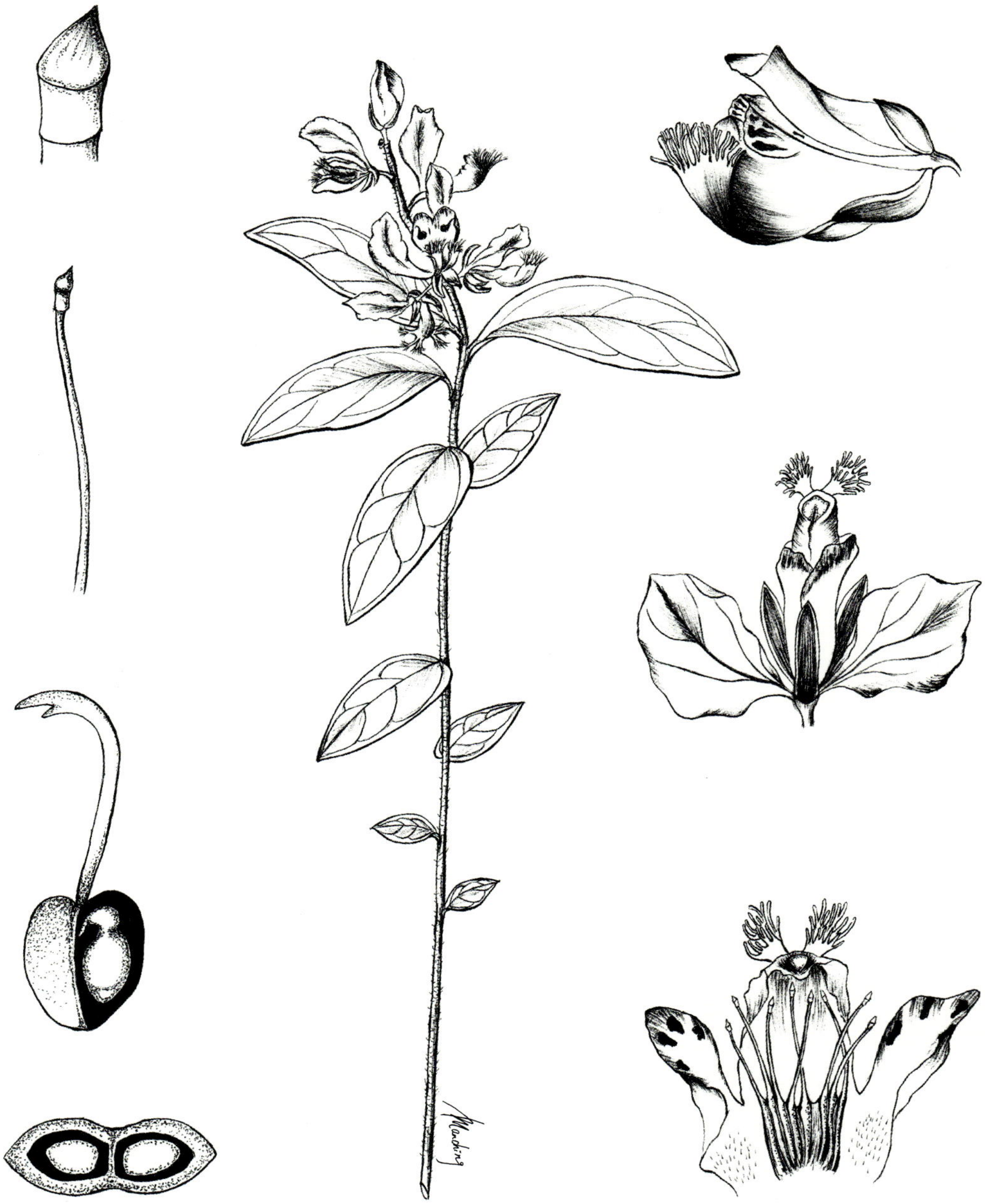

Polygala hongkongensis Hemsl. 香港遠志

Illustration 003 by M. C. Li
Authenticated by D. T. W. Lau

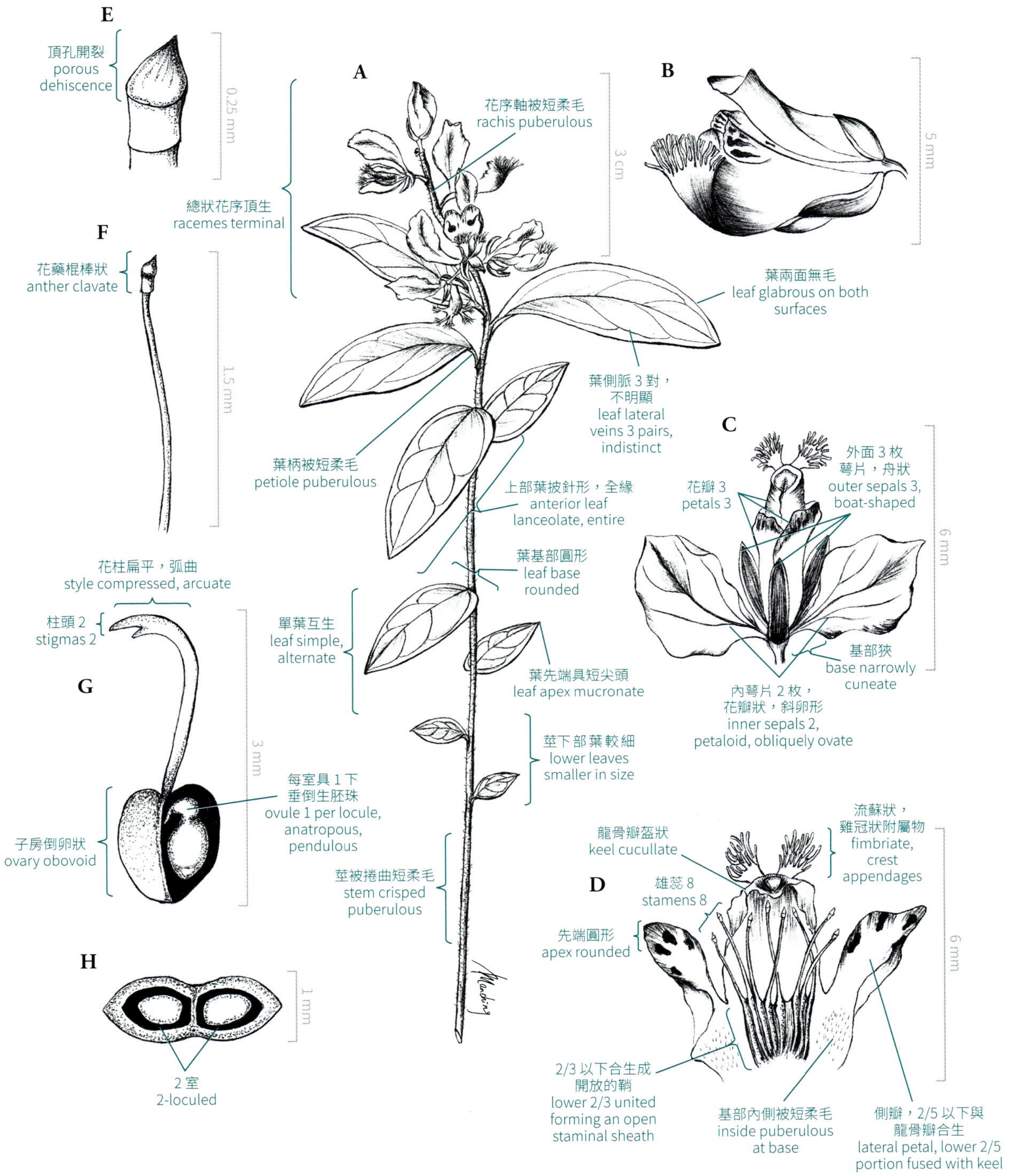

Polygala hongkongensis Hemsl. 香港遠志

Illustration 003 by M. C. Li
Authenticated by D. T. W. Lau

植物結構 PLANT STRUCTURES:
A. 花枝 Flowering branch **B.** 花，側面觀 Flower, lateral view **C.** 花，正面觀 Flower, frontal view **D.** 花解剖圖 Dissected flower **E.** 花藥 Anther **F.** 雄蕊 Stamen **G.** 雌蕊 Gynoecium **H.** 子房，橫切面 Ovary, in C. S.

參考標本 REFERENCE SPECIMENS: M. C. Li 024 (CUSLSH 2999); Hu & But 20444; Champion, Capt., #s.n. (K000592117)

參考文獻 REFERENCES:

Chen, S. K. (1997). Polygalaceae. In Delectis Florae Reipublicae Popularis Sinicae Agendae Academiae Sinicae Edita (Ed.), *Flora Reipublicae Popularis Sinicae* (Vol. 43(3), pp. 132–203). Science Press.
Chen, S. K., Ma, H. Y., & Parnell, J. A. N. (2008). Polygalaceae. In Z. Y. Wu, P. H. Raven & D. Y. Hong (Eds.), *Flora of China* (Vol. 11, pp. 139–159). Science Press & Missouri Botanical Garden Press.
Stevens, P. F. (2022). Fabales. Angiosperm Phylogeny Website. Retrieved March 17, 2022, from http://www.mobot.org/MOBOT/research/APweb/welcome.html
Wang, G. D. (2012). Polygalaceae. In Fairy Lake Botanical Garden, Shenzhen & Chinese Academy of Sciences (Eds.), *Flora of Shenzhen* (Vol. 3, pp. 22–29). China Forestry Publishing House.
Xia, N. H. (2008). Polygalaceae. In Hong Kong Herbarium, AFCD & South China Botanical Garden, CAS (Eds.), *Flora of Hong Kong* (Vol. 2, pp. 251–255). Agriculture, Fisheries and Conservation Department, HKSAR Government.

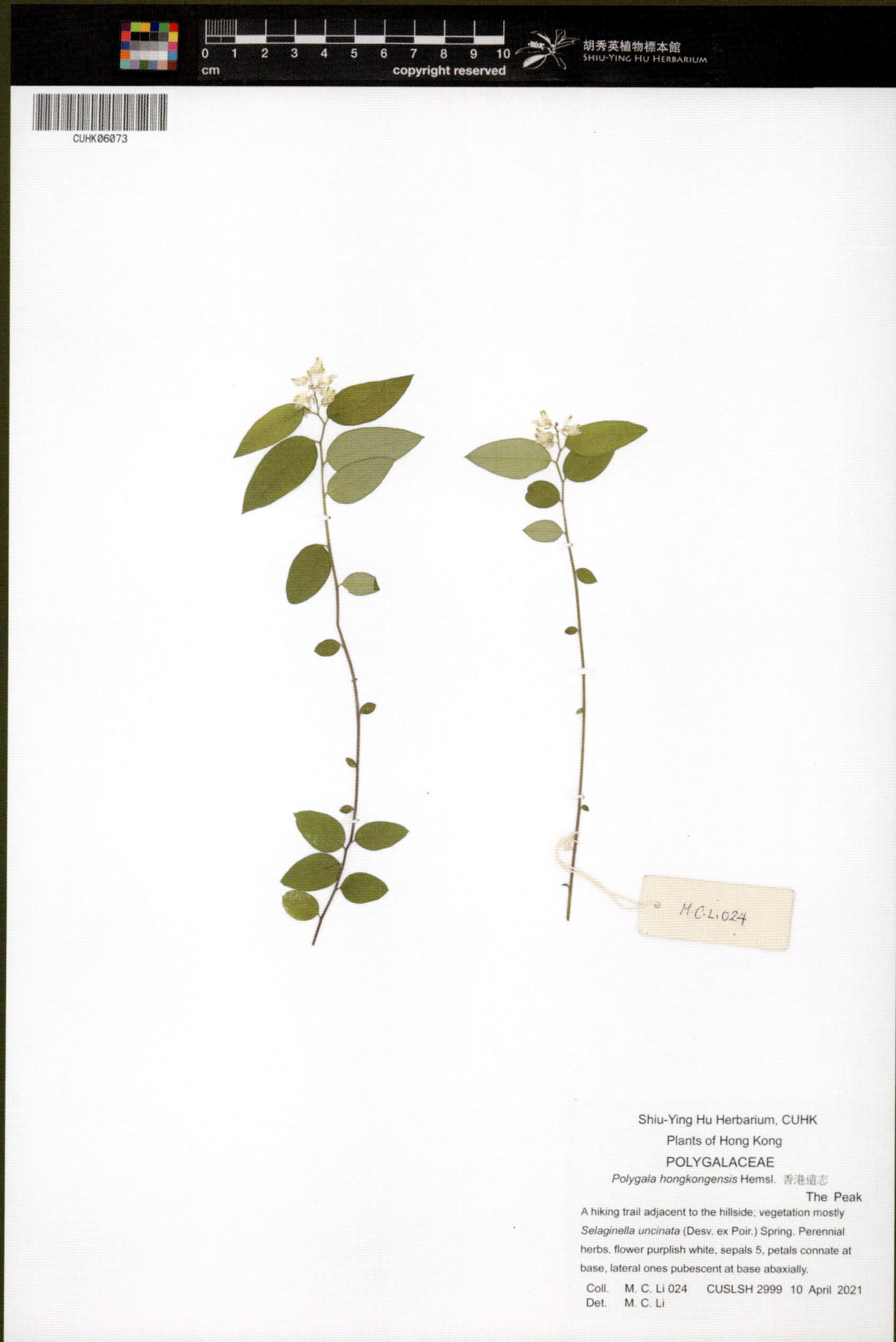

採集者編號 Collector no.
M. C. Li 024

採集年份 Year of collection
2021

外觀 Overview

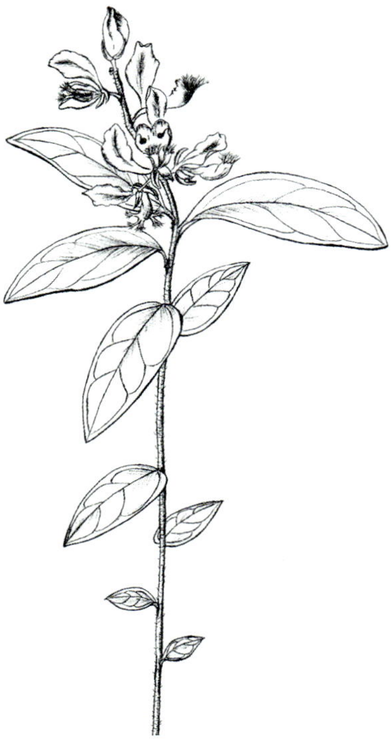

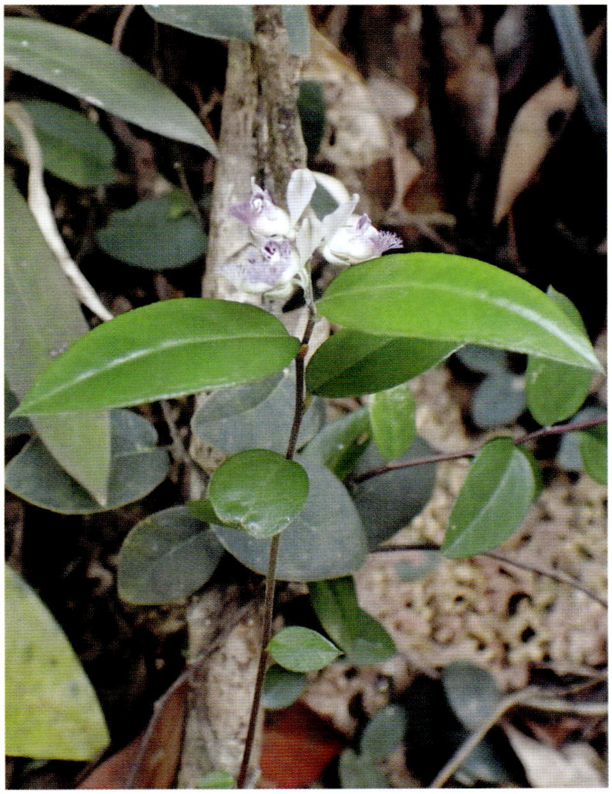

花枝 Flowering branch

葉 Leaf

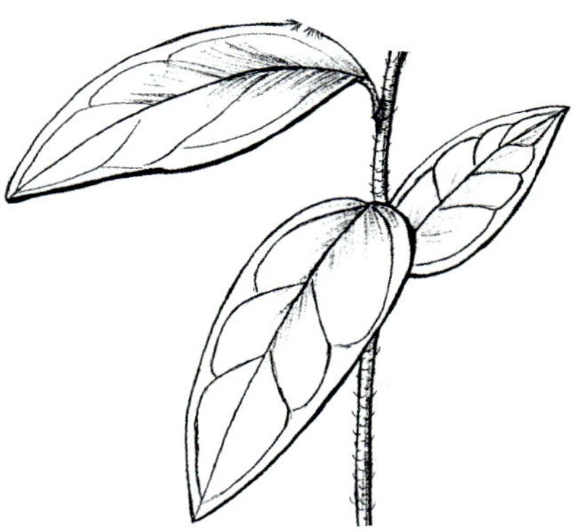

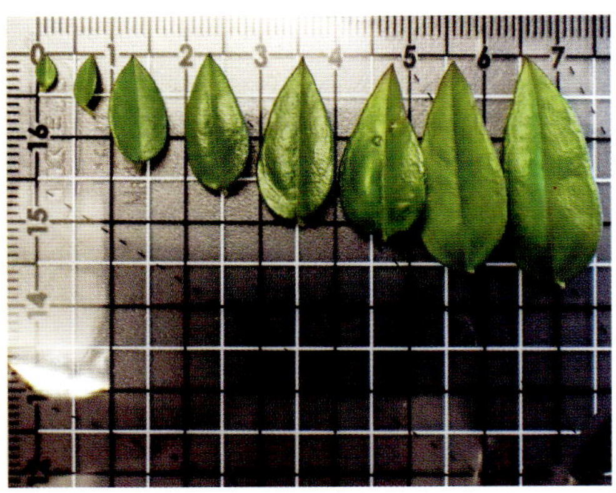

葉面 Leaves, adaxial

花 Flower

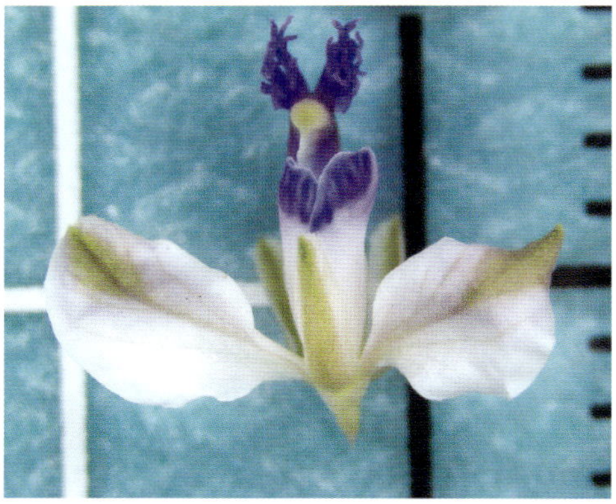

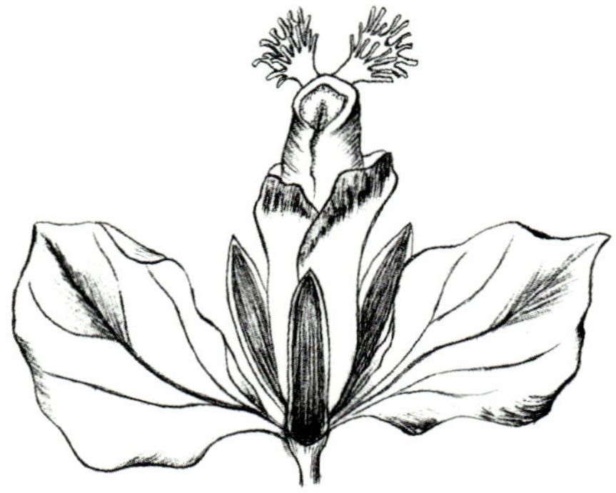

花，正面觀 Flower, frontal view

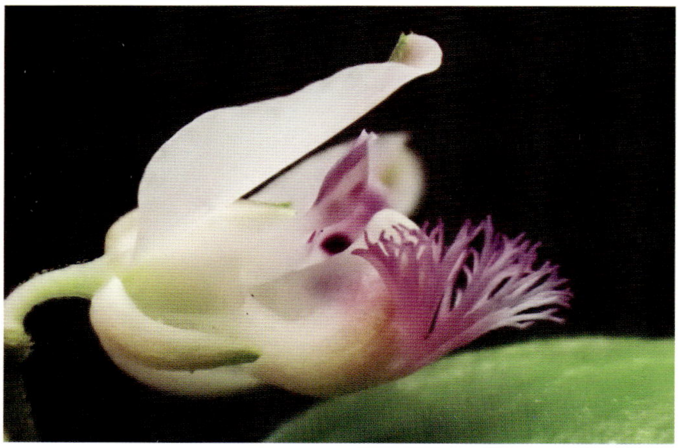

花，側面觀 Flower, lateral view

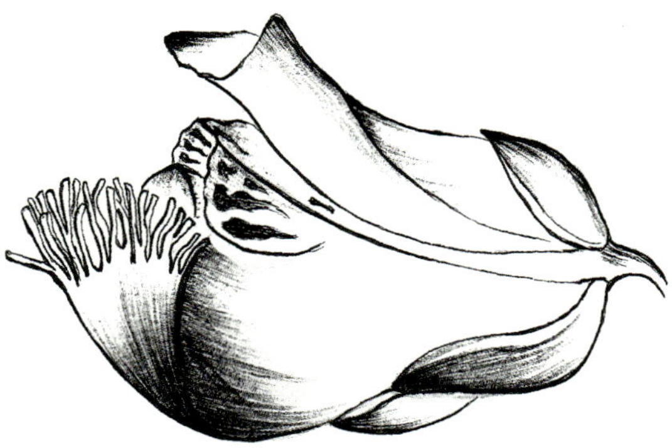

花 Flower

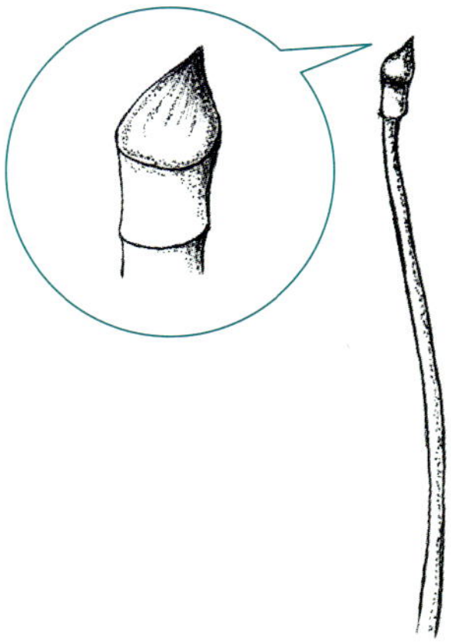

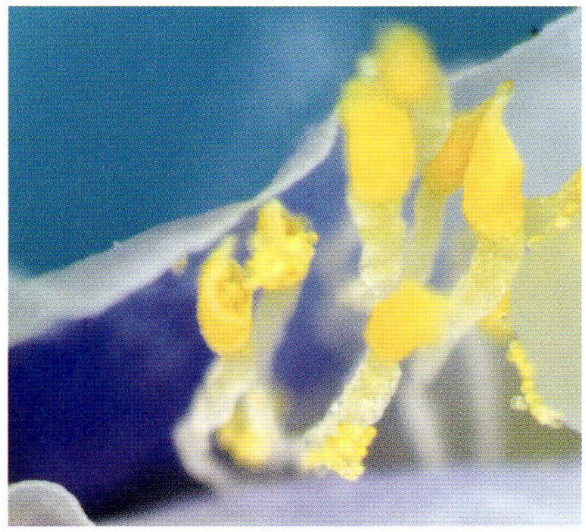

雄蕊 Stamen

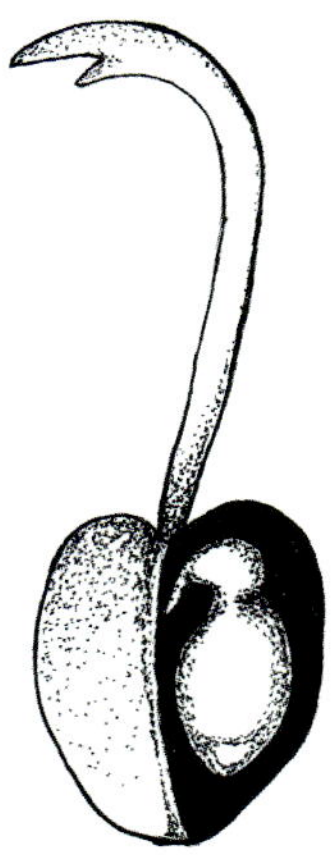

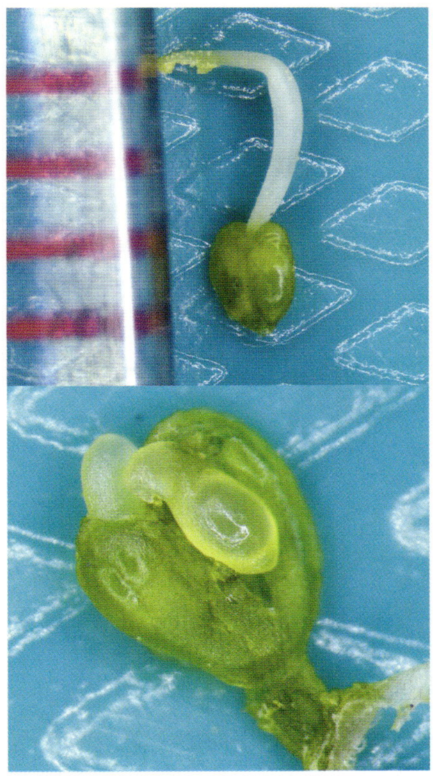

雌蕊 Gynoecium

Artocarpus hypargyreus Hance ex Benth.
白桂木

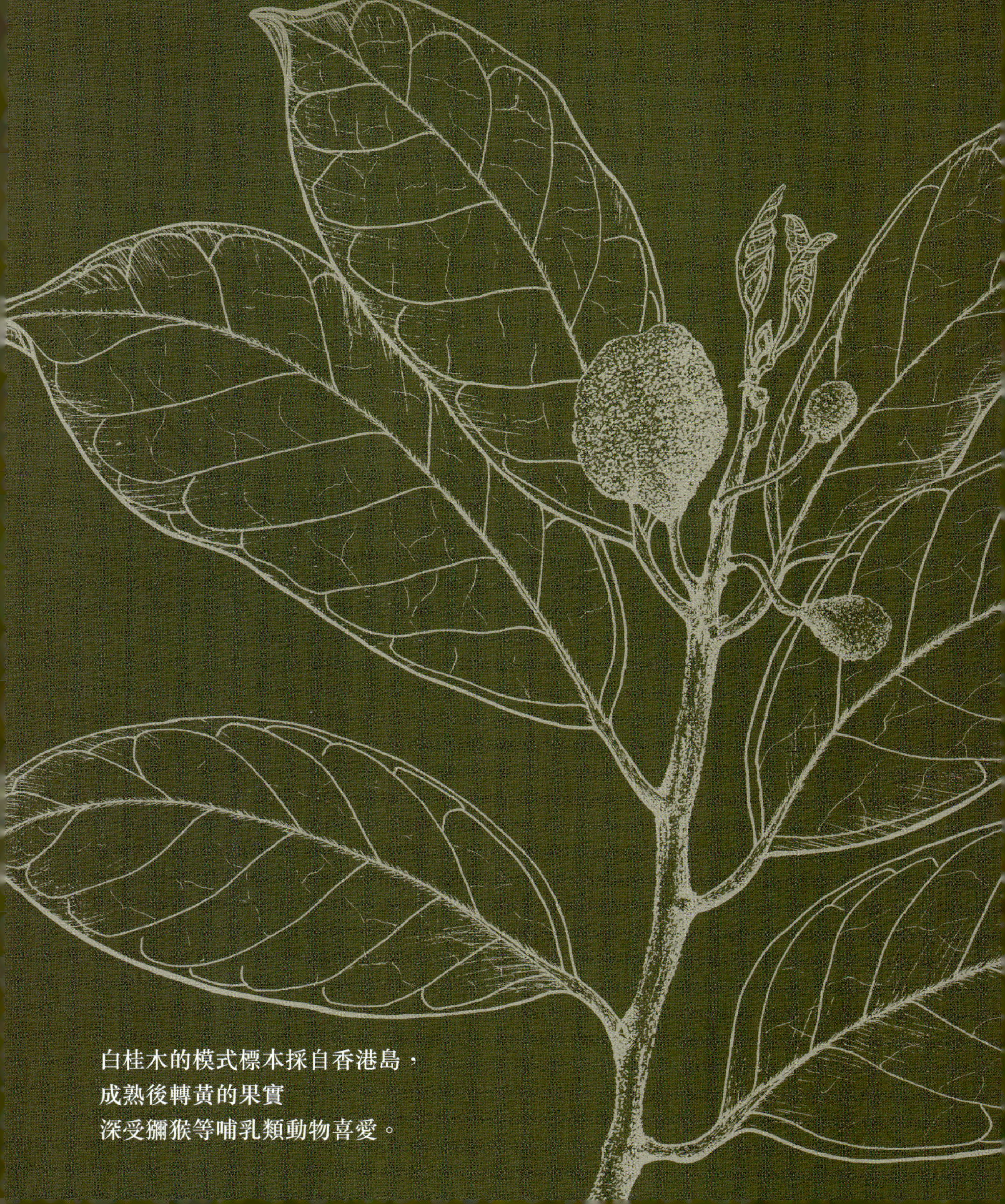

白桂木的模式標本採自香港島，
成熟後轉黃的果實
深受獼猴等哺乳類動物喜愛。

採集者編號
Collector no.
Shiu Ying Hu 13777

採集年份
Year of collection
1975

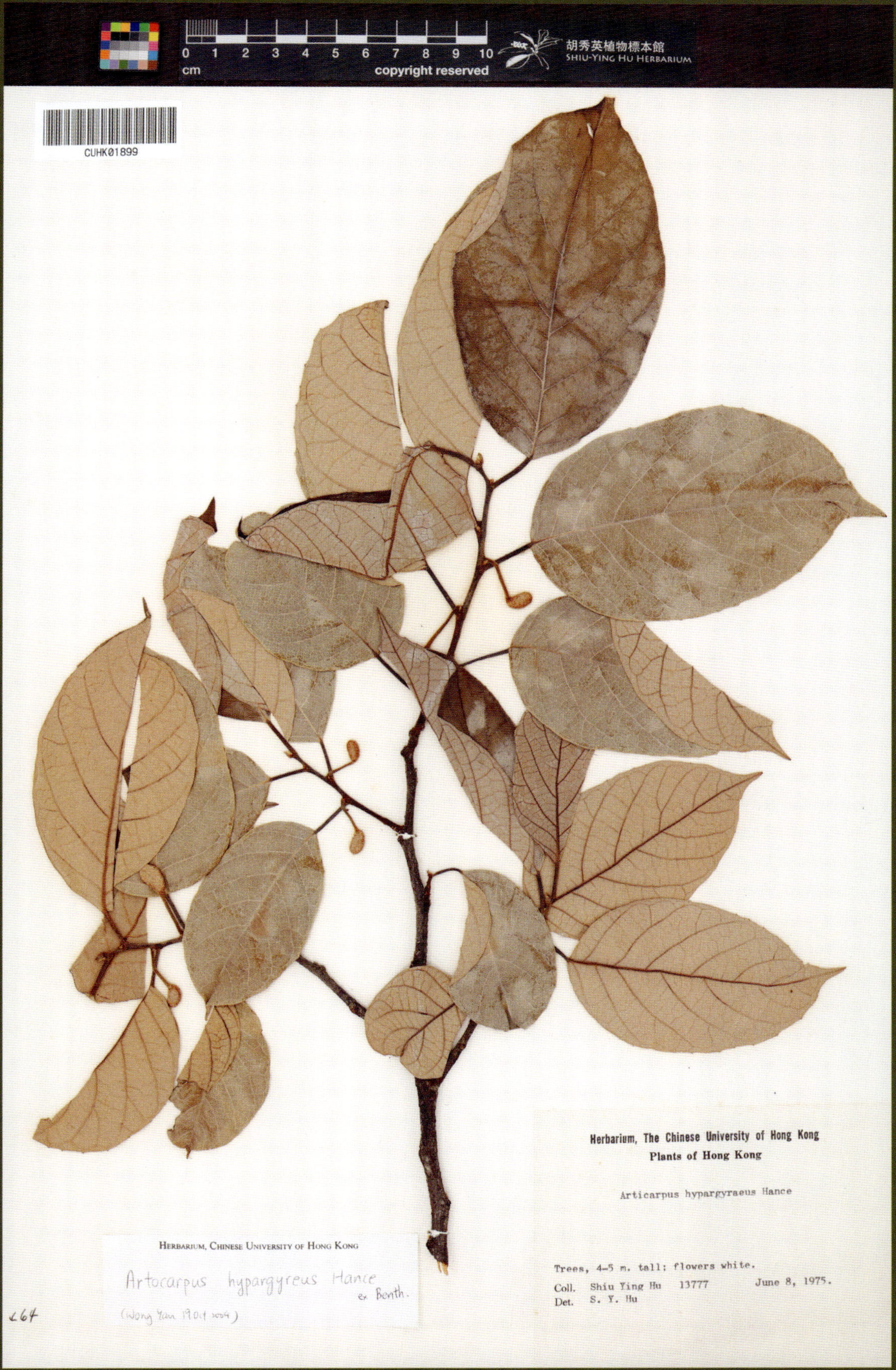

0 1 2 3 4 5 6 7 8 9 10
cm
copyright reserved
胡秀英植物標本館
SHIU-YING HU HERBARIUM
CUHK01899
Herbarium, The Chinese University of Hong Kong
Plants of Hong Kong
Articarpus hypargyraeus Hance
Trees, 4-5 m. tall; flowers white.
Coll. Shiu Ying Hu 13777 June 8, 1975.
Det. S. Y. Hu
HERBARIUM, CHINESE UNIVERSITY OF HONG KONG
Artocarpus hypargyreus Hance ex Benth.

白桂木

學名	*Artocarpus hypargyreus* Hance ex Benth.
中文名	白桂木
科名	桑科
本地分布狀態	原生植物
繪圖編號	M. C. Li 006

花期	花色	果期	果色
5至8月	雄花序：褐色 雌花序：白色	6至8月	橙黃色

白桂木為大型喬木，常見於常綠闊葉林，其模式標本採自香港島，由 H. F. Hance 於 1850 年前後採集。其成熟後轉黃的果實深受獼猴等哺乳類動物喜愛。

白桂木於《香港稀有及珍貴植物》中屬「近危」級別，並於國際自然保護聯盟瀕危物種紅色名錄中被列為「易危」物種。

《中國植物誌》(英文版) 的採用名稱

Artocarpus hypargyreus Hance in Bentham（科名：桑科）

《香港植物誌》的採用名稱

Artocarpus hypargyreus Hance ex Benth.（科名：桑科）

參考文獻 References

1. Corlett, R. T. (2011). Seed dispersal in Hong Kong, China: past, present and possible futures. *Integrative zoology 6*(2), 97–109. https://doi.org/10.1111/j.1749-4877.2011.00235.x
2. Hong Kong Herbarium. (2022). *HK Herbarium-Artocarpus hypargyreus*. HK Plant Database. Retrieved March 17, 2022, from https://herbarium.gov.hk/en/hk-plant-database/plant-detail/index.html?pType=species&oID=9207
3. IUCN. (1998). *Artocarpus hypargyreus*. The IUCN Red List of Threatened Species. Retrieved March 17, 2022, from https://www.iucnredlist.org/species/32383/9694685

Artocarpus hypargyreus Hance ex Benth.

Scientific name	*Artocarpus hypargyreus* Hance ex Benth.
Chinese name	白桂木
Family	Moraceae Gaudich.
Local distribution status	Native species
Illustration number	M. C. Li 006

FLOWERING PERIOD	COLOUR OF FLOWER	FRUITING PERIOD	COLOUR OF FRUIT
May to August	Male inflorescence: brown Female inflorescence: white	June to August	Yellowish orange

Artocarpus hypargyreus Hance ex Benth. is a large tree, commonly found in evergreen broad-leaved forests. The type specimen was collected from Hong Kong Island and was discovered by H. F. Hance around 1850. Its ripe, yellow fruits are attractive to mammals like macaques.

Artocarpus hypargyreus Hance ex Benth. is listed in the category of 'Near Threatened' in *Rare and Precious Plants of Hong Kong* and is classified as 'Vulnerable' in the IUCN Red List of Threatened Species.

Name in *Flora of China*
Artocarpus hypargyreus Hance in Bentham (Family: Moraceae)

Name in *Flora of Hong Kong*
Artocarpus hypargyreus Hance ex Benth. (Family: Moraceae)

4. Missouri Botanical Garden. (2022). *Name-Artocarpus hypargyreus*. Tropicos. Retrieved March 17, 2022, from https://www.tropicos.org/name/21302189
5. Zhou, Z. K., & Gilbert, M. G. (2003). Moraceae. In Z. Y. Wu, P. H. Raven & D. Y. Hong (Eds.), *Flora of China* (Vol. 5, pp. 21–73). Science Press & Missouri Botanical Garden Press.

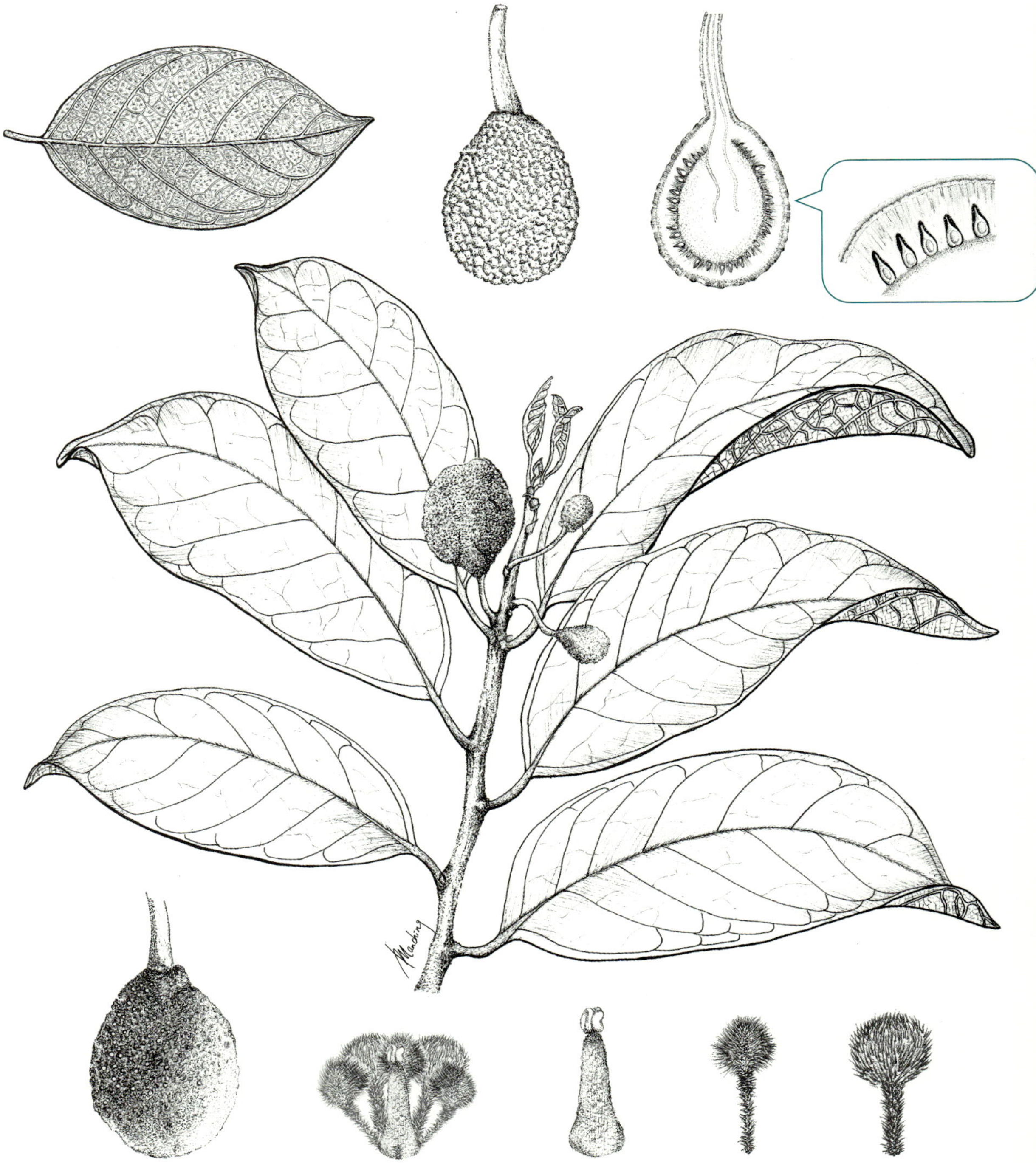

Artocarpus hypargyreus Hance ex Benth. 白桂木
Illustration 006 by M. C. Li
Authenticated by D. T. W. Lau

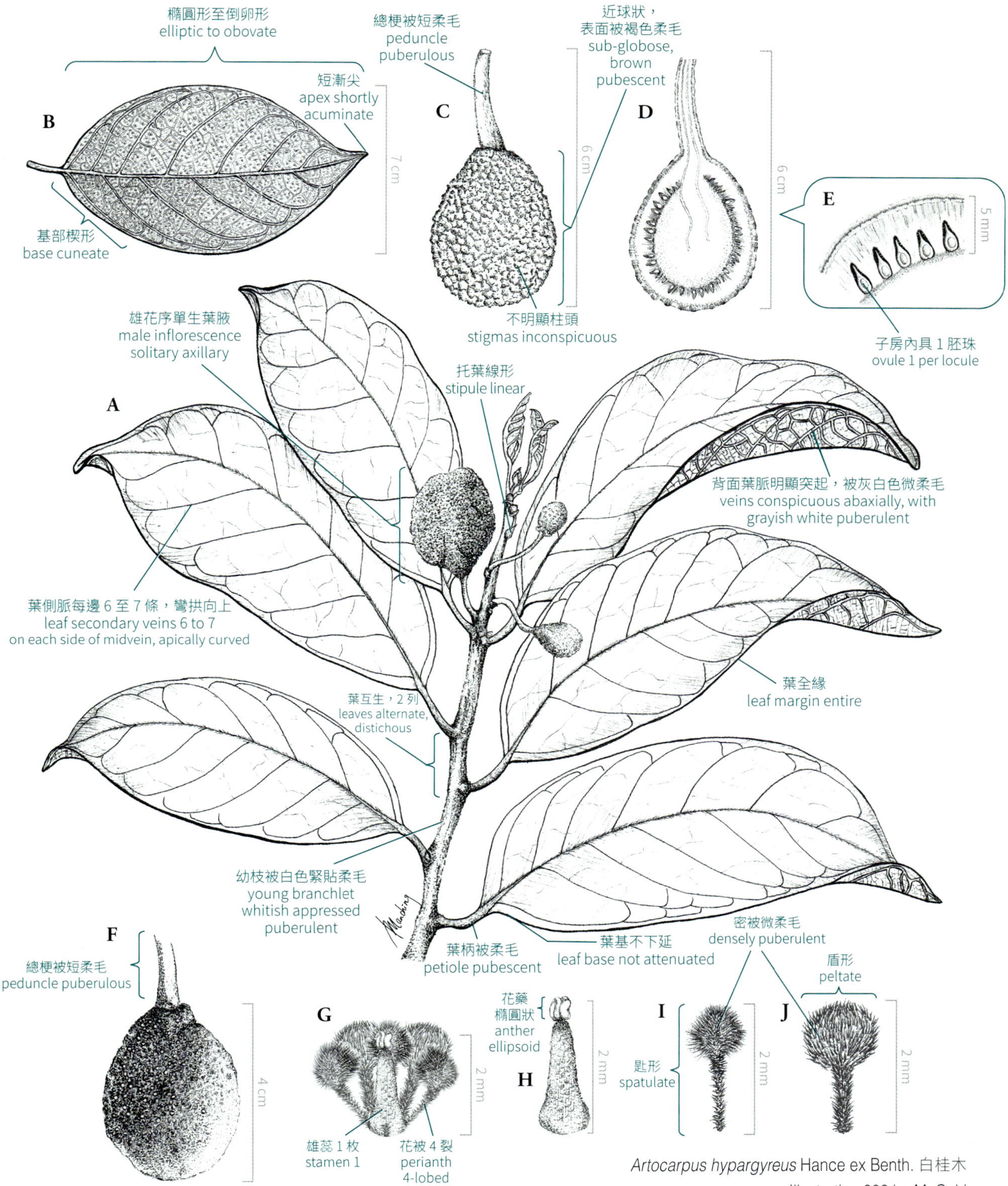

Artocarpus hypargyreus Hance ex Benth. 白桂木

Illustration 006 by M. C. Li

Authenticated by D. T. W. Lau

植物結構 PLANT STRUCTURES:

A. 花枝 Flowering branch **B.** 葉底 Leaf, abaxial **C.** 雌花序 Female inflorescence **D.** 雌花序，縱切面 Female inflorescence, in L. S. **E.** 雌花序，縱切面局部放大 Close-up of female inflorescence, in L. S. **F.** 雄花序 Male inflorescence **G.** 雄花 Male flower **H.** 雄蕊 Stamen **I.** 花被裂片 Perianth segment **J.** 苞片 Bract

參考標本 REFERENCE SPECIMENS: Rare and endangered plants (K. W. Lam, S. W. Shek, J. Y. Y. Lau, D. T. W. Lau) 15B; Shiu Ying Hu 13777

參考文獻 REFERENCES:

Chang, S. S., & Wu, C. Y. (1998). Moraceae. In Delectis Florae Reipublicae Popularis Sinicae Agendae Academiae Sinicae Edita (Ed.), *Flora Reipublicae Popularis Sinicae* (Vol. 23(1), pp. 1–224). Science Press.

Chen, F. H. (1987). Moraceae. In South China Botanical Garden, the Chinese Academy of Sciences (Ed.), *Flora of Guangdong* (Vol. 1, pp. 168–214). Guangdong Science and Technology Press.

Li, P. Q. (2017). Moraceae. In Fairy Lake Botanical Garden, Shenzhen & Chinese Academy of Sciences (Eds.), *Flora of Shenzhen* (Vol. 1, pp. 419–452). China Forestry Publishing House.

Stevens, P. F. (2022). Rosales. Angiosperm Phylogeny Website. Retrieved March 17, 2022, from http://www.mobot.org/MOBOT/research/APweb/welcome.html

Wu, D. L. (2007). Moraceae. In Hong Kong Herbarium, AFCD & South China Botanical Garden, CAS (Eds.), *Flora of Hong Kong* (Vol. 1, pp. 101–115). Agriculture, Fisheries and Conservation Department, HKSAR Government.

Zhou, Z. K., & Gilbert, M. G. (2003). Moraceae. In Z. Y. Wu, P. H. Raven & D. Y. Hong (Eds.), *Flora of China* (Vol. 5, pp. 21–73). Science Press & Missouri Botanical Garden Press.

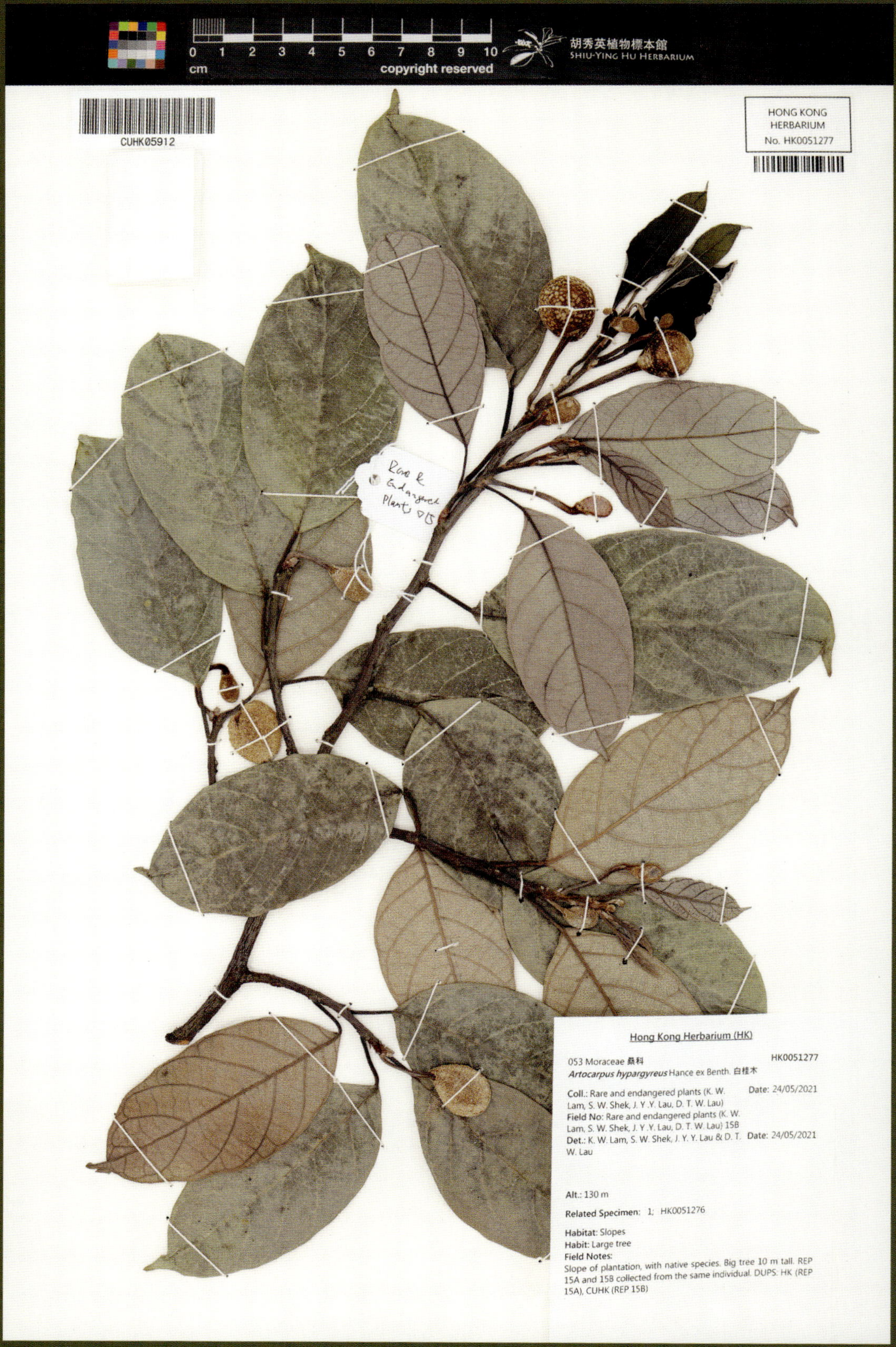

採集者編號 Collector no.

Rare and endangered plants (K. W. Lam, S. W. Shek, J. Y. Y. Lau, D. T. W. Lau) 15B

採集年份 Year of collection

2021

外觀 Overview

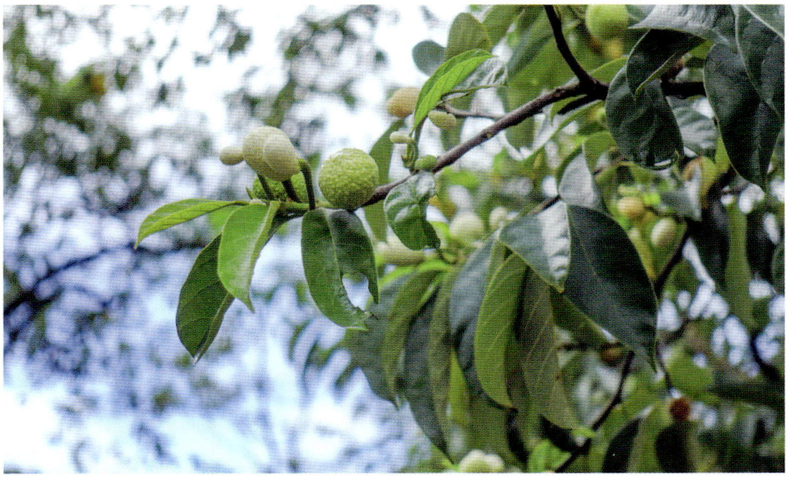

花枝 Flowering branch

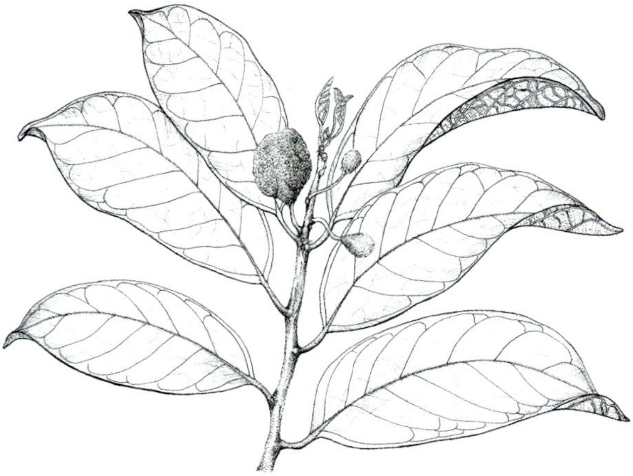

葉 Leaf

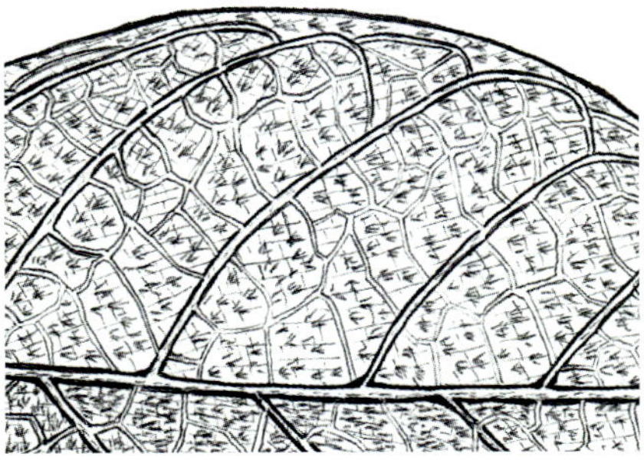

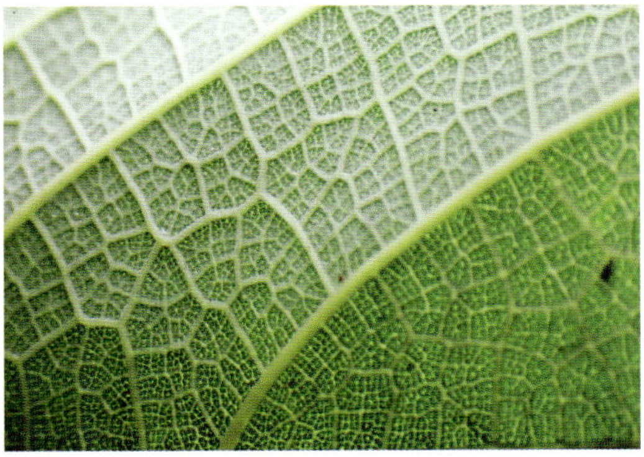

葉底 Leaf, abaxial

花 Flower

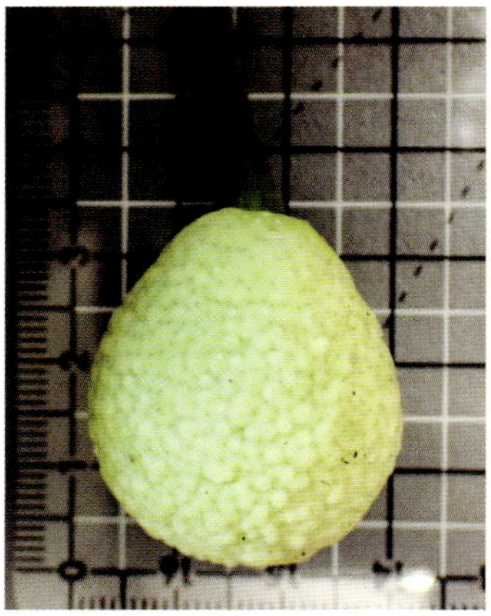

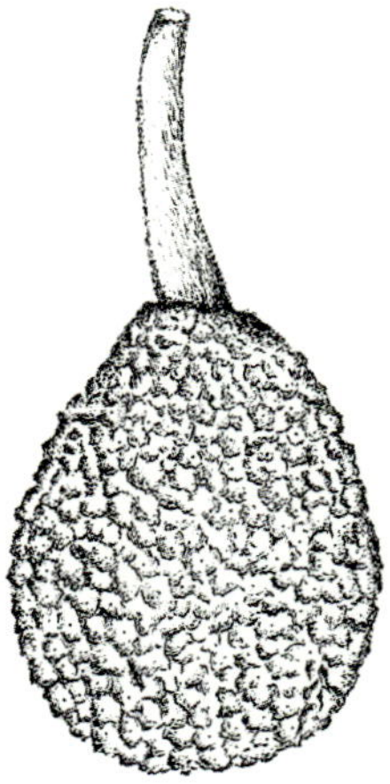

雌花序 Female inflorescence

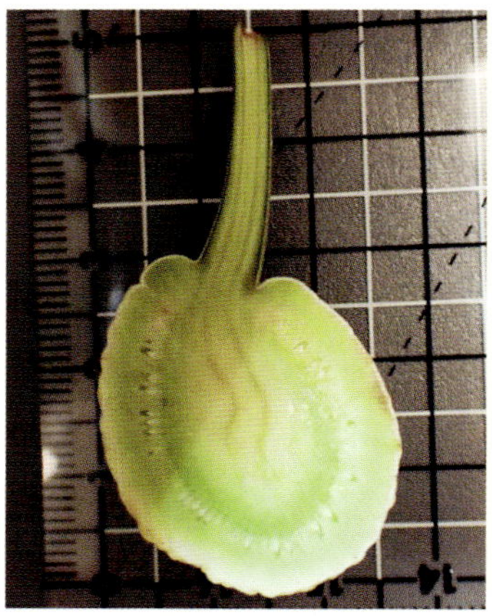

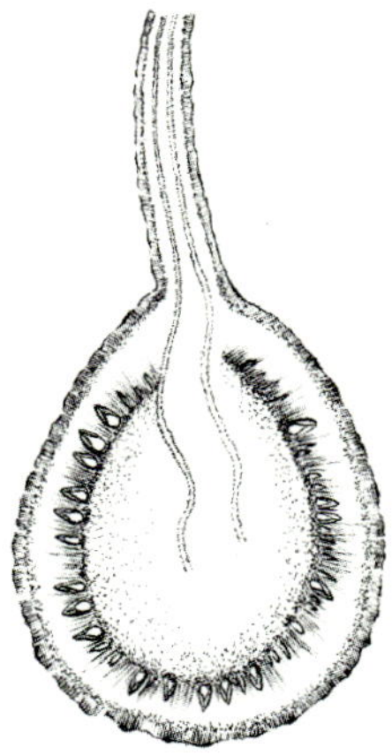

雌花序，縱切面
Female inflorescence, in L. S.

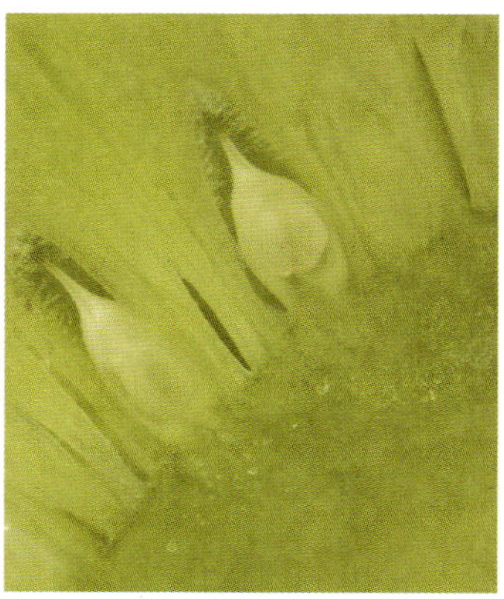

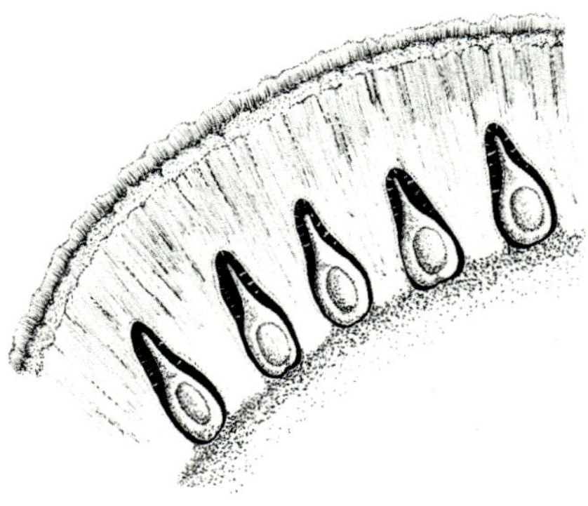

雌花序，縱切面局部放大
Close-up of female inflorescence, in L. S.

花 Flower

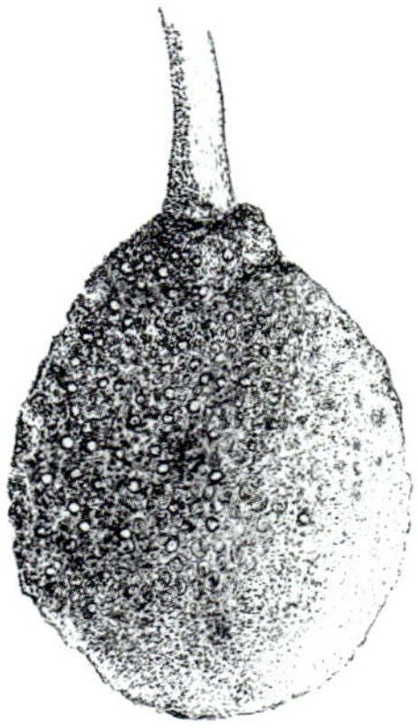

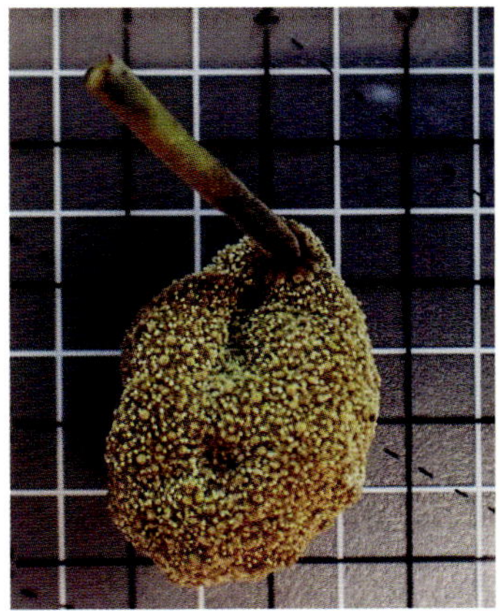

雄花序 Male inflorescence

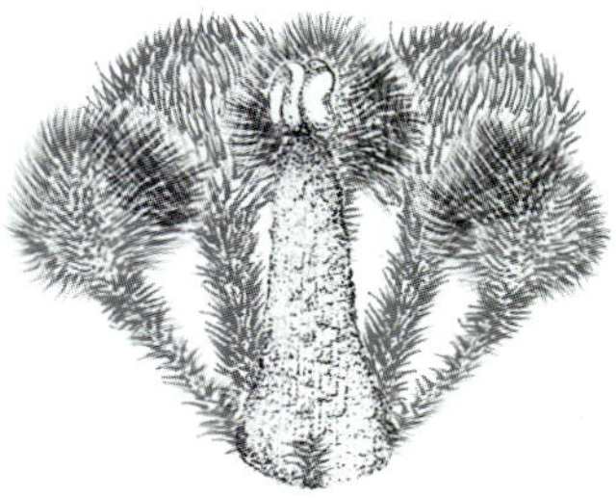

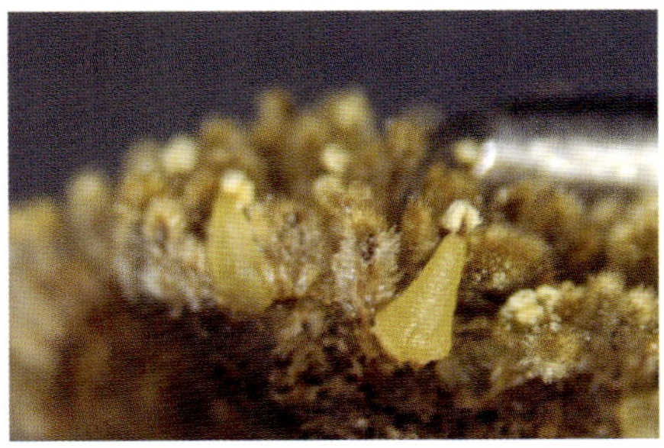

雄花 Male flower

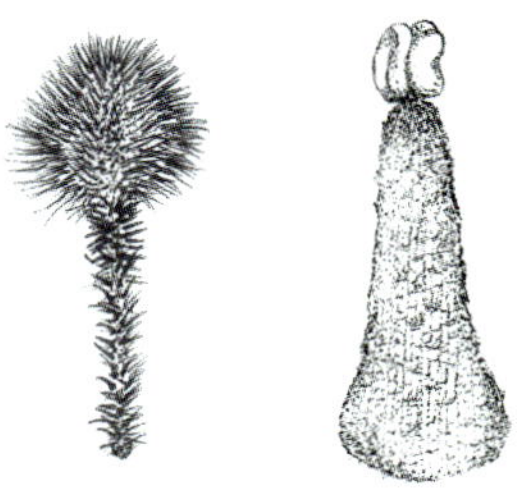

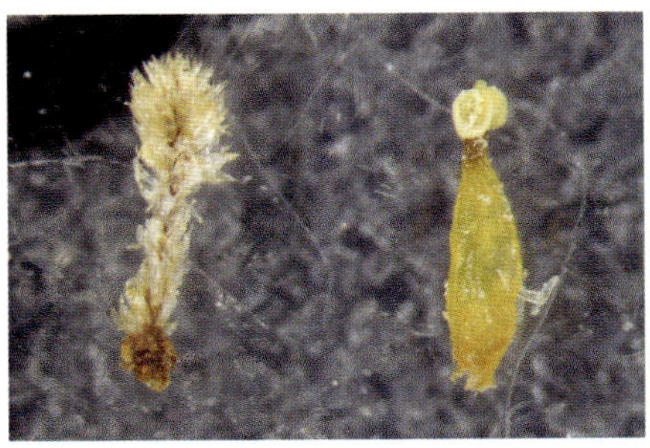

花被 Perianth 雄蕊 Stamen

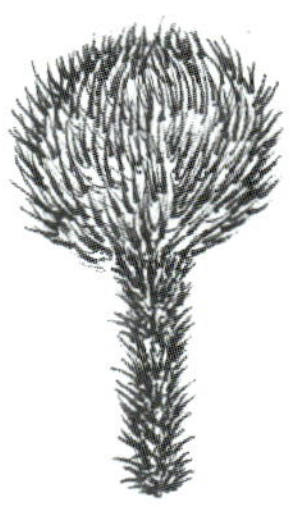

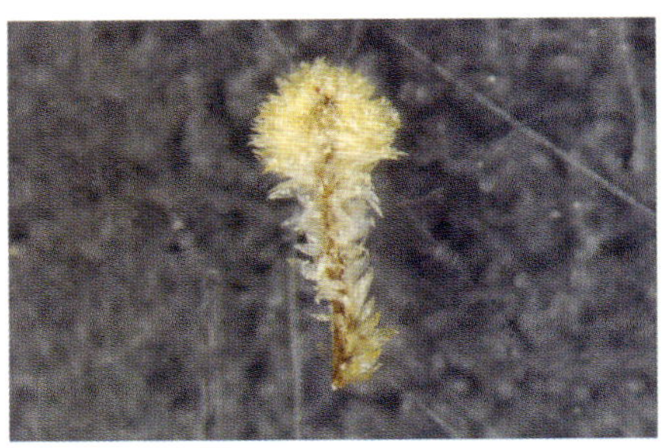

苞片 Bract

Sapindus saponaria L.
無患子

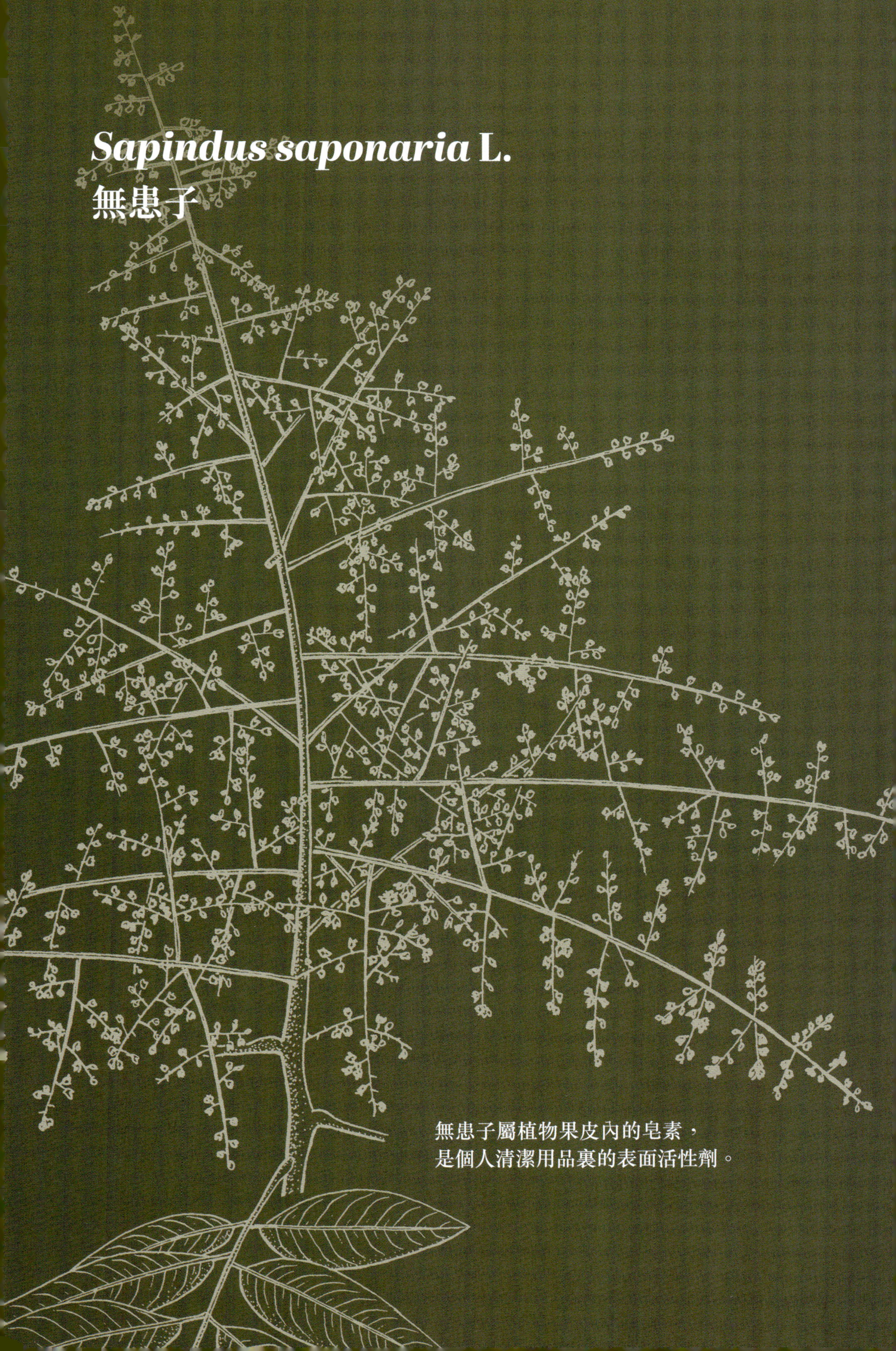

無患子屬植物果皮內的皂素，
是個人清潔用品裏的表面活性劑。

採集者編號
Collector no.
Hu & But 22184

採集年份
Year of collection
1993

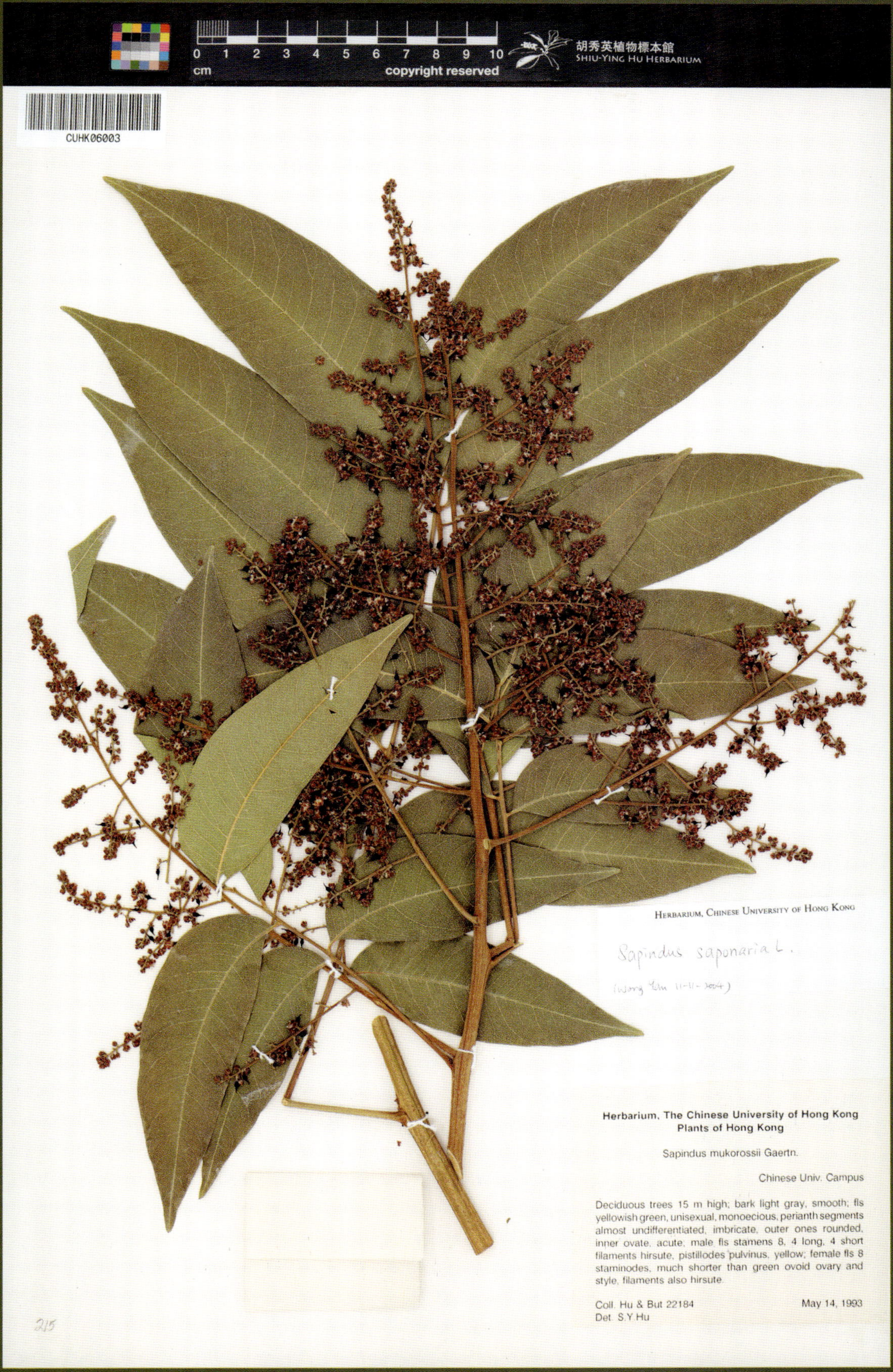
0 1 2 3 4 5 6 7 8 9 10
cm
copyright reserved
胡秀英植物標本館
SHIU-YING HU HERBARIUM
CUHK06003
HERBARIUM, CHINESE UNIVERSITY OF HONG KONG
Sapindus saponaria L.
Herbarium, The Chinese University of Hong Kong
Plants of Hong Kong
Sapindus mukorossii Gaertn.
Chinese Univ. Campus
Deciduous trees 15 m high; bark light gray, smooth; fls yellowish green, unisexual, monoecious, perianth segments almost undifferentiated, imbricate, outer ones rounded, inner ovate, acute; male fls stamens 8, 4 long, 4 short filaments hirsute, pistillodes pulvinus, yellow; female fls 8 staminodes, much shorter than green ovoid ovary and style, filaments also hirsute.
Coll. Hu & But 22184
May 14, 1993
Det. S.Y.Hu

無患子

學名	*Sapindus saponaria* L.
中文名	無患子
科名	無患子科
本地分布狀態	原生植物
繪圖編號	H. Y. Wong 008

花期	花色	果期	果色
3至5月	橙黃色	6至12月	黃褐色

無患子是大型的落葉喬木，可高達20米。

無患子屬植物有重要的經濟價值，其種子油是生物柴油的原料，而果皮內的皂素則是個人清潔用品裏的表面活性劑。在不同國家的傳統醫藥裏，無患子屬植物則有殺菌、驅蟲、止胃痛、治療皮膚炎等功用。

《中國植物誌》(英文版)的採用名稱

Sapindus saponaria Linnaeus (科名：無患子科)

《香港植物誌》的採用名稱

Sapindus saponaria L. (科名：無患子科)

參考文獻 References

1. Hong Kong Herbarium. (2021). *HK Herbarium-Sapindus saponaria*. HK Plant Database. Retrieved May 8, 2021, from https://www.herbarium.gov.hk/en/hk-plant-database/plant-detail/index.html?pType=species&oID=2877
2. Liu, J. M., Wang, L. C., Sun, C. W., Xi, B. Y., Li, D. D., Chen, Z., He, Q. Y., Weng, X. H., & Jia, L. M. (2021). Global distribution of soapberries (*Sapindus* L.) habitats under current and future climate scenarios. *Scientific Reports 11*(1). https://doi.org/10.1038/s41598-021-98389-8
3. Missouri Botanical Garden. (2021). *Name-Sapindus saponaria*. Tropicos. Retrieved May 8, 2021, from https://www.tropicos.org/name/28600572

Sapindus saponaria L.

Scientific name	*Sapindus saponaria* L.
Chinese name	無患子
Family	Sapindaceae Juss.
Local distribution status	Native species
Illustration number	H. Y. Wong 008

FLOWERING PERIOD	COLOUR OF FLOWER	FRUITING PERIOD	COLOUR OF FRUIT
March to May	Yellowish orange	June to December	Yellowish brown

Sapindus saponaria L. is a large deciduous tree, up to 20 m tall.

Plants under *Sapindus* are economically important. Their seed oil is the raw material of biodiesel, while the saponin in the pericarp is a surfactant in personal hygiene products. As a traditional medicine in different countries, the *Sapindus* species are used to kill bacteria, repel insects, relieve stomach ache, cure dermatitis, etc.

Name in *Flora of China*
Sapindus saponaria Linnaeus (Family: Sapindaceae)

Name in *Flora of Hong Kong*
Sapindus saponaria L. (Family: Sapindaceae)

4. Tsuzuki, J. K., Svidzinski, T. I. E., Shinobu, C. S., Silva, L. F. A., Rodrigues-Filho, E., Cortez, D. A. G., & Ferreira, I. C. P. (2007). Antifungal activity of the extracts and saponins from *Sapindus saponaria* L. *Anais da Academia Brasileira de Ciências 79*(4), 577–583. https://doi.org/10.1590/s0001-37652007000400002
5. Xia, N. H., & Gadek, P. A. (2007). Sapindaceae. In Z. Y. Wu, P. H. Raven & D. Y. Hong (Eds.), *Flora of China* (Vol. 12, pp. 5–24). Science Press & Missouri Botanical Garden Press.

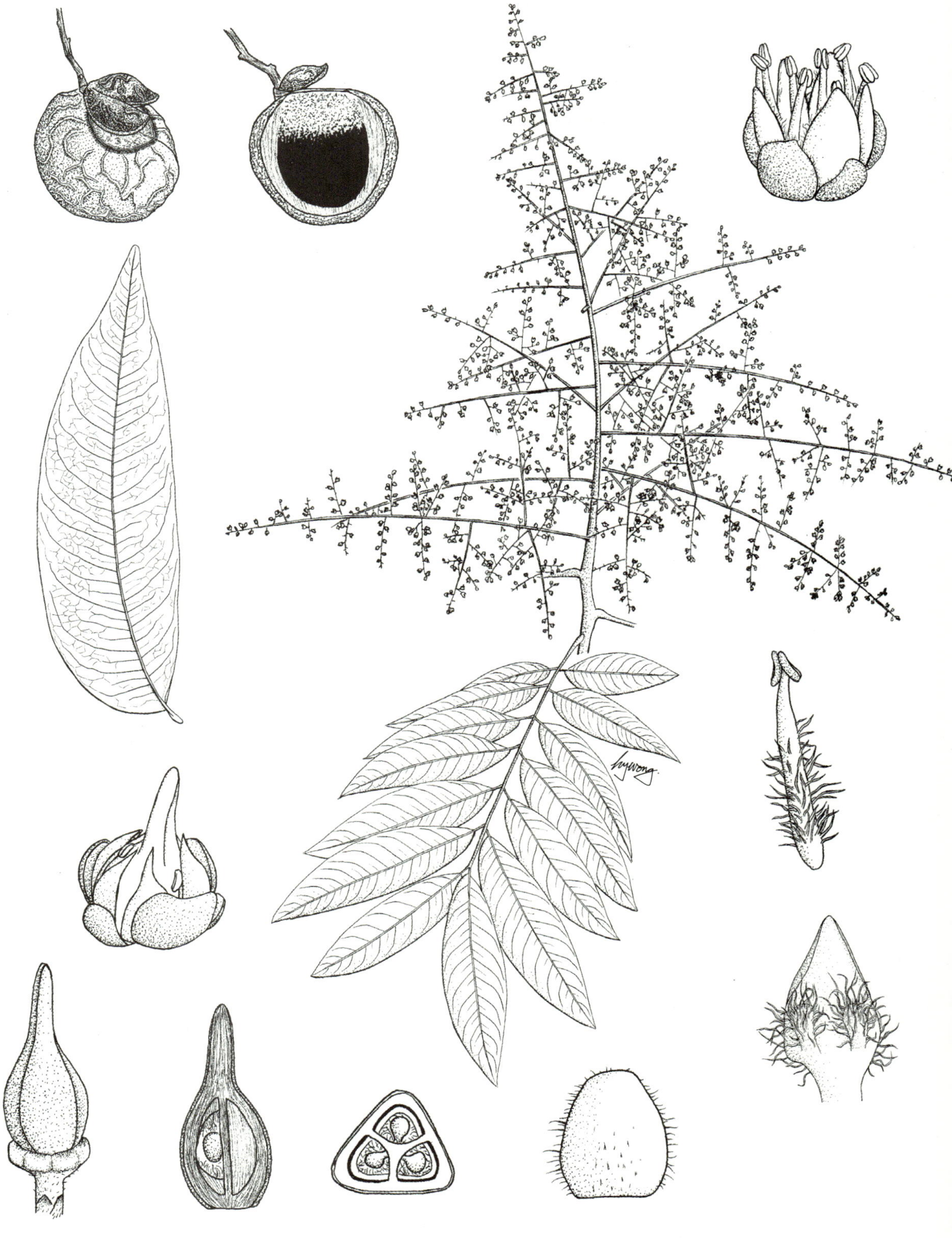

Sapindus saponaria L. 無患子
Illustration 008 by H. Y. Wong
Authentication by D. T. W. Lau

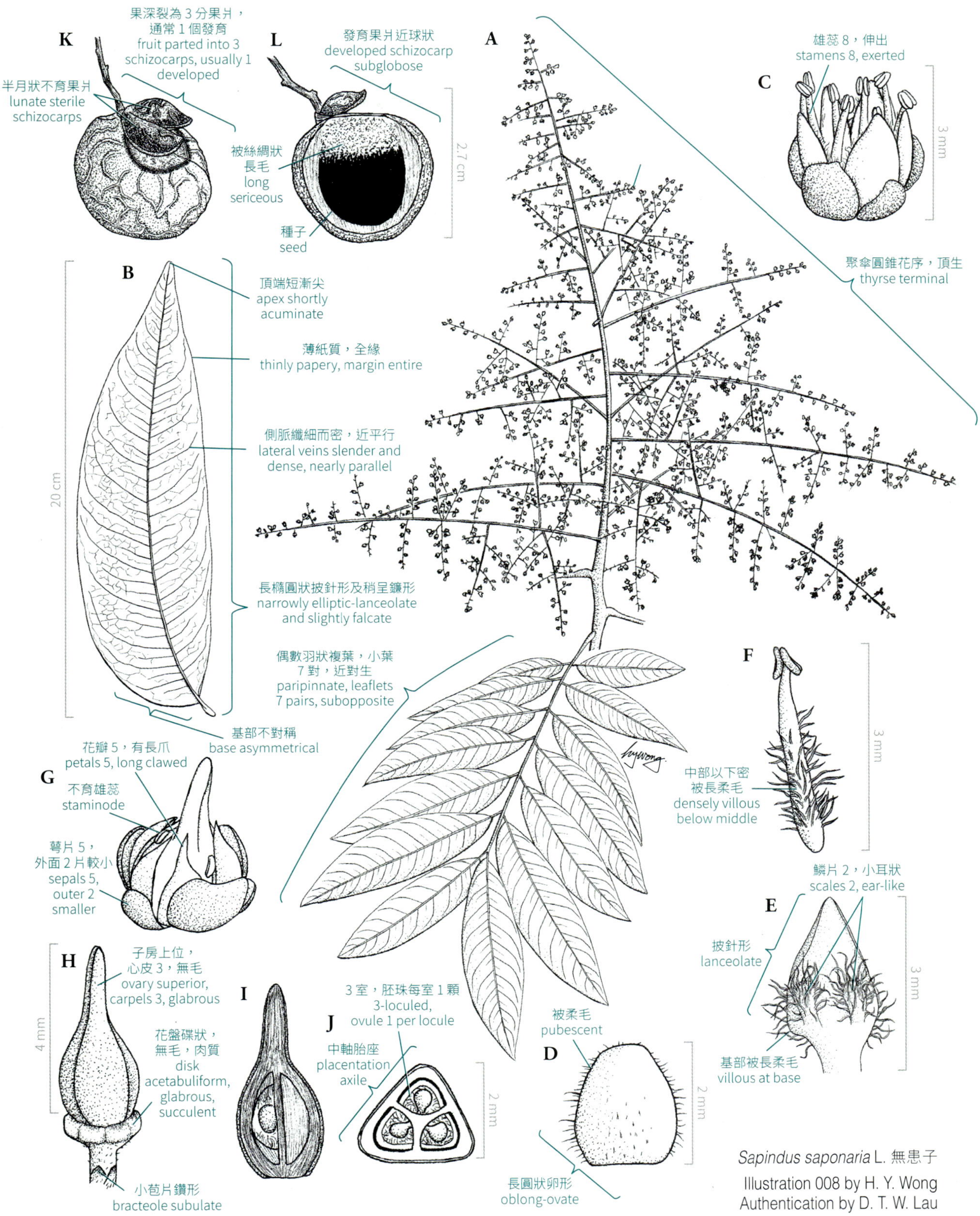

Sapindus saponaria L. 無患子
Illustration 008 by H. Y. Wong
Authentication by D. T. W. Lau

植物結構 PLANT STRUCTURES:
A. 具幼葉的花枝 Flowering branch with a young leaf **B.** 小葉葉底 Leaflet, abaxial **C.** 雄花 Male flower **D.** 雄花萼片 Sepal of male flower **E.** 雄花花瓣，向軸面 Petal of male flower, adaxial **F.** 雄蕊 Stamen **G.** 雌花 Female flower **H.** 雌蕊與花盤 Gynoecium and disk **I.** 子房，縱切面 Ovary, in L. S. **J.** 子房，橫切面 Ovary, in C. S. **K.** 果實 Fruit **L.** 果實，縱切面 Fruit, in L. S.

參考標本 REFERENCE SPECIMENS: H. Y. Wong 003 (CUSLSH3014); H. Y. Wong 009 (CUSLSH3120); Hu & But 22184

參考文獻 REFERENCES:

Liao, W. B., & Luo, L. (2012). Sapindaceae. In Fairy Lake Botanical Garden, Shenzhen & Chinese Academy of Sciences (Eds.), *Flora of Shenzhen* (Vol. 3, pp. 33–39). China Forestry Publishing House.
Lo, H. S. (1987). Sapindaceae. In South China Botanical Garden, the Chinese Academy of Sciences (Ed.), *Flora of Guangdong* (Vol. 1, pp. 236–256). Guangdong Science and Technology Press.
Lo, H. S., & Chen, T. C. (1985). Sapindaceae. In Delectis Florae Reipublicae Popularis Sinicae Agendae Academiae Sinicae Edita (Ed.), *Flora Reipublicae Popularis Sinicae* (Vol. 47(1), pp. 1–72). Science Press.
Stevens, P. F. (2021). Sapindales. Angiosperm Phylogeny Website. Retrieved May 7, 2021, from http://www.mobot.org/MOBOT/research/APweb/welcome.html
Xia, N. H. (2009). Sapindaceae. In Hong Kong Herbarium, AFCD & South China Botanical Garden, CAS (Eds.), *Flora of Hong Kong* (Vol. 2, pp. 257–261). Agriculture, Fisheries and Conservation Department, HKSAR Government.
Xia, N. H., & Gadek, P. A. (2007). Sapindaceae. In Z. Y. Wu, P. H. Raven & D. Y. Hong (Eds.), *Flora of China* (Vol. 12, pp. 5–24). Science Press & Missouri Botanical Garden Press.

憑證標本
Voucher Specimen

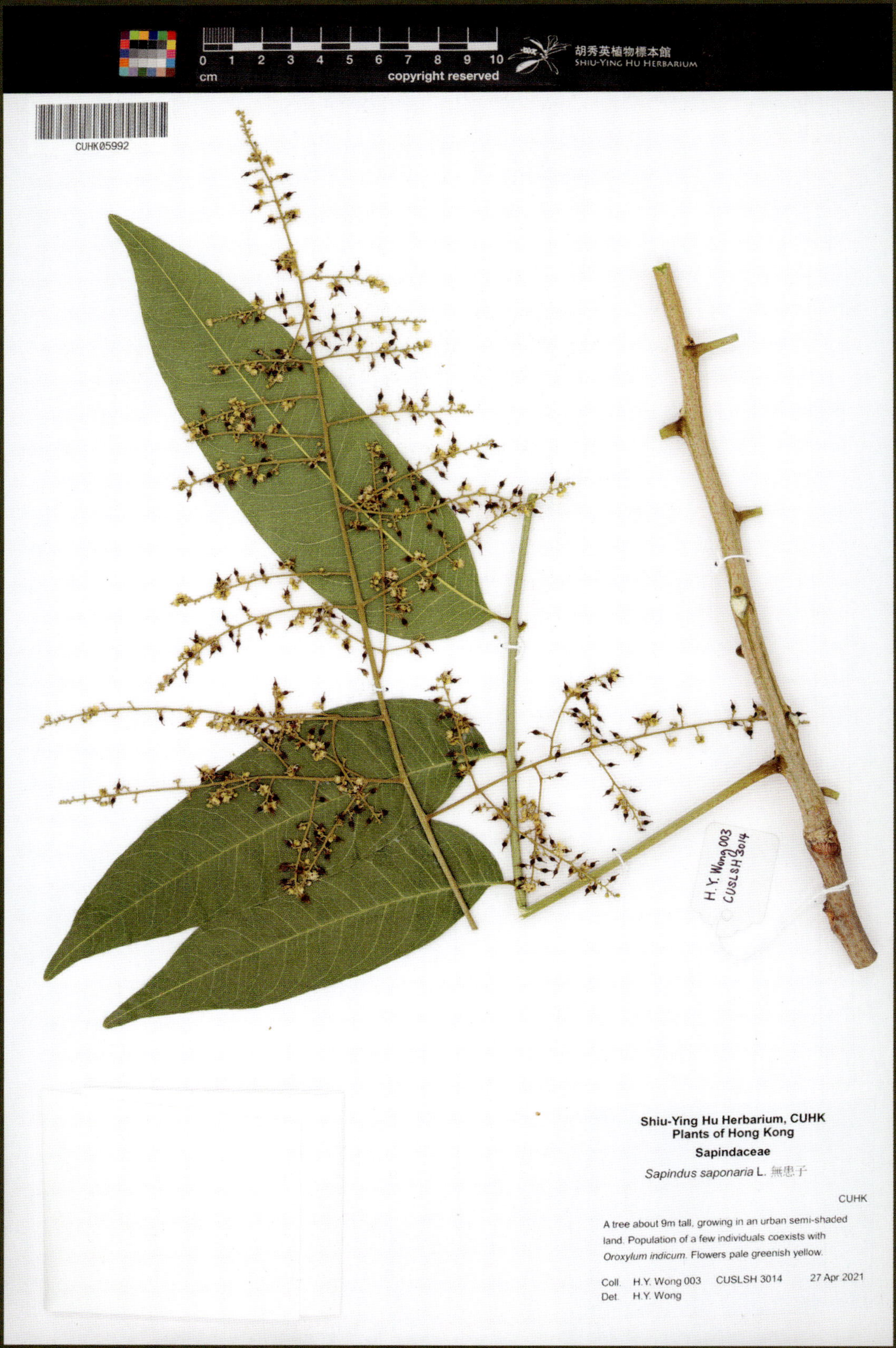

採集者編號 Collector no.
H. Y. Wong 003
採集年份 Year of collection
2021

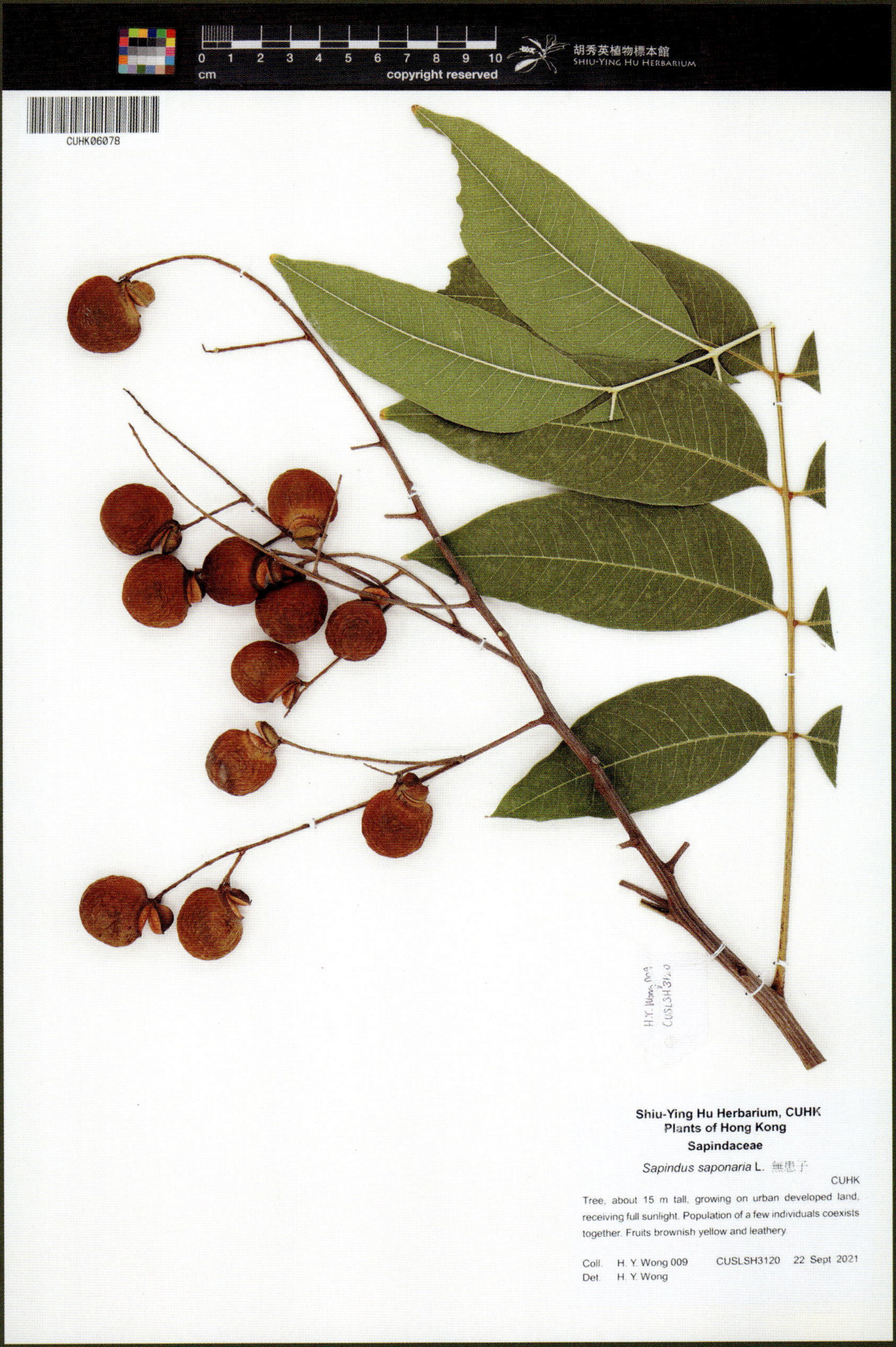

採集者編號 Collector no.
H. Y. Wong 009

採集年份 Year of collection
2021

外觀 Overview

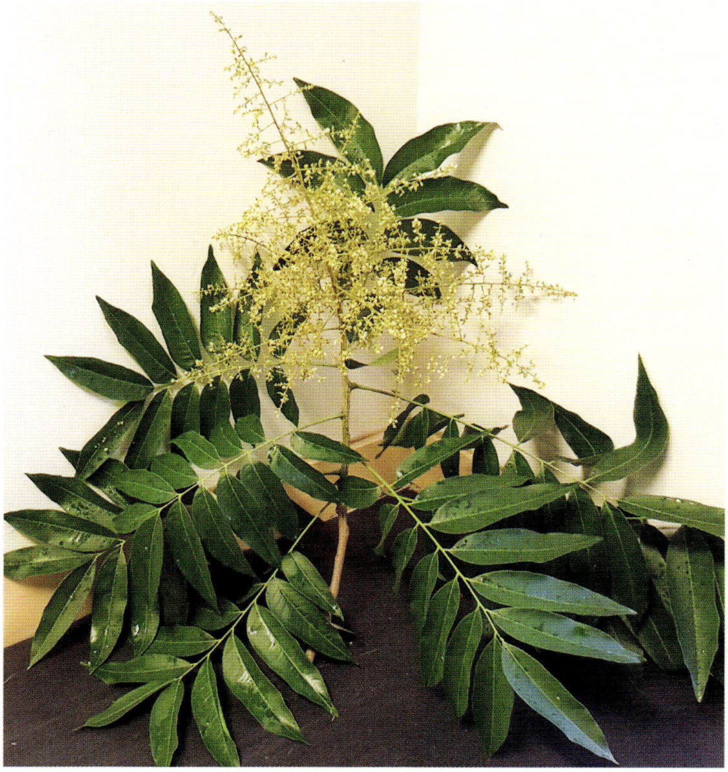

花枝 Flowering branch

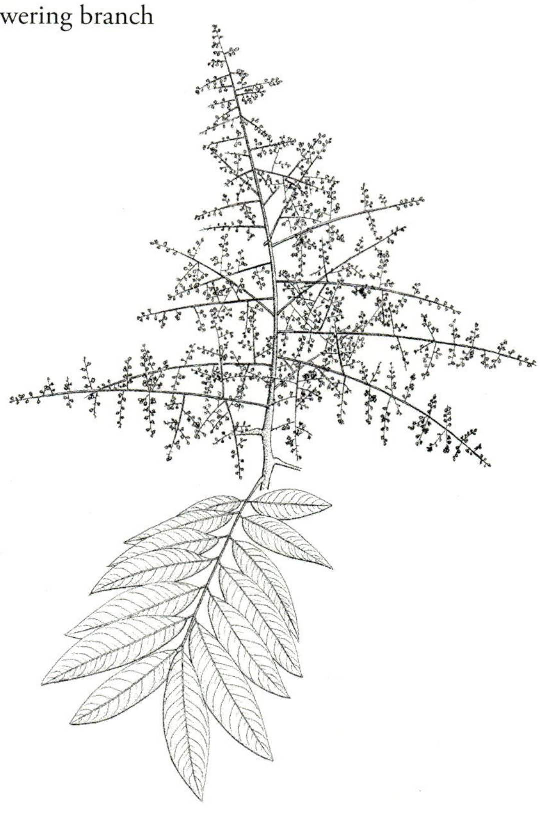

葉 Leaf

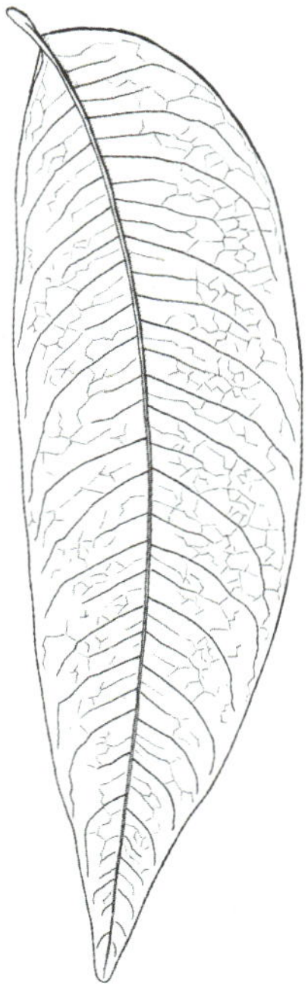

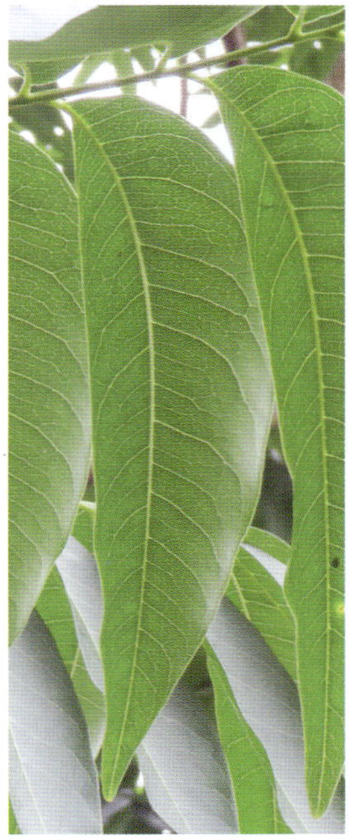

小葉葉底
Leaflet, abaxial

花 Flower

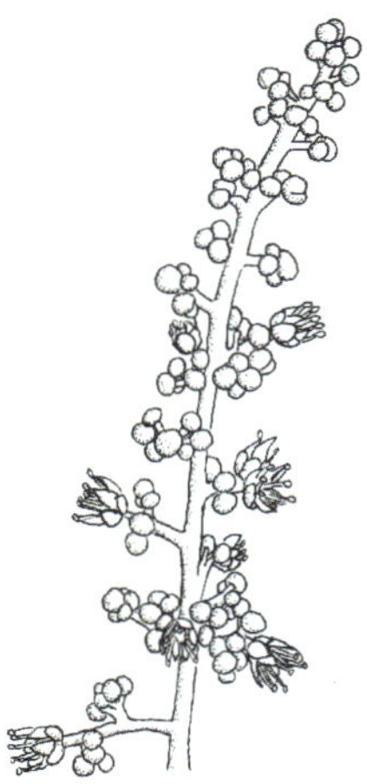

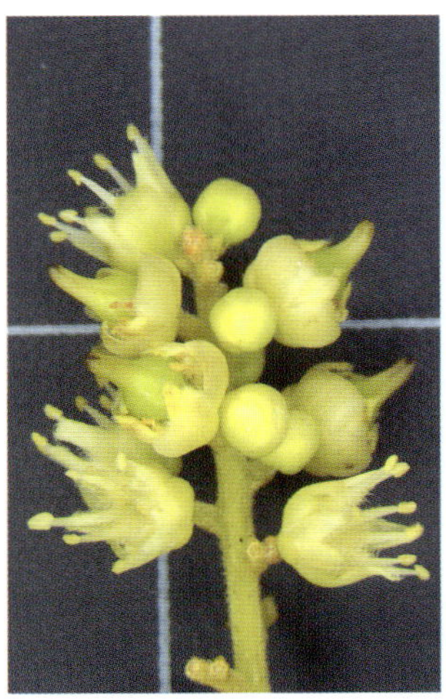

花序局部放大，
展示數朵雄花及雌花
Close-up of an inflorescence, showing a few male and female flowers

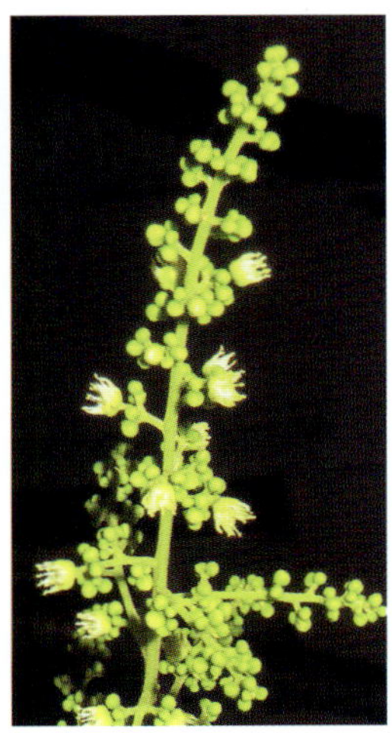

部分的花序
Part of an inflorescence

花 Flower

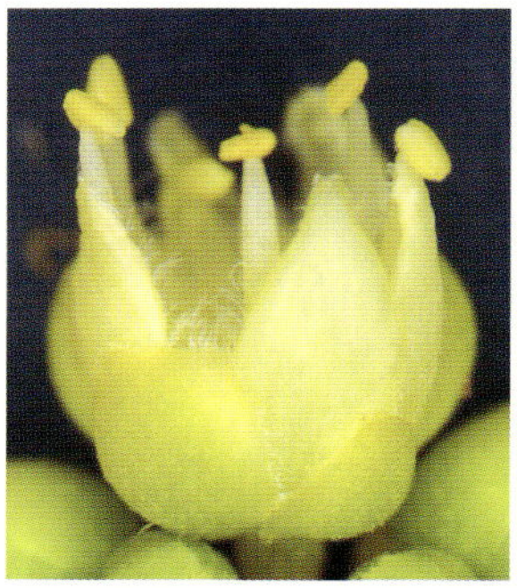

雄花 Male flower

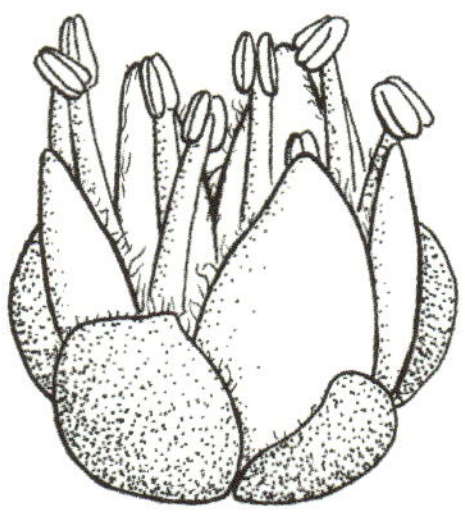

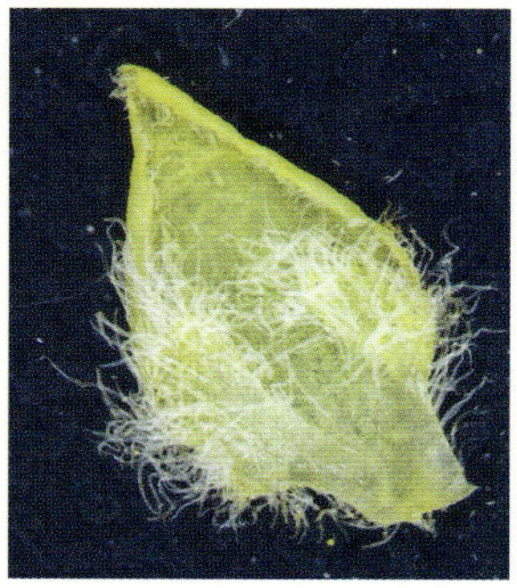

雄花花瓣，向軸面
Petal of male flower, adaxial

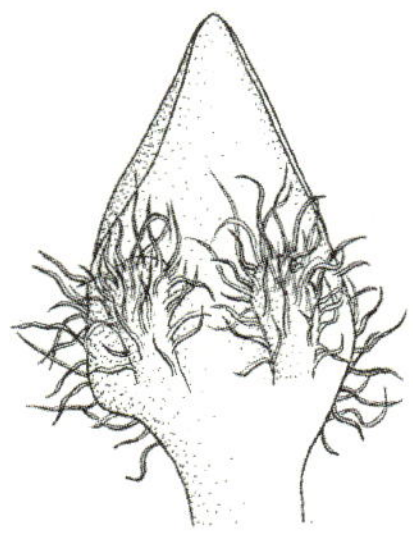

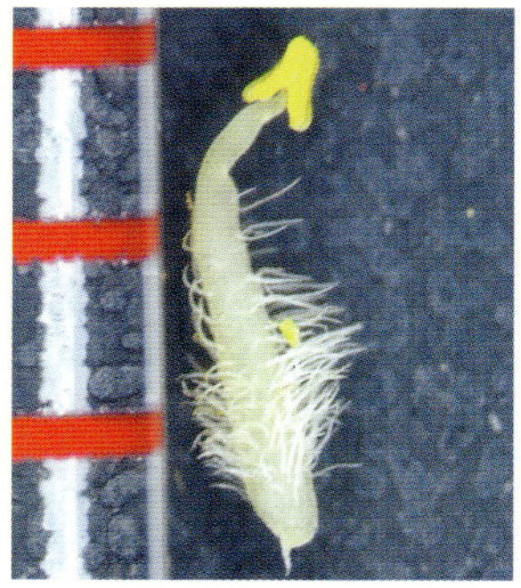

雄蕊
Stamen

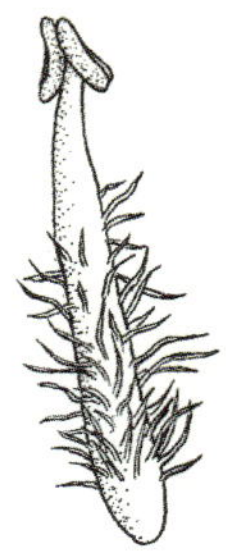

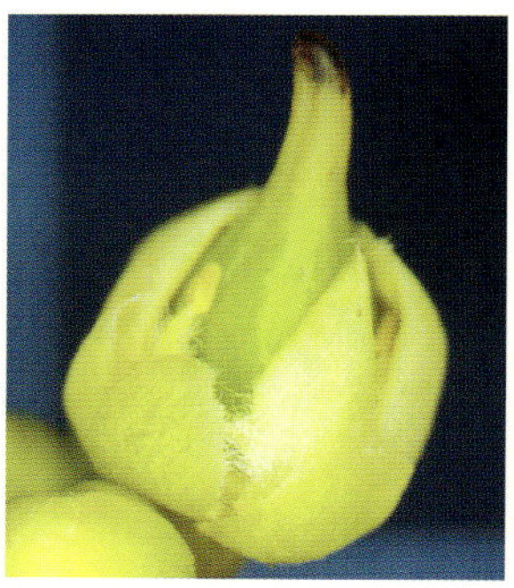

雌花 Female flower

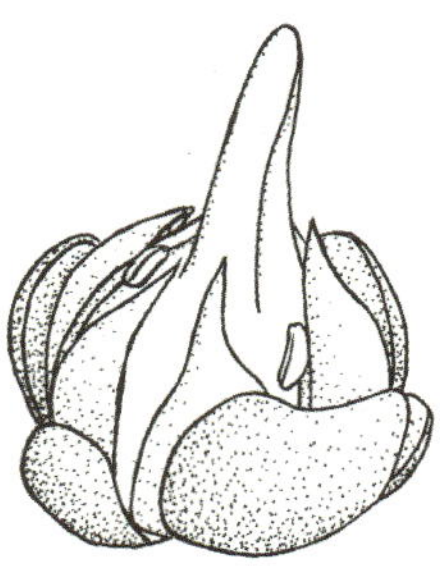

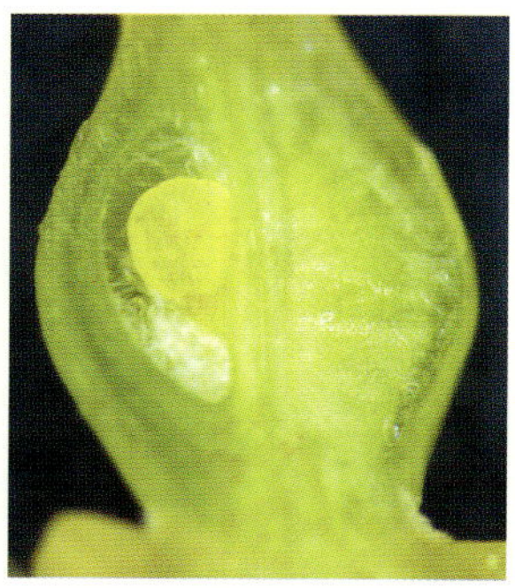

子房，縱切面
Ovary, in L. S.

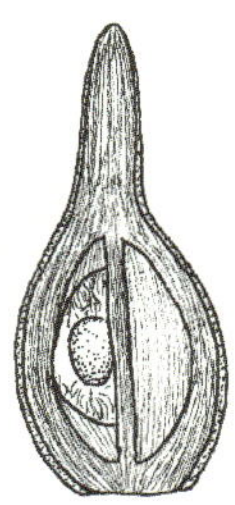

果 Fruit

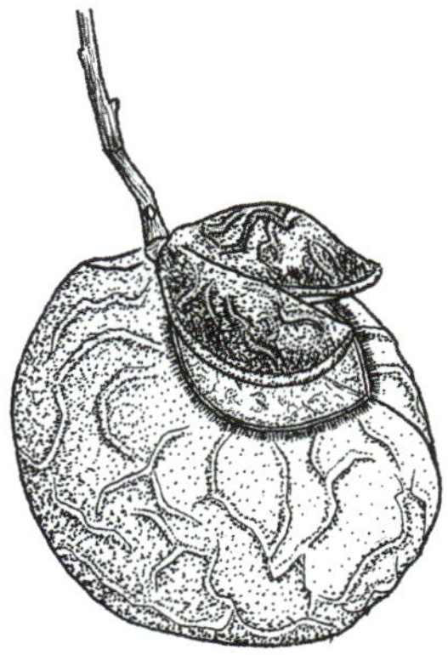
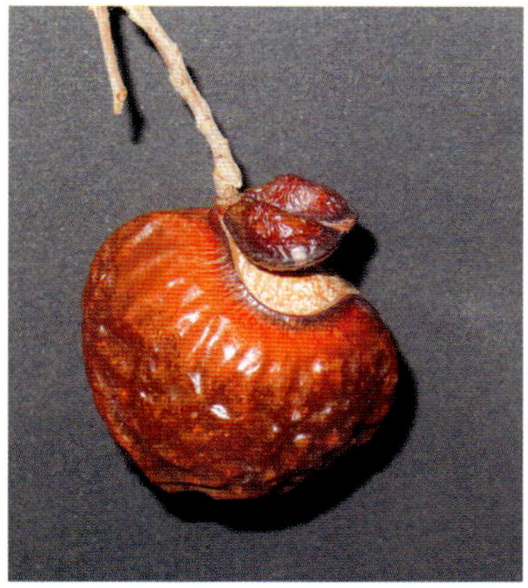

分果 Schizocarps

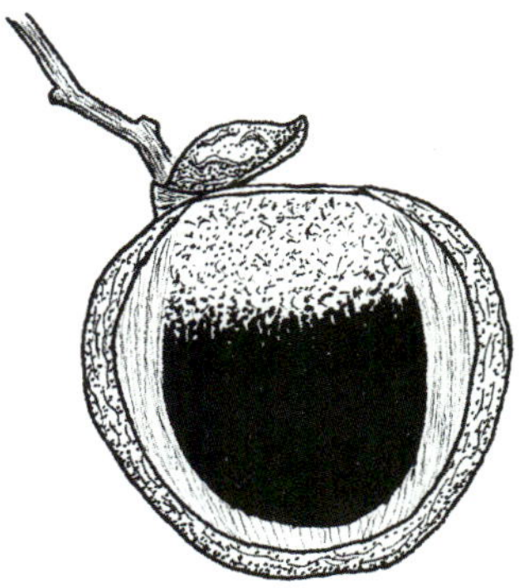
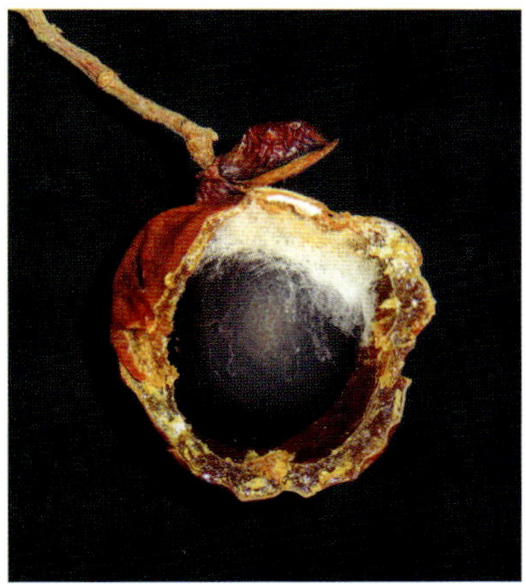

分果，縱切面
Schizocarp, in L. S.

Aquilaria sinensis (Lour.) Spreng.
土沉香

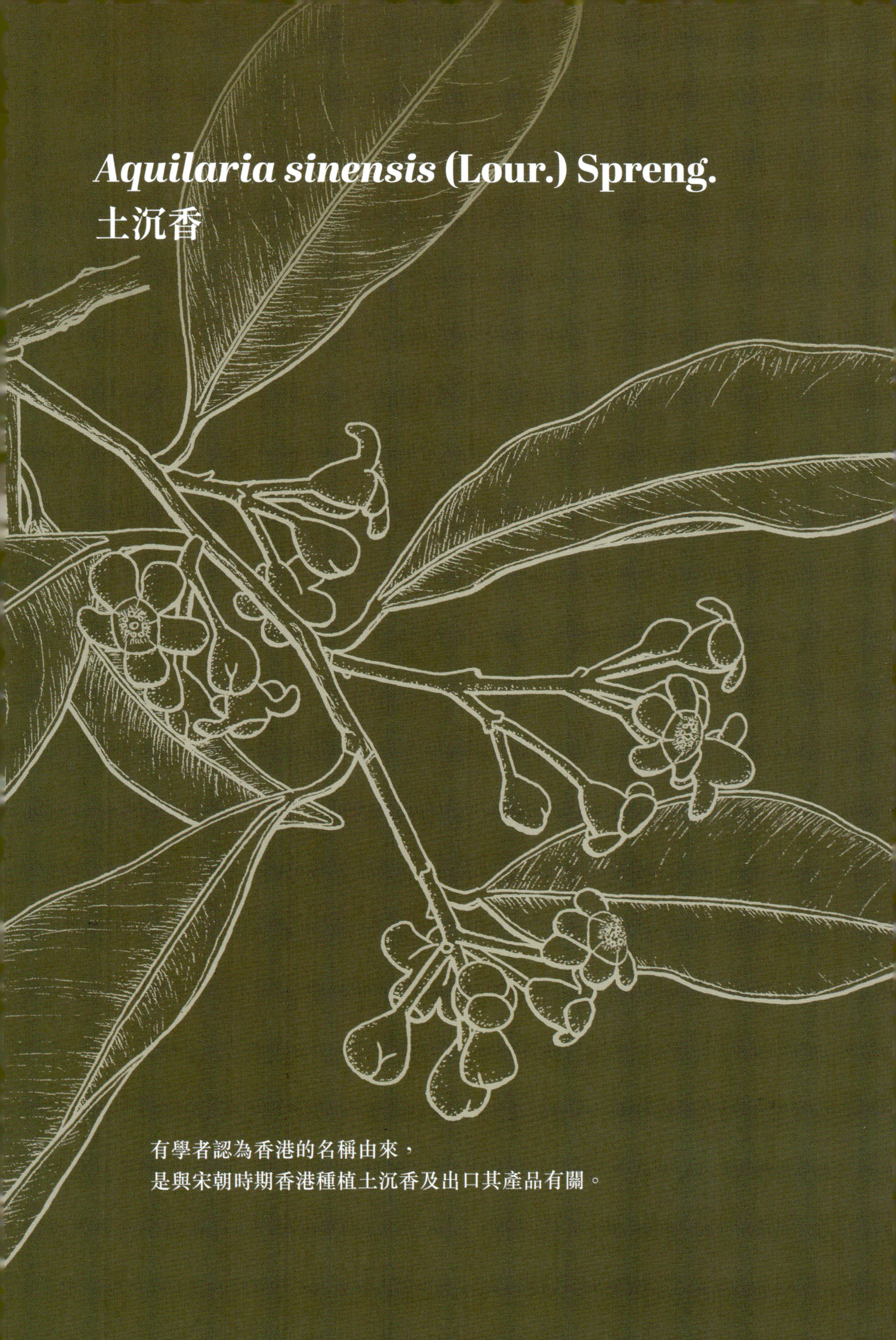

有學者認為香港的名稱由來，
是與宋朝時期香港種植土沉香及出口其產品有關。

採集者編號
Collector no.
Shiu Ying Hu 5263

採集年份
Year of collection
1968

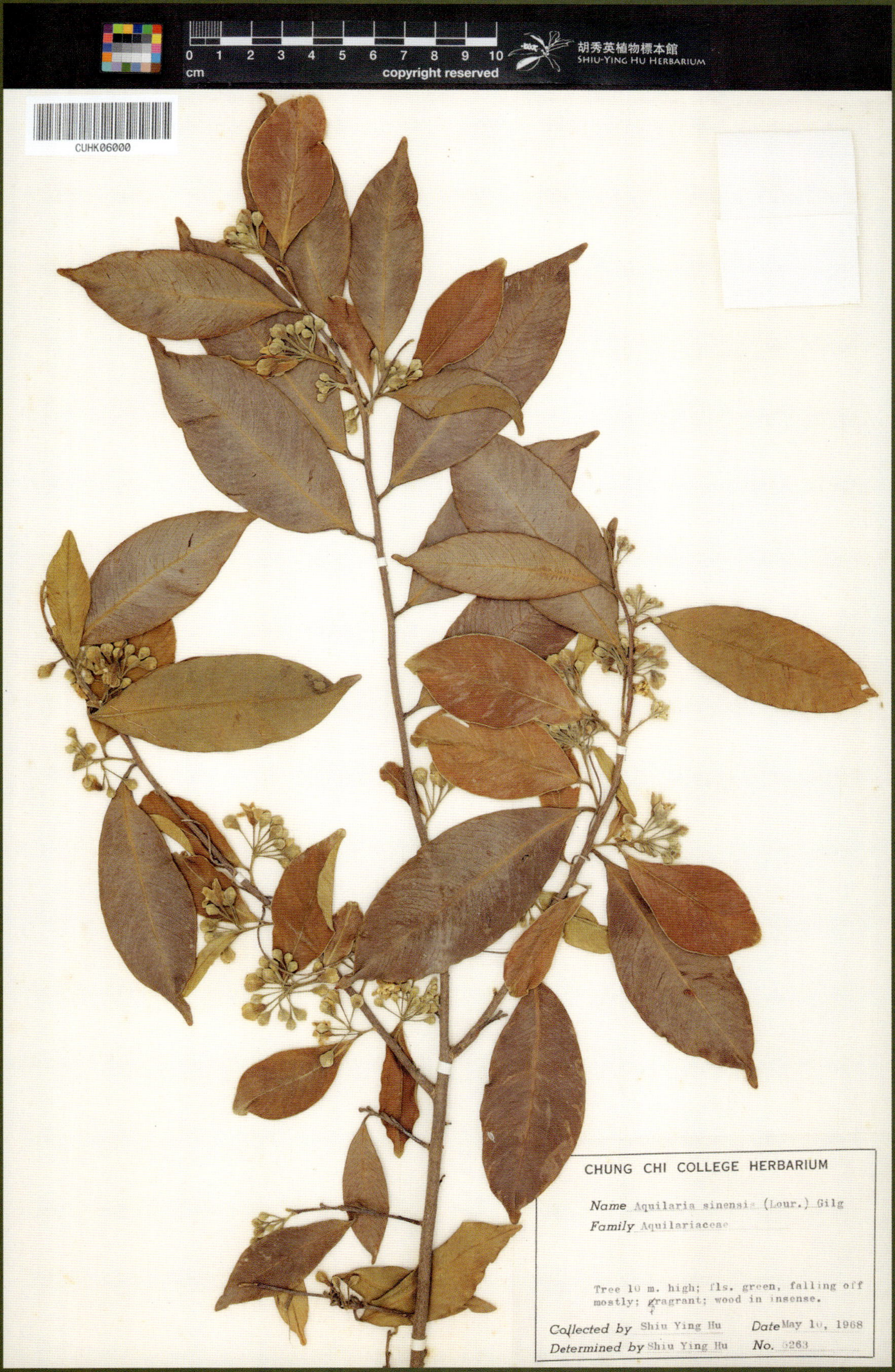
0 1 2 3 4 5 6 7 8 9 10
cm
copyright reserved
胡秀英植物標本館
SHIU-YING HU HERBARIUM
CUHK06000
CHUNG CHI COLLEGE HERBARIUM
Name Aquilaria sinensis (Lour.) Gilg
Family Aquilariaceae
Tree 10 m. high; fls. green, falling off mostly; gragrant; wood in insense.
Collected by Shiu Ying Hu
Date May 10, 1968
Determined by Shiu Ying Hu
No. 5263

土沉香

學名	*Aquilaria sinensis* (Lour.) Spreng.
中文名	土沉香
科名	瑞香科
本地分布狀態	原生植物
繪圖編號	H. Y. Wong 006

花期	花色	果期	果色
3 至 5 月	黃綠色	9 至 10 月	黃綠色

土沉香為常綠喬木，常見於低地森林。

有學者認為香港的名稱由來，是與宋朝時期香港種植土沉香及出口其產品有關。由於土沉香受真菌感染後，會產出具藥用價值及可作香料的名貴樹脂，導致其野生群落不斷被非法砍伐。現時香港及中國內地均有土沉香的人工培植及研究，以作保育、綠化及經濟用途。

土沉香於《香港稀有及珍貴植物》中屬「近危」，在《中國植物紅皮書》為「漸危」，而在國際自然保護聯盟瀕危物種紅色名錄中則為「易危」。

土沉香受香港法例第 586 章《保護瀕危動植物物種條例》規管，亦列於《瀕危野生動植物種國際貿易公約》附錄II。

《中國植物誌》(英文版) 的採用名稱

Aquilaria sinensis (Loureiro) Sprengel (科名：瑞香科)

《香港植物誌》的採用名稱

Aquilaria sinensis (Lour.) Spreng. (科名：瑞香科)

參考文獻 References

1. Hong Kong Herbarium. (2021a). *Botanical Nomenclature-Aquilaria sinensis and origin of the name of Hong Kong*. Special Topics. Retrieved June 1, 2021, from https://herbarium.gov.hk/en/special-topics/botanical-nomenclature/botanical-nomenclature-detail/index-id-1.html
2. Hong Kong Herbarium. (2021b). *Aquilaria sinensis*. Rare and Precious Plants of Hong Kong (Online Version). Retrieved June 1, 2021, from https://www.herbarium.gov.hk/en/publications/books/book2/text/aquilaria-sinensis/index.html
3. Hong Kong Herbarium. (2021c). *HK Herbarium-Aquilaria sinensis*. HK Plant Database. Retrieved May 15, 2021, from https://herbarium.gov.hk/tc/hk-plant-database/plant-detail/index.html?pType=species&oID=6214
4. IUCN. (2018). *Aquilaria sinensis*. The IUCN Red List of Threatened Species. Retrieved June 1, 2021, from https://www.iucnredlist.org/species/32382/2817115

Aquilaria sinensis (Lour.) Spreng.

Scientific name	*Aquilaria sinensis* (Lour.) Spreng.
Chinese name	土沉香
Family	Thymelaeaceae Juss.
Local distribution status	Native species
Illustration number	H. Y. Wong 006

FLOWERING PERIOD	COLOUR OF FLOWER	FRUITING PERIOD	COLOUR OF FRUIT
March to May	Yellowish green	September to October	Yellowish green

Aquilaria sinensis (Lour.) Spreng. is an evergreen tree, usually found in lowland forests. Some scholars believe the origin of the name 'Hong Kong' is related to the fact that Hong Kong had *Aquilaria sinensis* (Lour.) Spreng. plantations and exportation of the derived products during the Song dynasty. Fungal infection of *Aquilaria sinensis* (Lour.) Spreng. will produce valuable resin which is used as medicine and incense. This has led to illegal logging of its populations. Currently, Hong Kong and mainland China have propagation and research on the species, for conservation, greening and commercial purposes.

Aquilaria sinensis (Lour.) Spreng. is listed as 'Near Threatened' in *Rare and Precious Plants of Hong Kong*, as 'Vulnerable' in the *China Plant Red Data Book* and as 'Vulnerable' in the IUCN Red List of Threatened Species. It is regulated under Hong Kong's Protection of Endangered Species of Animals and Plants Ordinance (Cap. 586) and is also included in CITES Appendix II.

Name in *Flora of China*
Aquilaria sinensis (Loureiro) Sprengel (Family: Thymelaeaceae)

Name in *Flora of Hong Kong*
Aquilaria sinensis (Lour.) Spreng. (Family: Thymelaeaceae)

5. Lin, B., Guo, Y. Y., Zeng, X. Q., & Du, Z. H. (2019). Current situation and development suggestions of *Aquilaria Sinensis* breeding in Hainan Province. *Tropical Forestry 47*(2), 44–47. https://doi.org/10.3969/j.issn.1672-0938.2019.02.012
6. Missouri Botanical Garden. (2021). *Name-Aquilaria sinensis.* Tropicos. Retrieved June 7, 2021, from https://www.tropicos.org/name/50175642
7. The CITES Secretariat. (n.d.). *Aquilaria sinensis.* Checklist of CITES Species. Retrieved June 7, 2021, from https://checklist.cites.org/#/en
8. Wang, Y. Z., Gilbert, M. G., Mathew, B., Brickell, C. D., & Nevling, L. I. (2007). Thymelaeaceae. In Z. Y. Wu & P. H. Raven (Eds.), *Flora of China* (Vol. 13, pp. 213–250). Science Press & Missouri Botanical Garden Press.

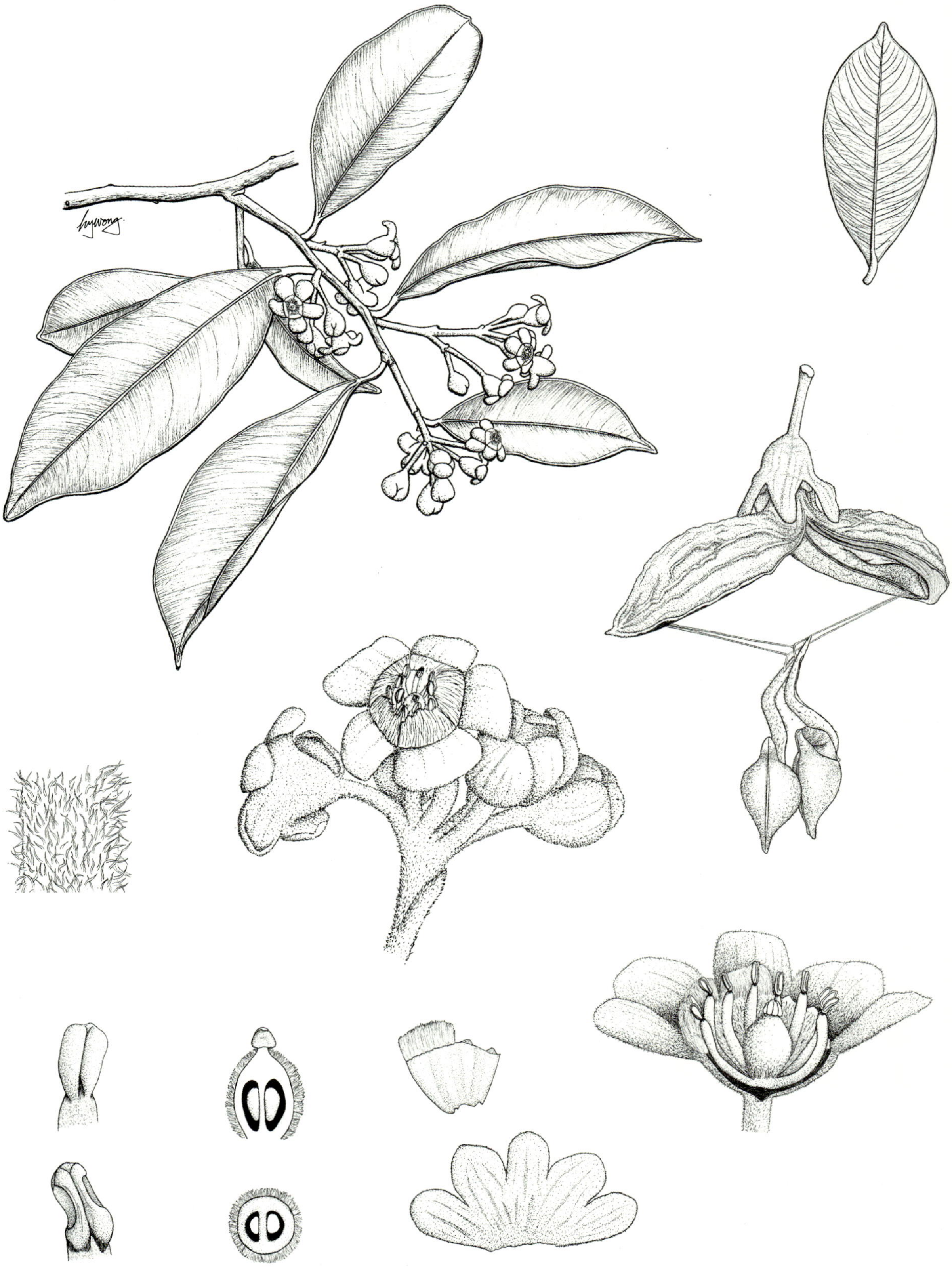

Aquilaria sinensis (Lour.) Spreng. 土沉香
Illustration 006 by H. Y. Wong
Authenticated by D. T. W. Lau

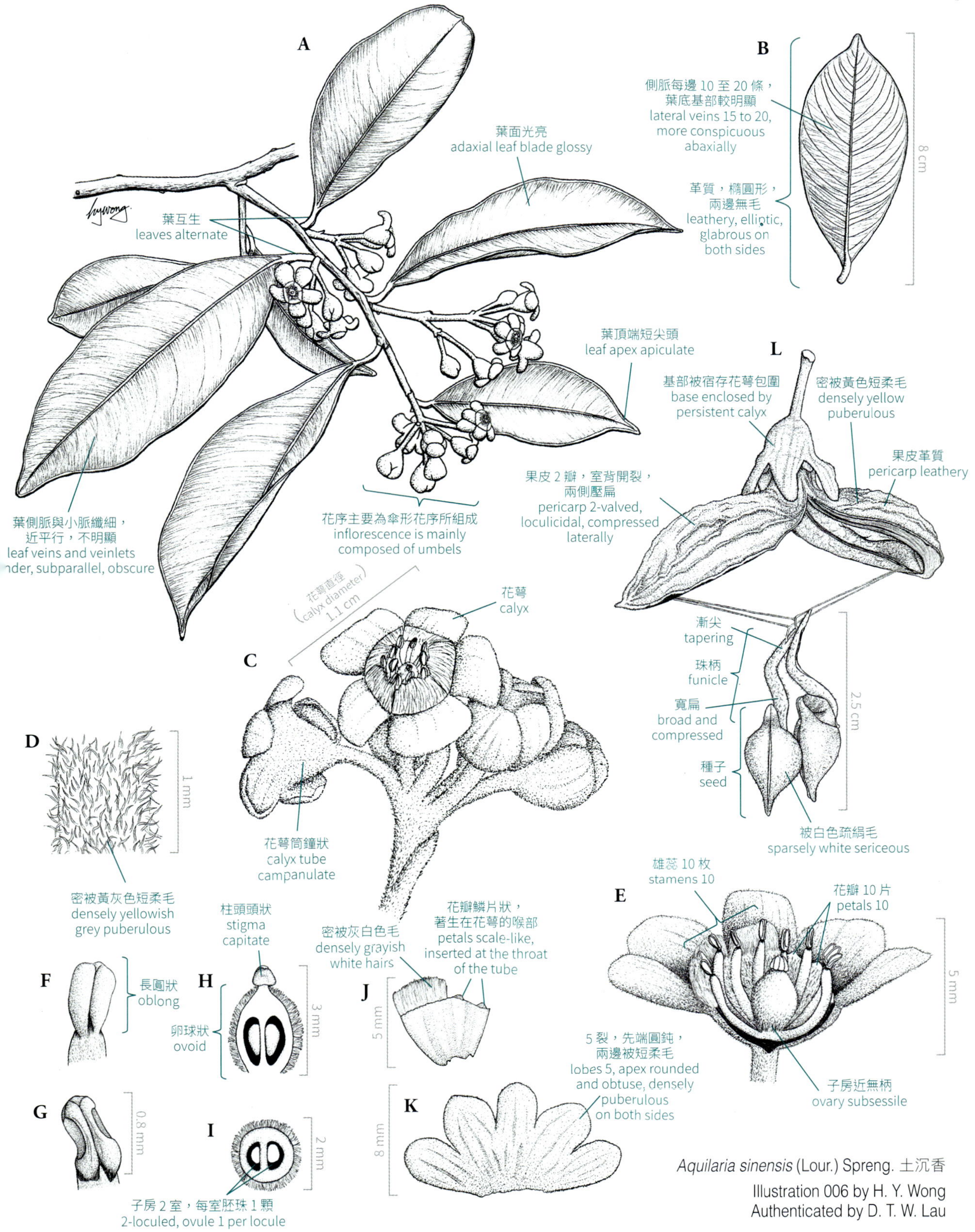

Aquilaria sinensis (Lour.) Spreng. 土沉香

Illustration 006 by H. Y. Wong
Authenticated by D. T. W. Lau

植物結構 PLANT STRUCTURES:

A. 花枝 Flowering branch **B.** 葉底 Leaf, abaxial **C.** 花序 Inflorescence **D.** 花梗的被毛 Hairs on pedicel **E.** 花解剖圖，展示花萼、2 片花瓣、雄蕊及雌蕊 Dissected flower, showing calyx, 2 petals, androecium and gynoecium **F.** 花藥，背軸面 Anther, abaxial **G.** 花藥，向軸面 Anther, adaxial **H.** 子房，縱切面 Ovary, in L. S. **I.** 子房，橫切面 Ovary, in C. S. **J.** 部分的花萼，除去 2 片花瓣上的被毛，展示 2 片花瓣 Part of the calyx tube, hairs on 2 petals removed, showing 2 petals **K.** 花萼，背軸面 Calyx tube, abaxial **L.** 果實 Fruit

參考標本 REFERENCE SPECIMENS: Rare and endangered plants (S. W. Shek, K. W. Lam, D. T. W. Lau, J. Y. Y. Lau) 10B; H. Y. Wong 005 (CUSLSH3055); S. Y. Hu 5263

參考文獻 REFERENCES:

Huang, S. C., & Zhang, Z. R. (1999). Thymelaeaceae. In Delectis Florae Reipublicae Popularis Sinicae Agendae Academiae Sinicae Edita (Ed.), *Flora Reipublicae Popularis Sinicae* (Vol. 52(1), pp. 287–400). Science Press.

Stevens, P. F. (2021). Malvales. Angiosperm Phylogeny Website. Retrieved June 8, 2021, from http://www.mobot.org/MOBOT/research/APweb/welcome.html

Wang, Y. Z., Gilbert, M. G., Mathew, B., Brickell, C. D., & Nevling, L. I. (2007). Thymelaeaceae. In Z. Y. Wu & P. H. Raven (Eds.), *Flora of China* (Vol. 13, pp. 213–250). Science Press & Missouri Botanical Garden Press.

Wei, Z. F. (1995). Thymelaeaceae. In South China Botanical Garden, the Chinese Academy of Sciences (Ed.), *Flora of Guangdong* (Vol. 3, pp. 84–92). Guangdong Science and Technology Press.

Xia, N. H. (2008). Thymelaeaceae. In Hong Kong Herbarium, AFCD & South China Botanical Garden, CAS (Eds.), *Flora of Hong Kong* (Vol. 2, pp. 132–135). Agriculture, Fisheries and Conservation Department, HKSAR Government.

Zhang, Y. X. (2010). Thymelaeaceae. In Fairy Lake Botanical Garden, Shenzhen & Chinese Academy of Sciences (Eds.), *Flora of Shenzhen* (Vol. 2, pp. 404–409). China Forestry Publishing House.

憑證標本
Voucher Specimen

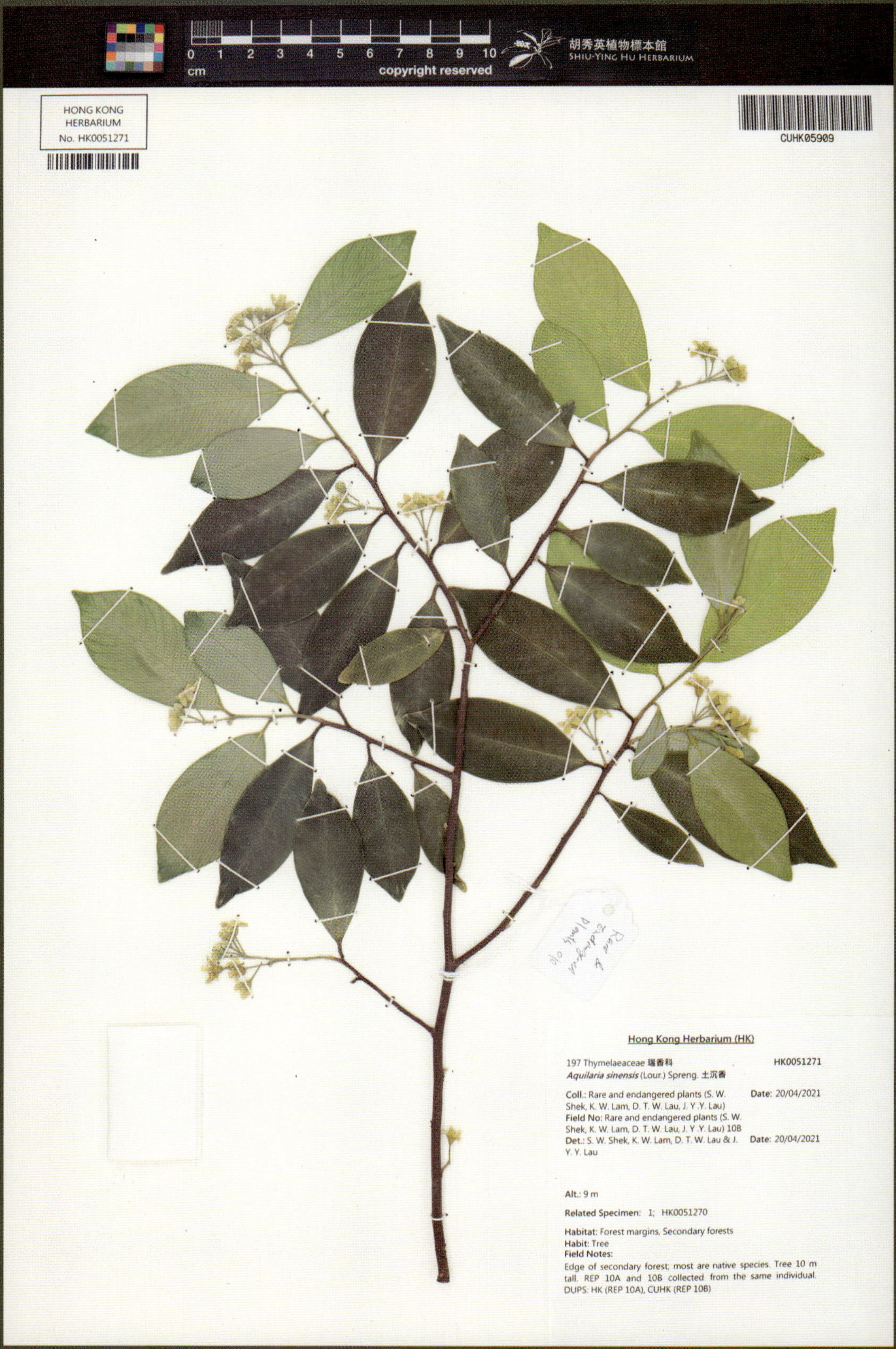

採集者編號 Collector no.

Rare and endangered plants (S. W. Shek, K. W. Lam, D. T. W. Lau, J. Y. Y. Lau) 10B

採集年份 Year of collection

2021

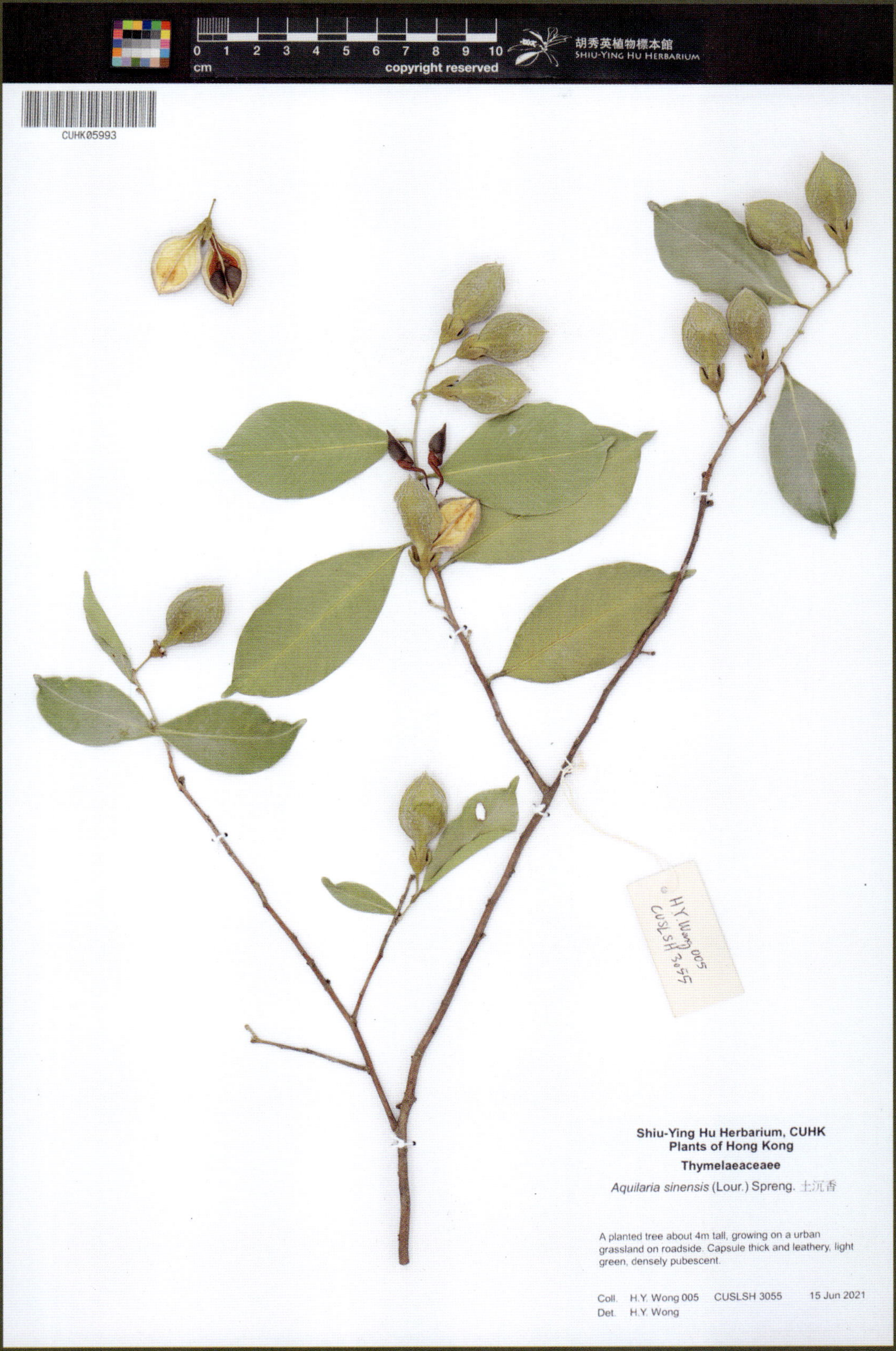

採集者編號 Collector no.
H. Y. Wong 005

採集年份 Year of collection
2021

外觀 Overview

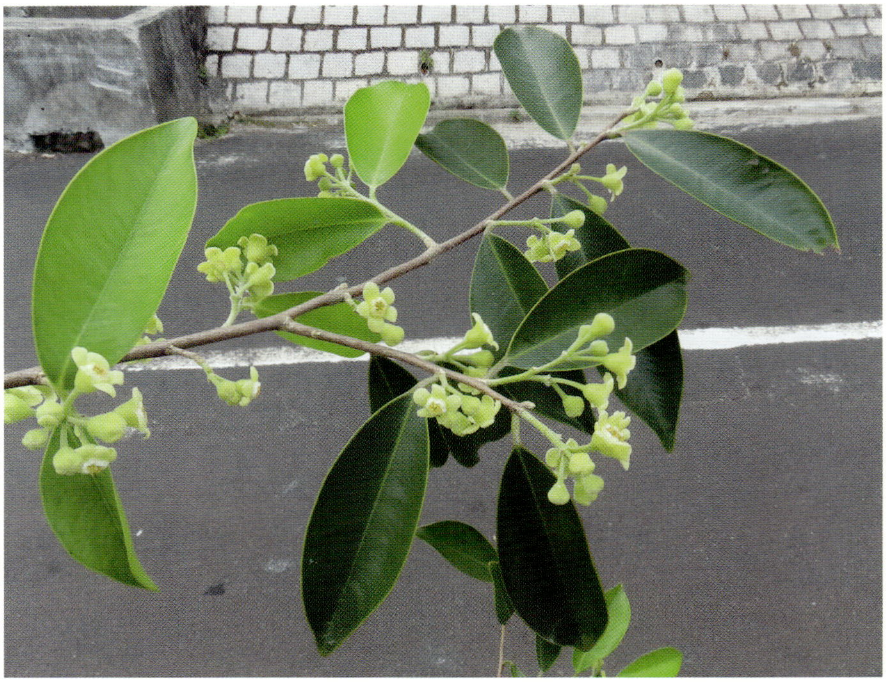

花枝 Flowering branch

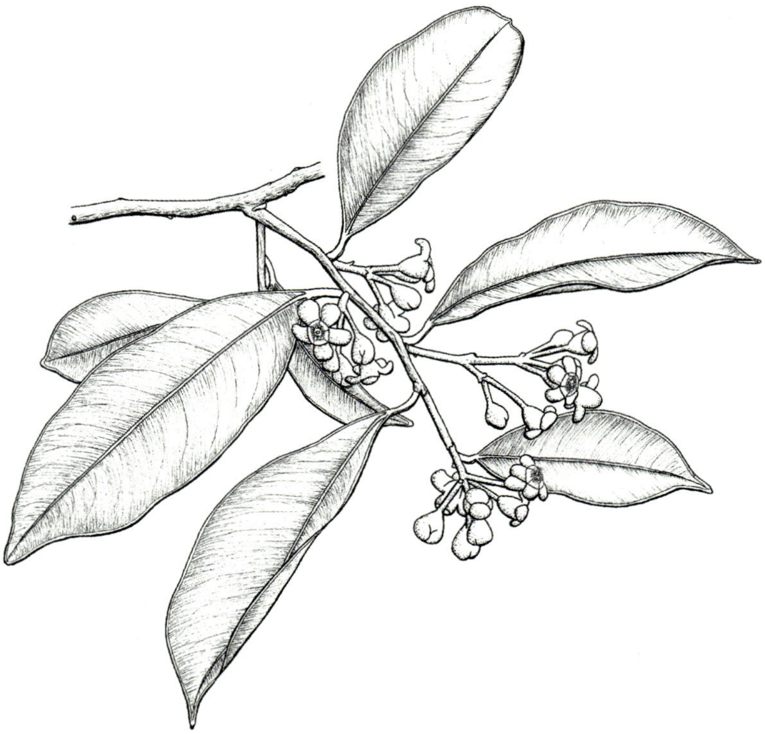

葉 Leaf

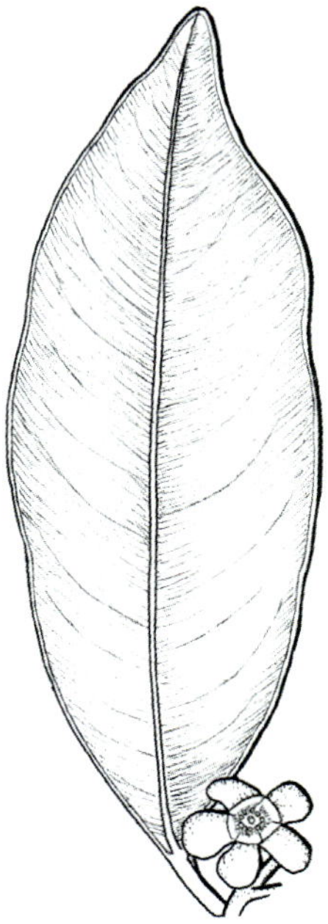

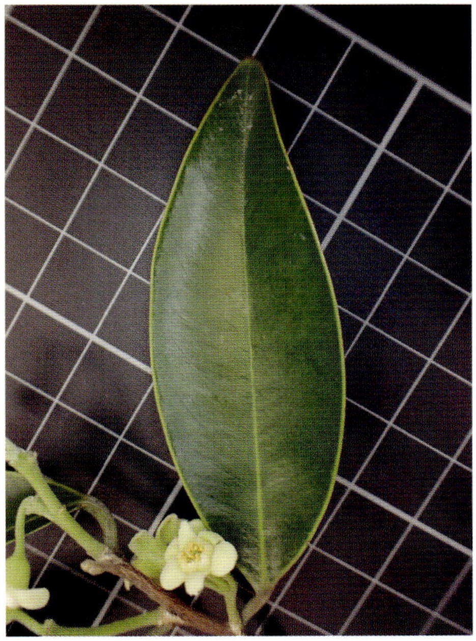

葉面 Leaf, adaxial

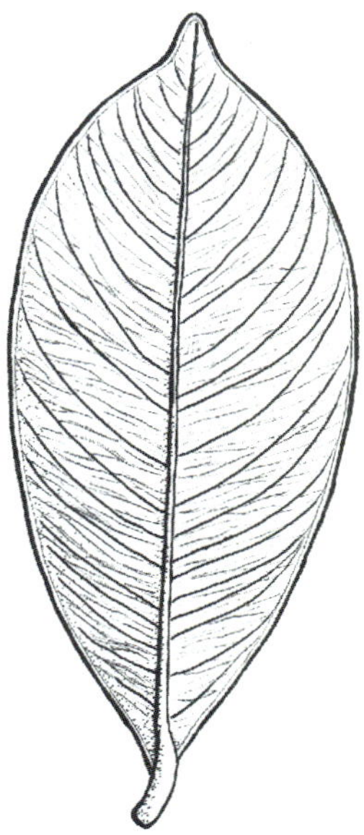

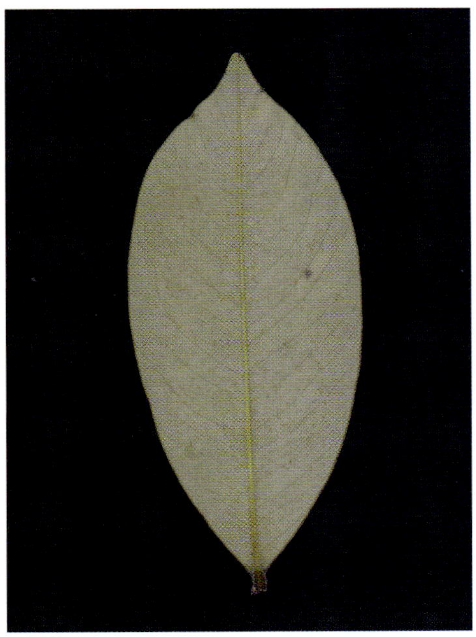

葉底 Leaf, abaxial

花 Flower

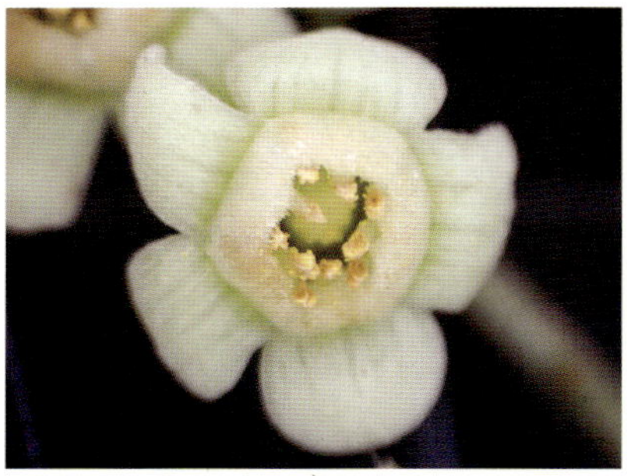
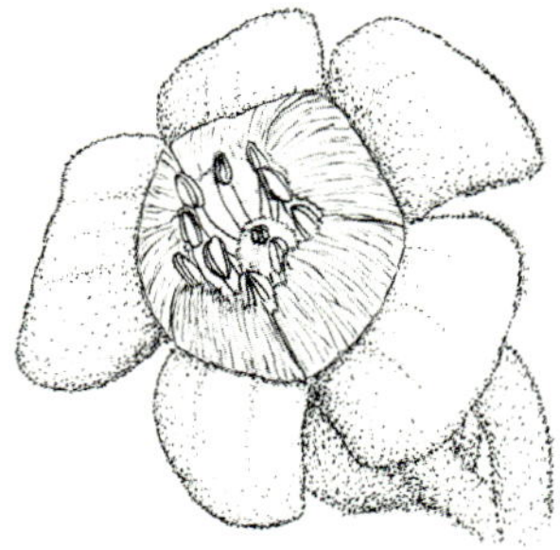

花，正面觀 Flower, frontal view

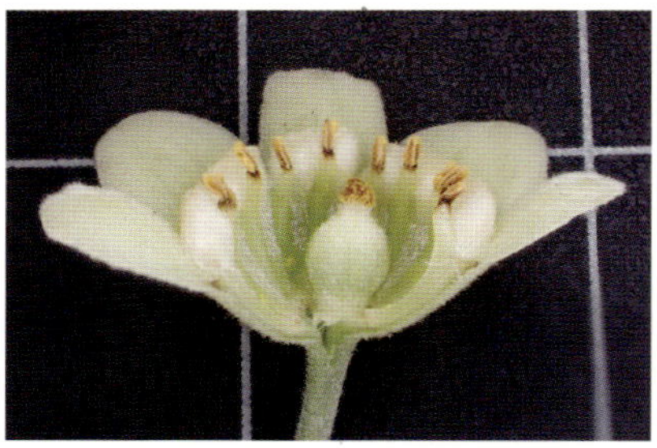
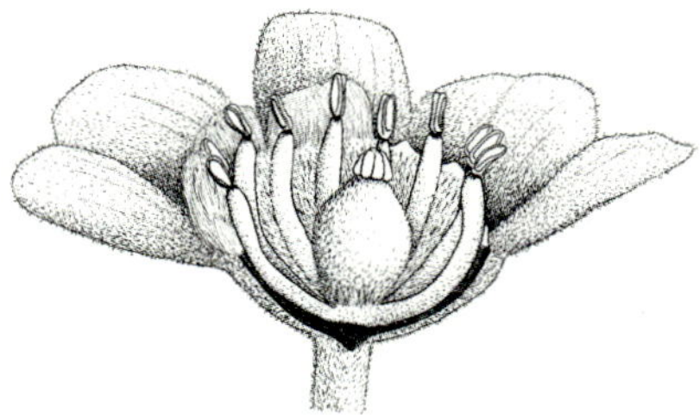

花解剖圖，展示花萼、
雄蕊及雌蕊
Dissected flower, showing calyx, androecium and gynoecium

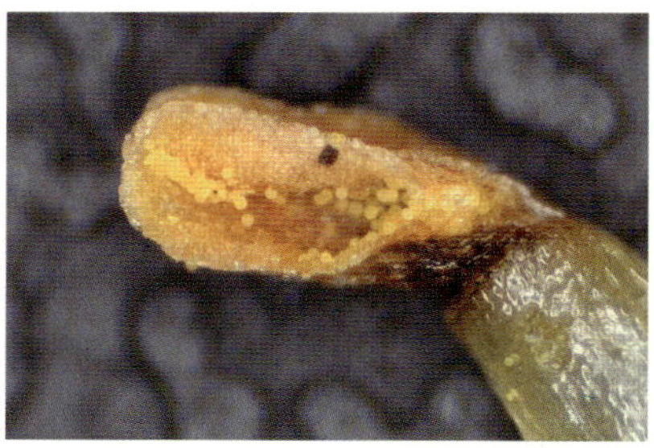
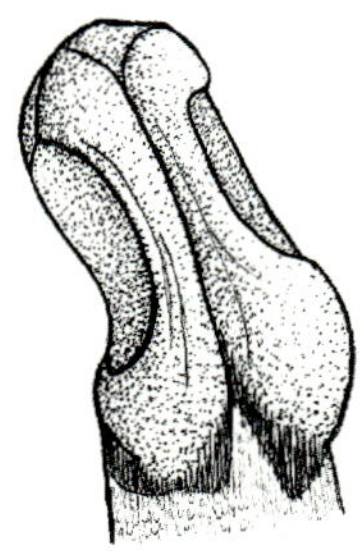

花藥，側面觀 Anther, lateral view

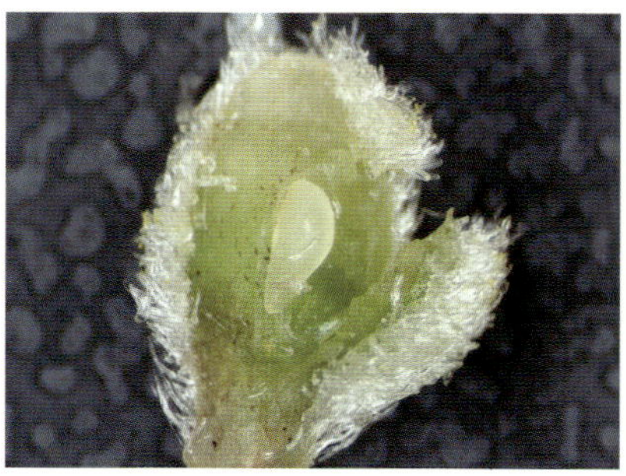
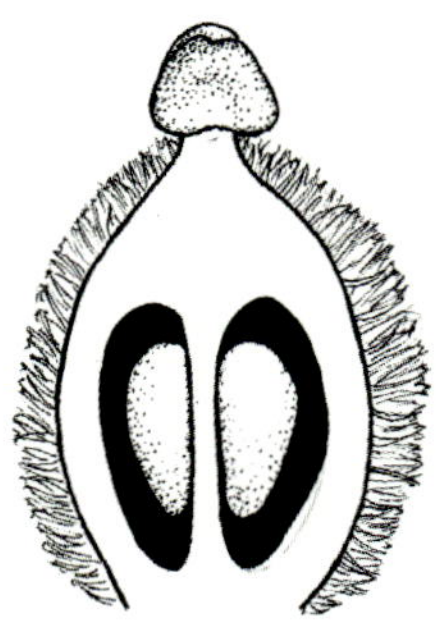

子房，縱切面 Ovary, in L. S.

果 Fruit

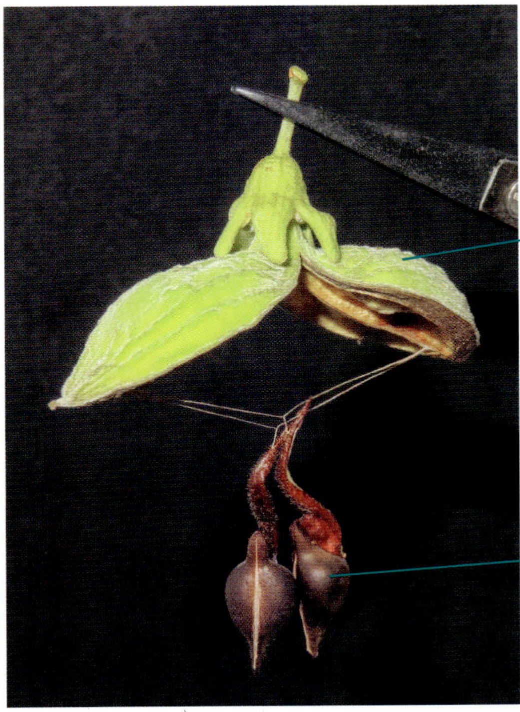

蒴果 Capsule

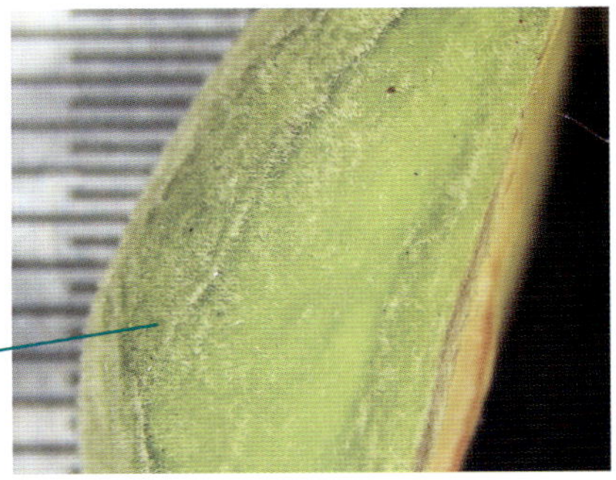

果皮 Pericarp

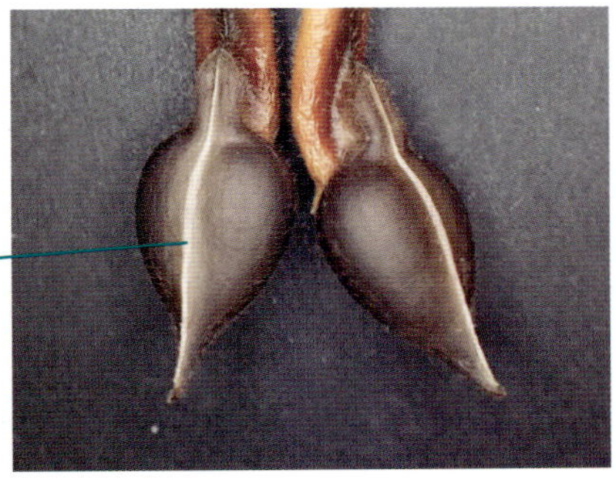

種子 Seeds

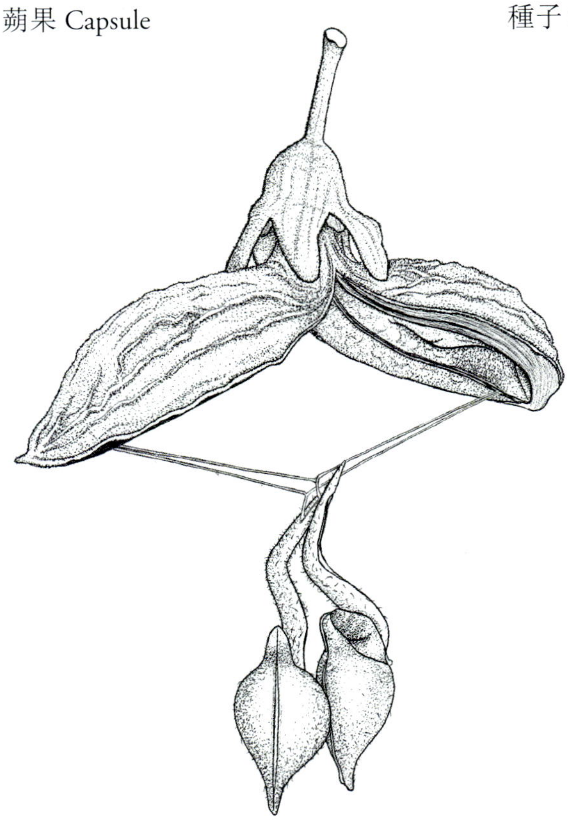

Lysimachia alpestris Champ. ex Benth.
香港過路黃

香港過路黃的模式標本是由
J. G. Champion於1850年前後在太平山採集。

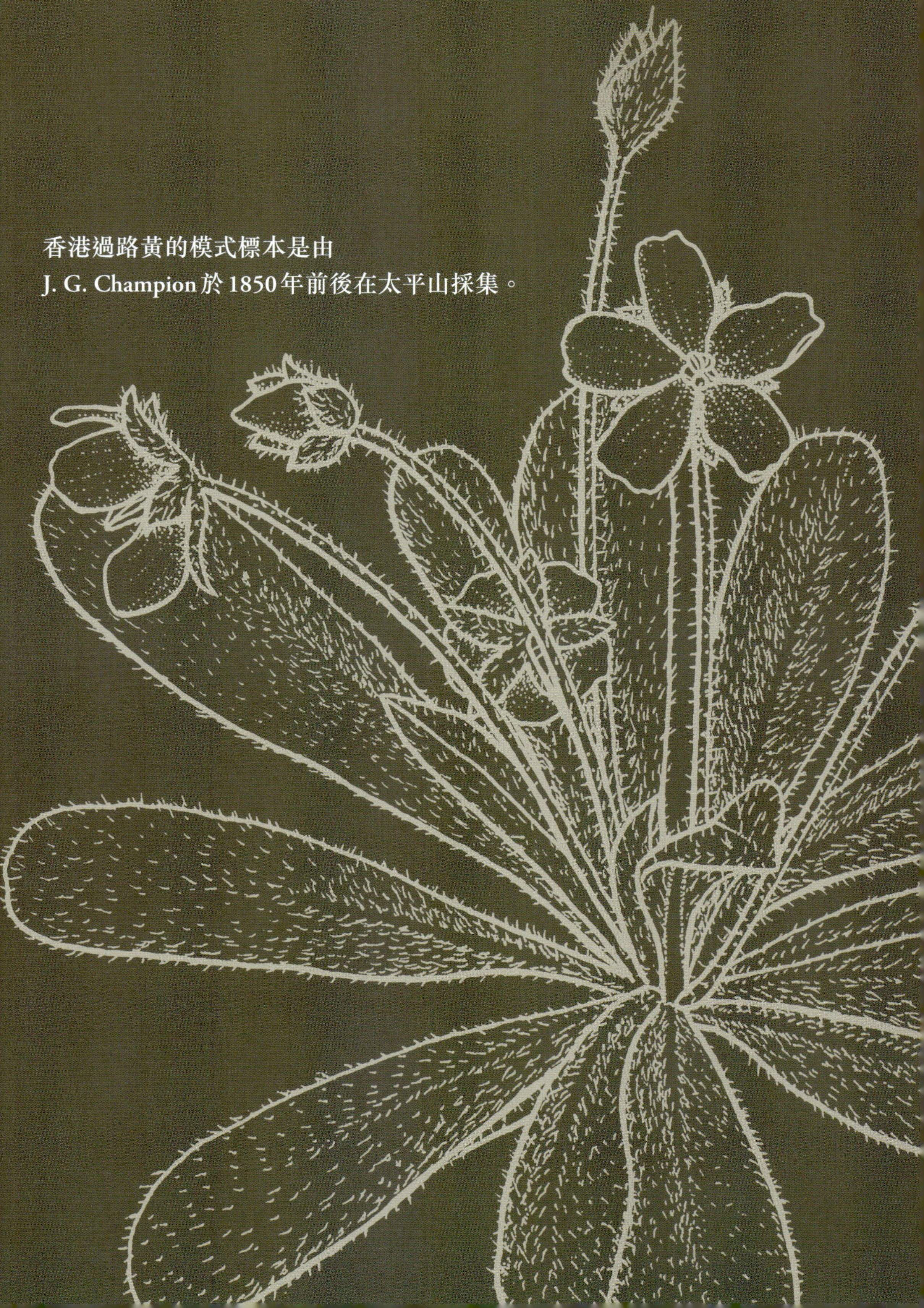

胡秀英教授採集的標本
Specimen Collected by Professor Shiu-Ying Hu

採集者編號
Collector no.
Hu & But 20443

採集年份
Year of collection
1992

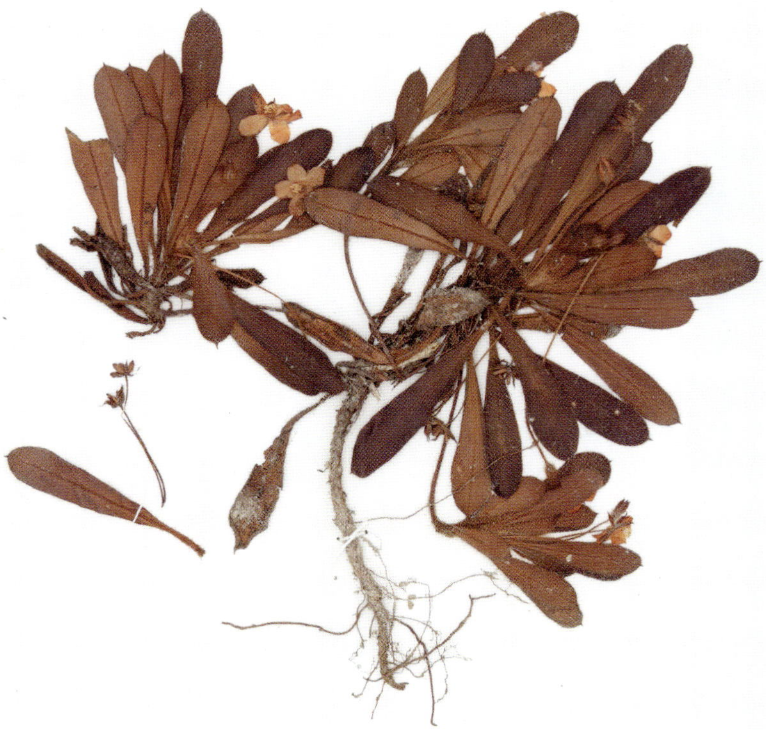

Herbarium, The Chinese University of Hong Kong
Plants of Hong Kong

Lysinachia alpestris Champ. ex Bentham

Acaulescent herbs stoloniferous; fls yellow.

Coll. Hu & But 20443 Apr. 21, 1992
Det. S.Y. Hu

295

香港過路黃

學名 *Lysimachia alpestris* Champ. ex Benth.

中文名 香港過路黃

科名 報春花科

本地分布狀態 原生植物

繪圖編號 H. Y. Wong 004

花期	花色	果期	果色
4至5月	黃色	未有記錄	未有記錄

香港過路黃是多年生的草本植物，可見於溪邊、次生林及灌木叢。其模式標本是由J. G. Champion於1850年前後在太平山採集。

香港過路黃的地區分布狹窄，僅見於香港局部地區（極少量見亦於廣東新會古兜山），於《香港稀有及珍貴植物》中屬「極危」。

《中國植物誌》（英文版）的採用名稱

Lysimachia alpestris Champion ex Bentham（科名：報春花科）

《香港植物誌》的採用名稱

Lysimachia alpestris Champ. ex Benth.（科名：報春花科）

參考文獻 References

1. Hong Kong Herbarium. (2021a). *Lysimachia alpestris*. Rare and Precious Plants of Hong Kong (Online Version). Retrieved May 2, 2021, from https://www.herbarium.gov.hk/en/publications/books/book2/text/lysimachia-alpestris/index.html
2. Hong Kong Herbarium. (2021b). *HK Herbarium-Lysimachia alpestris*. HK Plant Database. Retrieved May 2, 2021, from https://www.herbarium.gov.hk/tc/hk-plant-database/plant-detail/index.html?pType=species&oID=2703

Lysimachia alpestris Champ. ex Benth.

Scientific name	*Lysimachia alpestris* Champ. ex Benth.
Chinese name	香港過路黃
Family	Primulaceae Batsch
Local distribution status	Native species
Illustration number	H. Y. Wong 004

FLOWERING PERIOD	COLOUR OF FLOWER	FRUITING PERIOD	COLOUR OF FRUIT
April to May	Yellow	Not recorded	Not recorded

Lysimachia alpestris Champ. ex Benth. is a perennial herb that can be found by streams or in secondary forests and shrublands. Its type specimen was collected by J. G. Champion in Victoria Peak around 1850.

The distribution of *Lysimachia alpestris* Champ. ex Benth. is restricted to some areas of Hong Kong (and a very small number in Gudou Shan, Guangdong). It is listed as 'Critically Endangered' in *Rare and Precious Plants of Hong Kong*.

Name in *Flora of China*
Lysimachia alpestris Champion ex Bentham (Family: Primulaceae)

Name in *Flora of Hong Kong*
Lysimachia alpestris Champ. ex Benth. (Family: Primulaceae)

3. Hu, C. M., & Kelso, S. (1996). Primulaceae. In Z. Y. Wu & P. H. Raven (Eds.), *Flora of China* (Vol. 15, pp. 39–189). Science Press & Missouri Botanical Garden Press.
4. Missouri Botanical Garden. (2021). *Name-Lysimachia alpestris*. Tropicos. Retrieved May 4, 2021, from https://www.tropicos.org/name/26400815

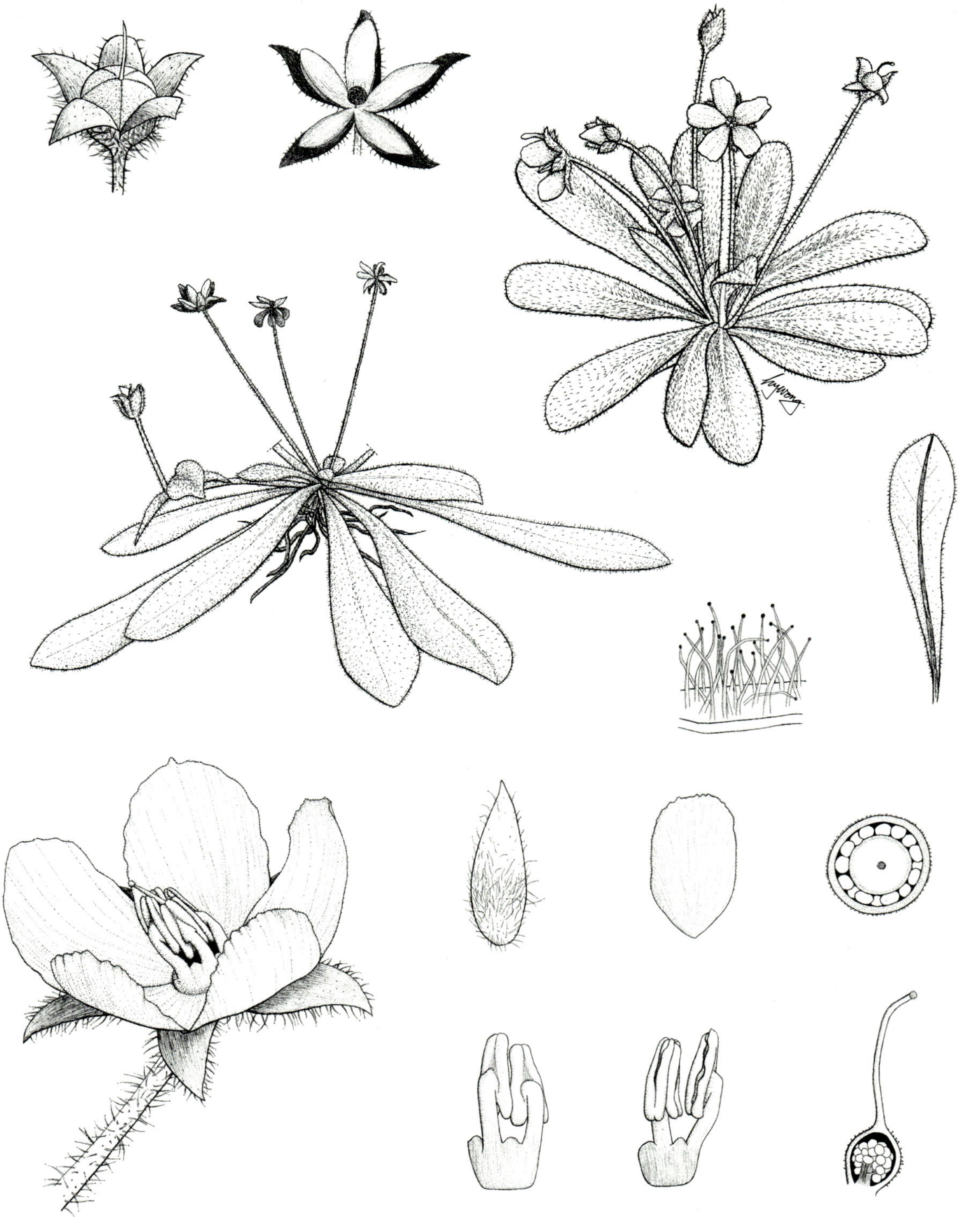

Lysimachia alpestris Champ. ex Benth. 香港過路黃

Illustration 004 by H. Y. Wong

Authentication by D. T. W. Lau

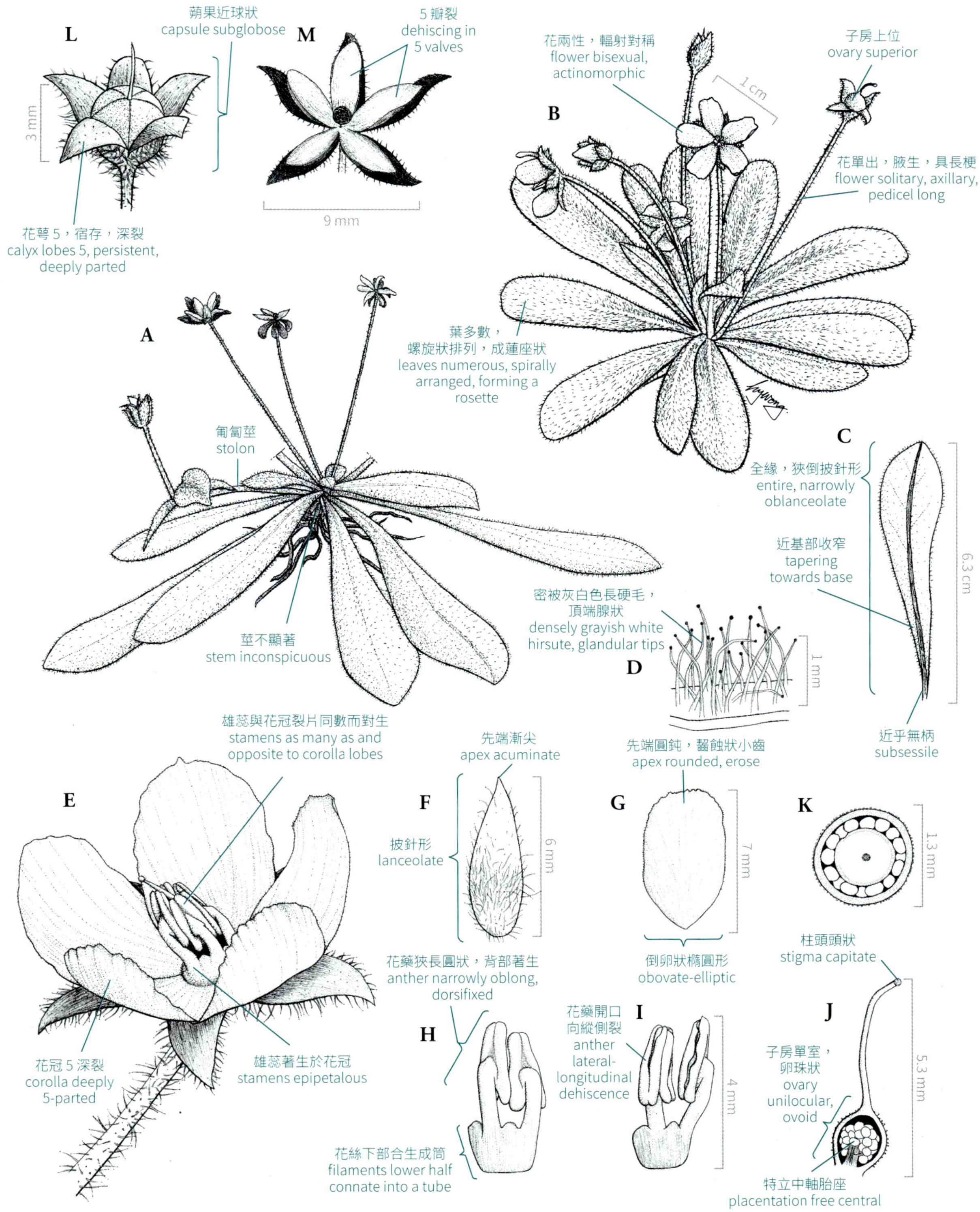

Lysimachia alpestris Champ. ex Benth. 香港過路黃

Illustration 004 by H. Y. Wong
Authentication by D. T. W. Lau

植物結構 PLANT STRUCTURES:

A. 具果的全株 Whole plant with fruits **B.** 具花的全株 Whole plant with flowers **C.** 葉底 Leaf, abaxial **D.** 葉底的被毛 Hairs on abaxial leaf **E.** 花 Flower **F.** 花萼裂片 Calyx lobe **G.** 花冠裂片 Corolla lobe **H.** 雄蕊，背軸面 Stamens, abaxial **I.** 雄蕊，向軸面 Stamens, adaxial **J.** 子房，縱切面 Ovary, in L. S. **K.** 子房，橫切面 Ovary, in C. S. **L.** 果實 Fruit **M.** 蒴果的殘存裂片及萼片 Residual valves and calyx of capsule

參考標本 REFERENCE SPECIMENS: Rare and endangered plants (K. W. Lam, S. W. Shek, J. Y. Y. Lau, D. T. W. Lau) 8B; Shiu Ying Hu 13773; Shiu Ying Hu 13262; Hu & But 20443

參考文獻 REFERENCES:

Chen, F. H., & Hu, Q. M. (1987). Primulaceae. In South China Botanical Garden, the Chinese Academy of Sciences (Ed.), *Flora of Guangdong* (Vol. 1, pp. 541–556). Guangdong Science and Technology Press.

Chen, F. H., Hu, C. M., Fang, Y. I., Cheng, C. Z., Yang, Y. C., & Huang, R. F. (1989). Primulaceae. In Delectis Florae Reipublicae Popularis Sinicae Agendae Academiae Sinicae Edita (Ed.), *Flora Reipublicae Popularis Sinicae* (Vol. 59(1), pp. 1–201). Science Press.

Hu, C. M., & Kelso, S. (1996). Primulaceae. In Z. Y. Wu & P. H. Raven (Eds.), *Flora of China* (Vol. 15, pp. 39–189). Science Press & Missouri Botanical Garden Press.

Hu, Q. M. (2007). Primulaceae. In Hong Kong Herbarium, AFCD & South China Botanical Garden, CAS (Eds.), *Flora of Hong Kong* (Vol. 1, pp. 306–308). Agriculture, Fisheries and Conservation Department, HKSAR Government.

Stevens, P. F. (2021). Ericales. Angiosperm Phylogeny Website. Retrieved May 4, 2021, from http://www.mobot.org/MOBOT/research/APweb/welcome.html

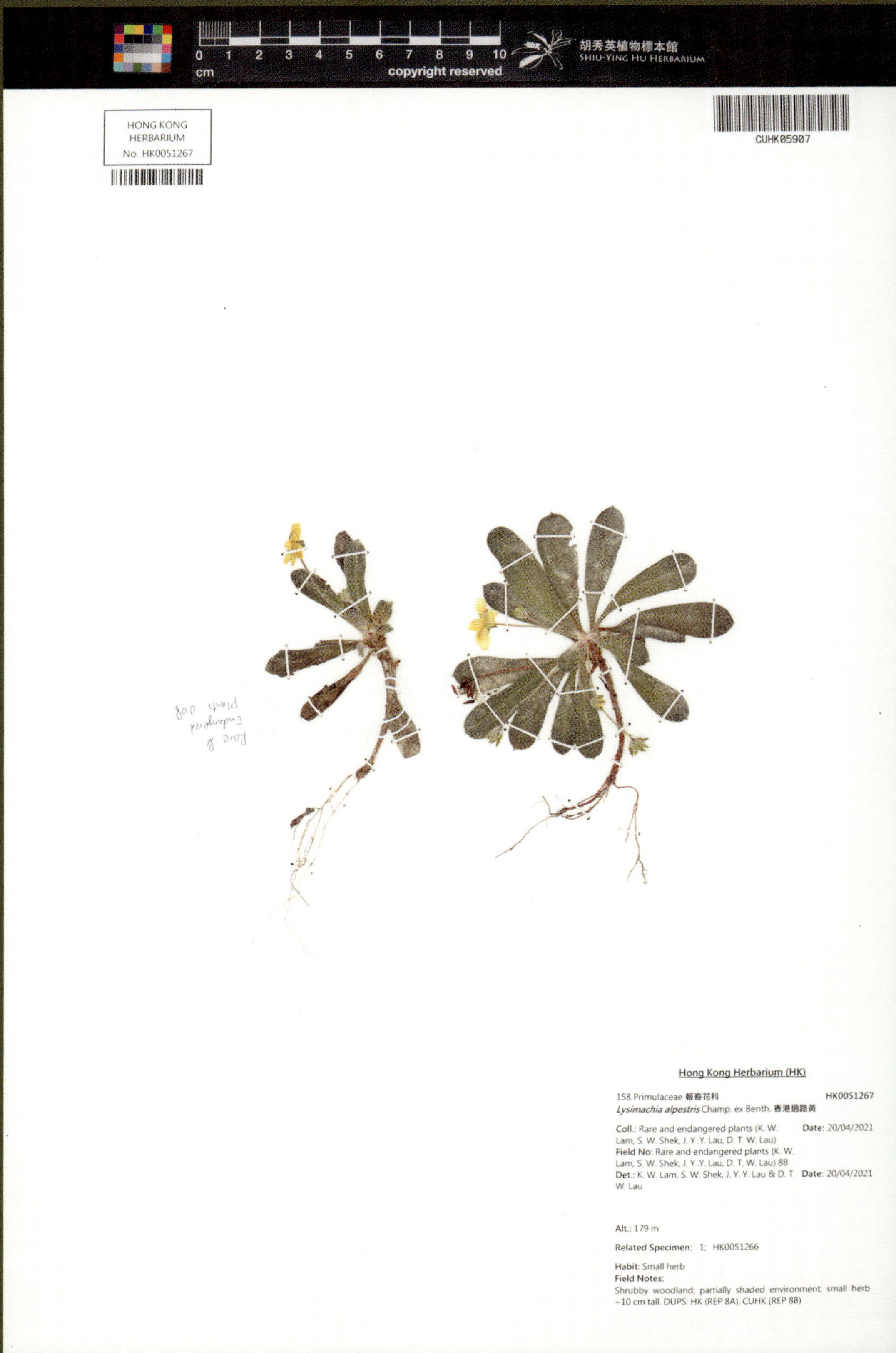

採集者編號 Collector no.

Rare and endangered plants (K. W. Lam, S. W. Shek, J. Y. Y. Lau, D. T. W. Lau) 8B

採集年份 Year of collection

2021

外觀 Overview

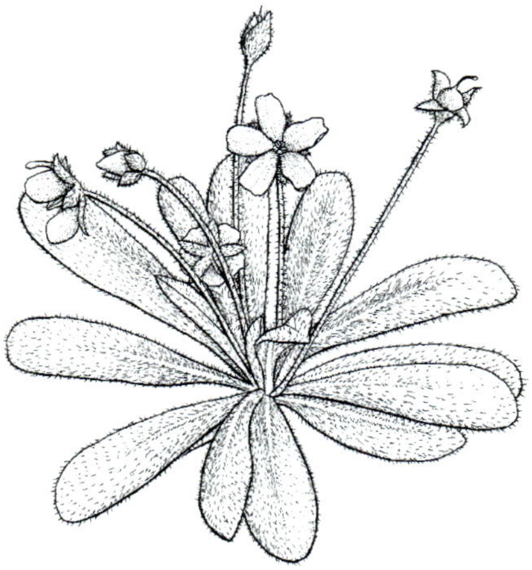

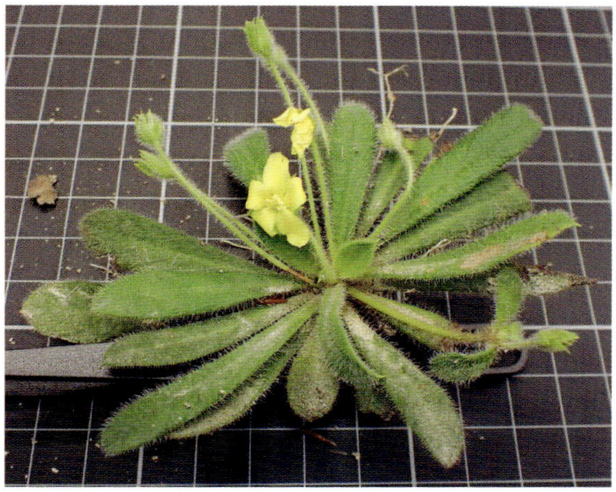

具花的全株
Whole plant with flowers

葉 Leaf

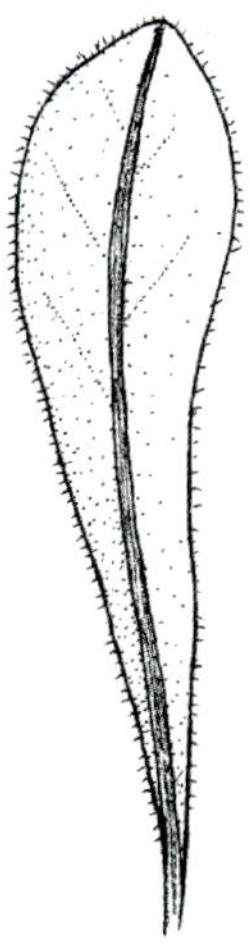

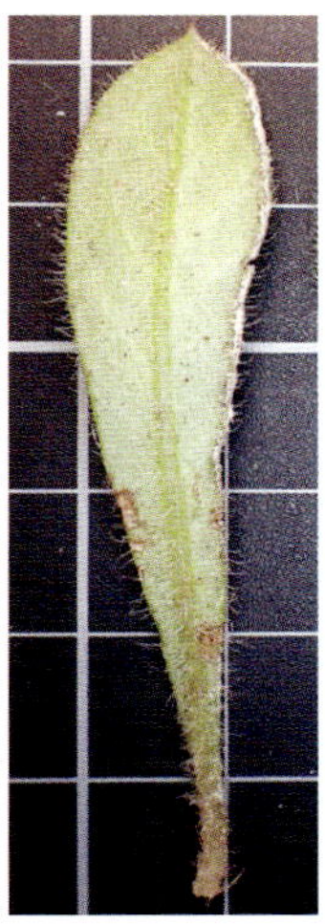

葉底
Leaf, abaxial

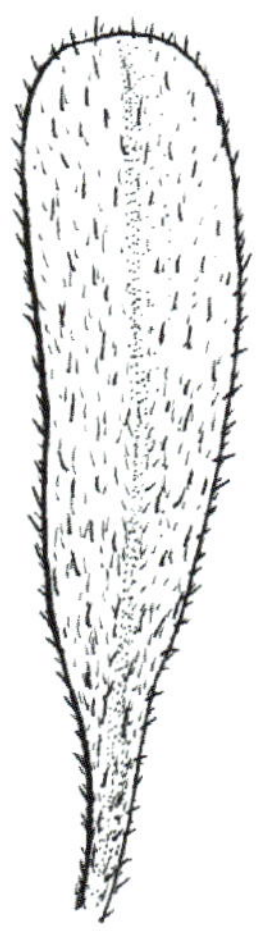

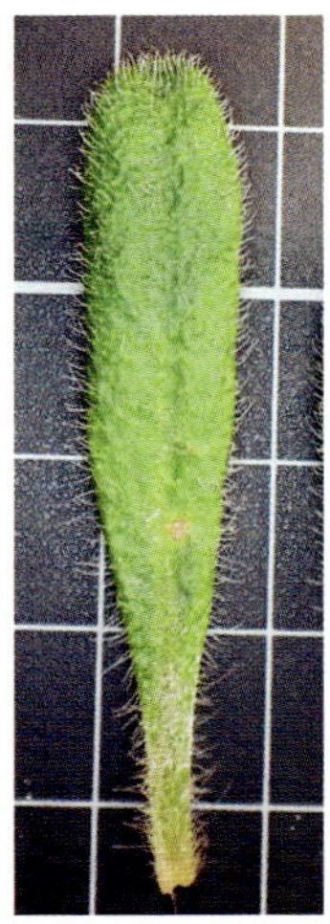

葉面
Leaf, adaxial

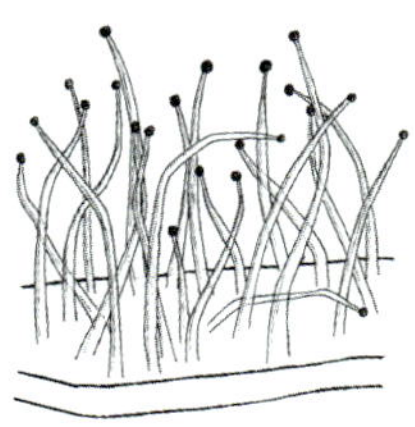

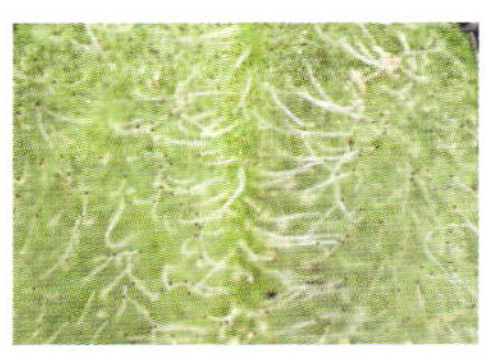

葉底的被毛
Hairs on abaxial leaf

花 Flower

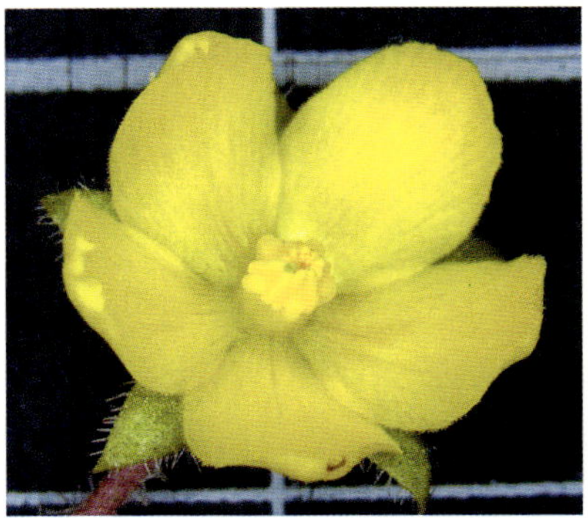
花 Flower

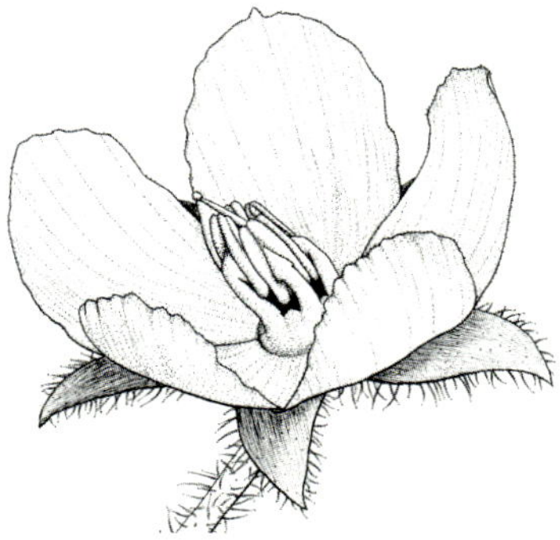

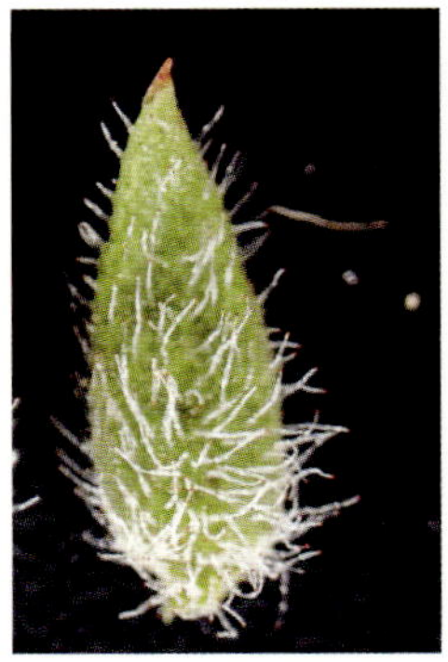
花萼裂片
Calyx lobe

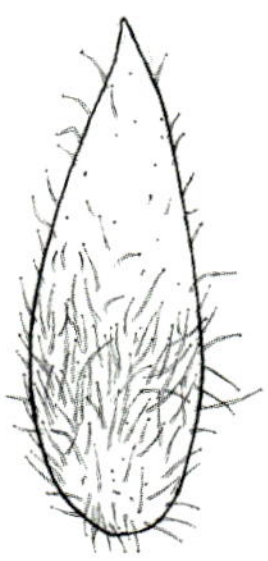

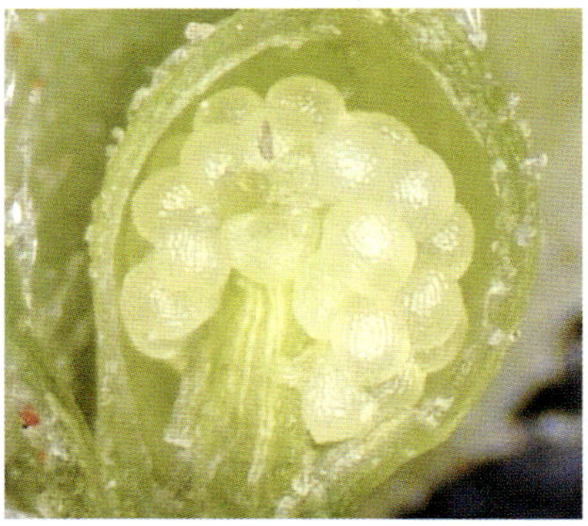
子房，縱切面 Ovary in L. S.

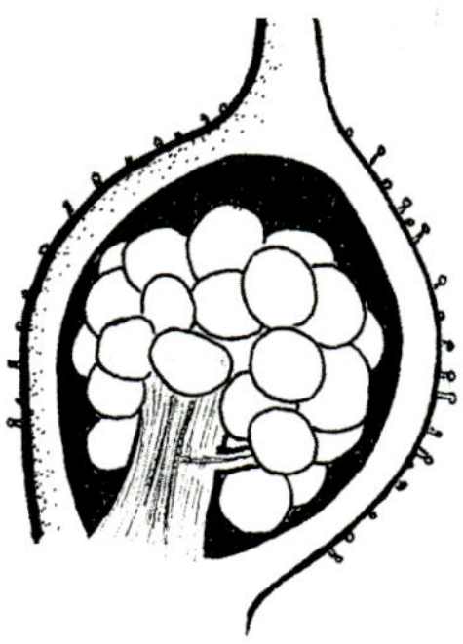

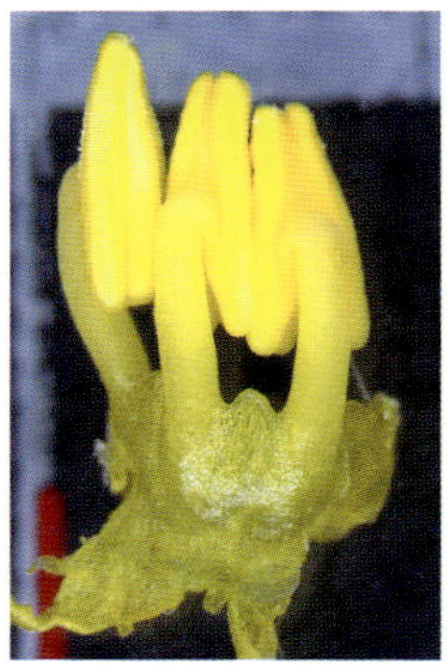
雄蕊，背軸面
Stamens, abaxial

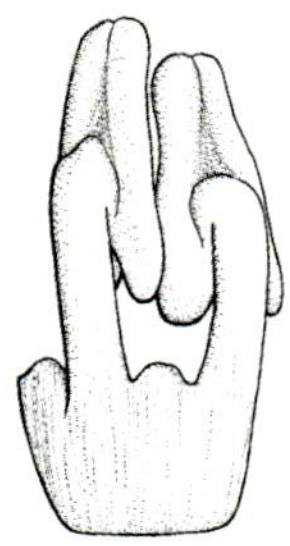

果 Fruit

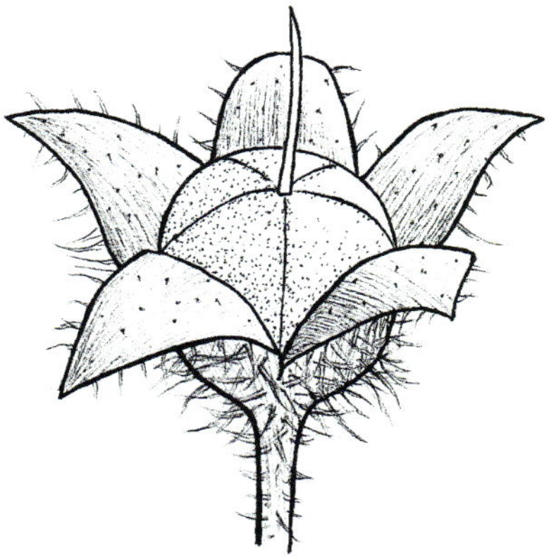

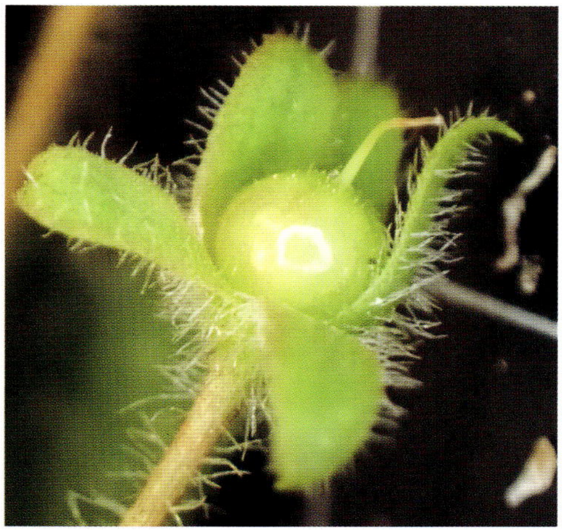

蒴果 Capsule

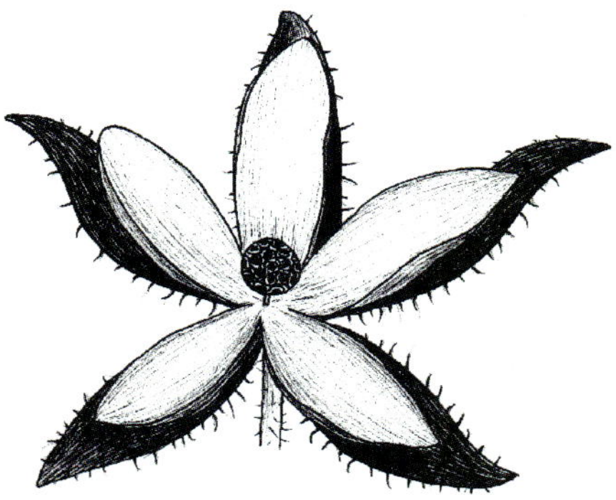

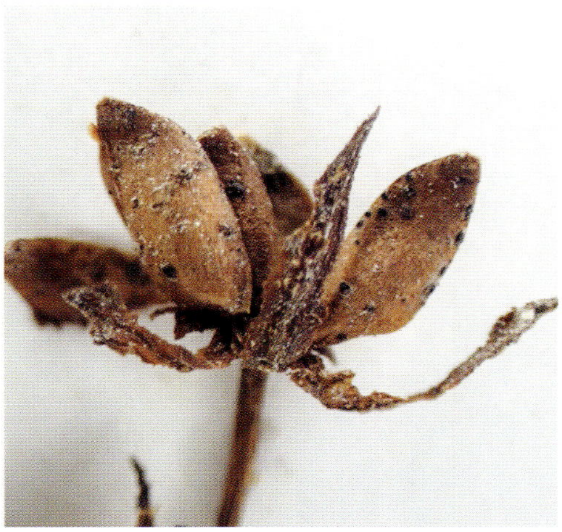

蒴果的殘存裂片及萼片
Residual valves and calyx of capsule

Camellia oleifera C. Abel
油茶

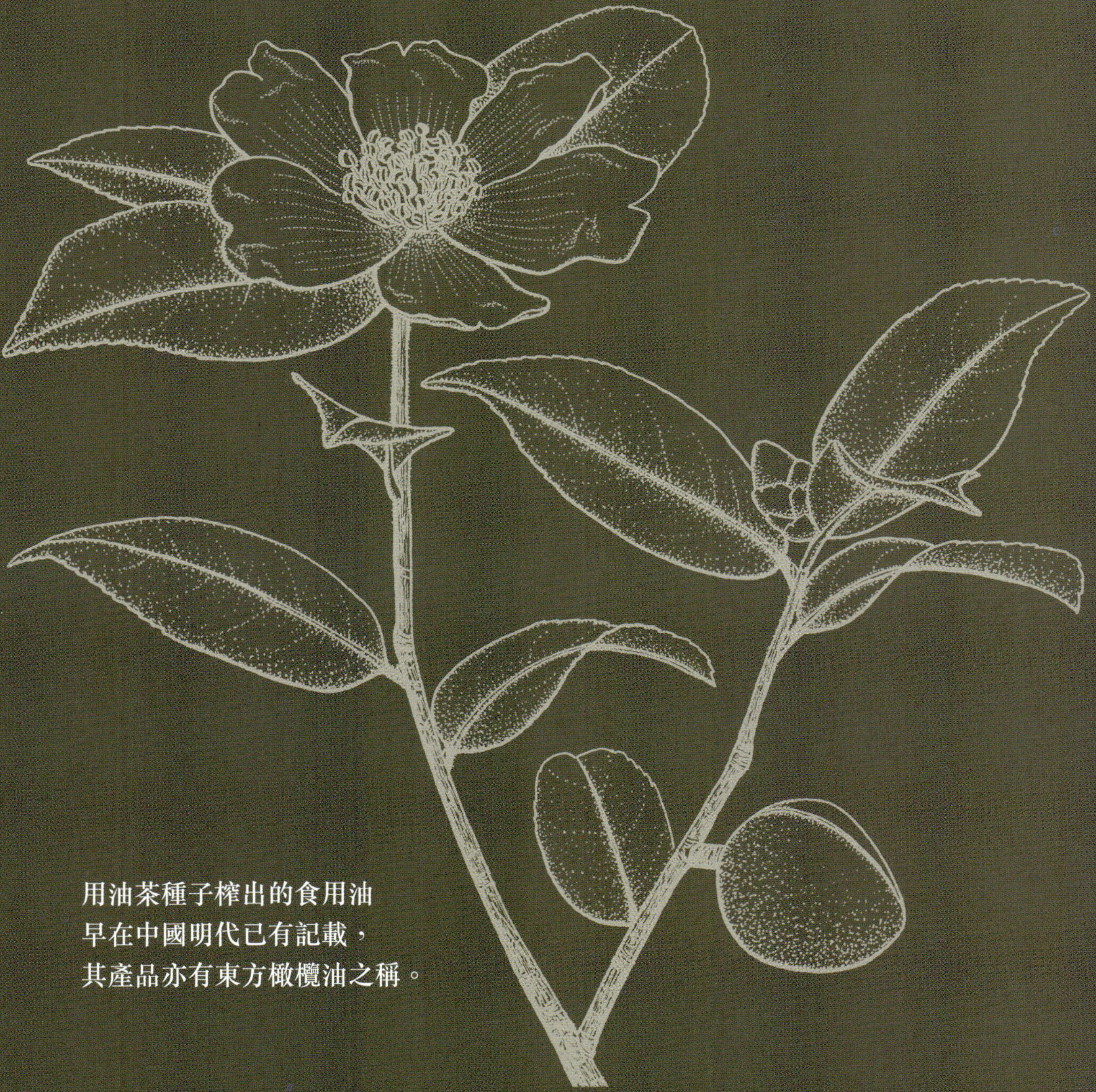

用油茶種子榨出的食用油
早在中國明代已有記載，
其產品亦有東方橄欖油之稱。

採集者編號
Collector no.
Shiu Ying Hu 11241

採集年份
Year of collection
1971

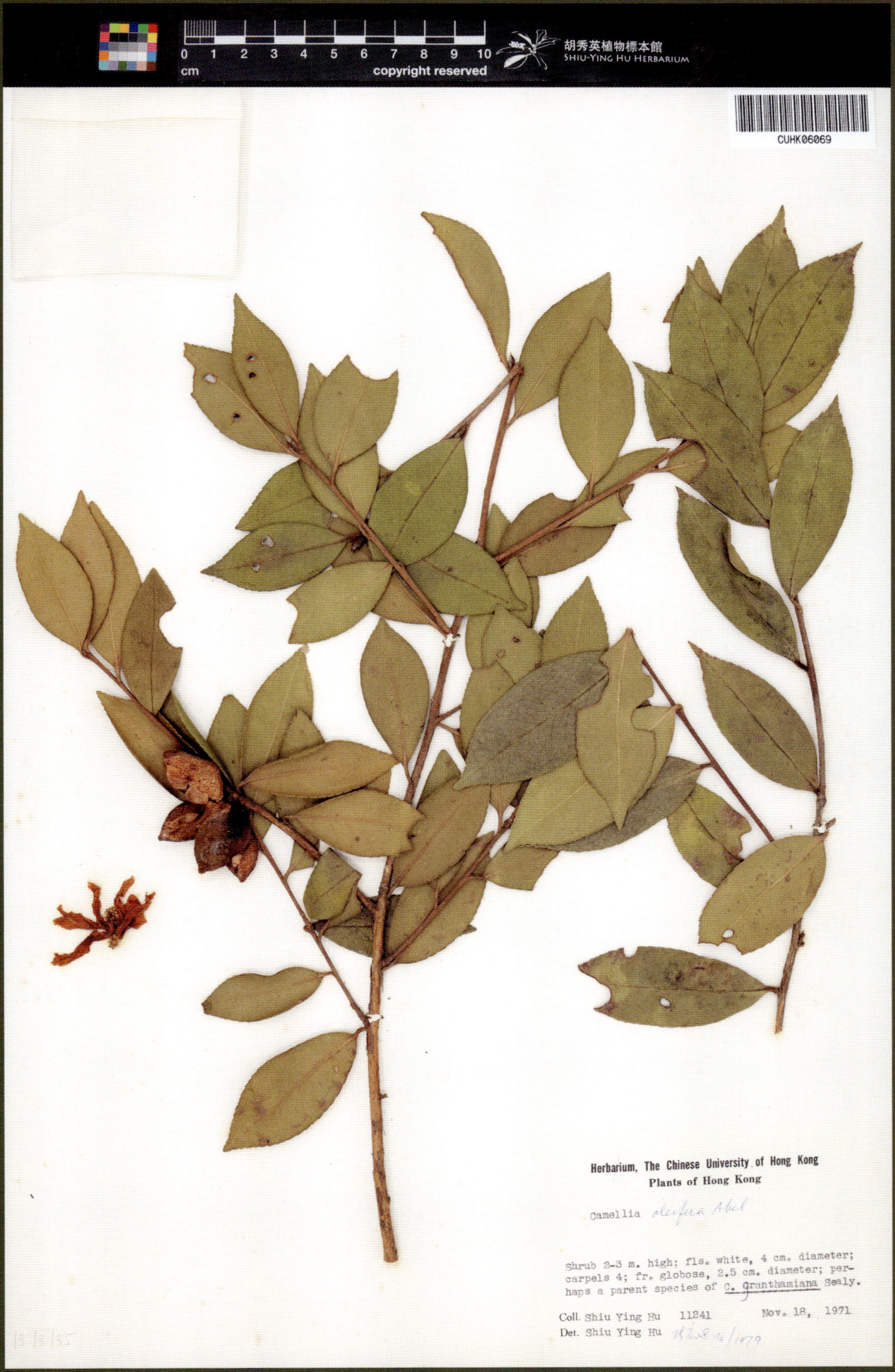
0 1 2 3 4 5 6 7 8 9 10
cm
copyright reserved
胡秀英植物標本館
SHIU-YING HU HERBARIUM
CUHK06069
Herbarium, The Chinese University of Hong Kong
Plants of Hong Kong
Camellia oleifera Abel
Shrub 2-3 m. high; fls. white, 4 cm. diameter; carpels 4; fr. globose, 2.5 cm. diameter; perhaps a parent species of C. Granthamiana Sealy.
Coll. Shiu Ying Hu 11241 Nov. 18, 1971
Det. Shiu Ying Hu

油茶

學名 *Camellia oleifera* C. Abel
中文名 油茶
科名 山茶科
本地分布狀態 原生植物
繪圖編號 H. Y. Wong 009

花期	花色	果期	果色
12 至翌年 1 月	白色	9 至 10 月	紅褐色

油茶為灌木或小喬木，可見於森林地區。

油茶是重要的經濟作物，用油茶種子榨出的食用油早在中國明代已有記載，其產品亦有東方橄欖油之稱。油茶在中國被廣泛栽培，衍生出不同的栽培種，而野生群落的基因多樣性則有助培育更良好的栽培種。

油茶受香港法例第 96A章《林務規例》的規管，而在國際自然保護聯盟瀕危物種紅色名錄中，則被列為「無危」物種。

《中國植物誌》(英文版) 的採用名稱
Camellia oleifera C. Abel (科名：山茶科)

《香港植物誌》的採用名稱
Camellia oleifera Abel (科名：山茶科)

參考文獻 References

1. Dong, B., Wu, B., Hong, W. H., Li, X. P., Li, Z., Xue, L., & Huang, Y. F. (2017). Transcriptome analysis of the tea oil camellia (*Camellia oleifera*) reveals candidate drought stress genes. *PLoS One 12* (7), Article e0181835. https://doi.org/10.1371/journal.pone.0181835
2. Hong Kong Herbarium. (2021). *HK Herbarium-Camellia oleifera*. HK Plant Database. Retrieved December 20, 2021, from https://www.herbarium.gov.hk/en/hk-plant-database/plant-detail/index.html?pType=species&oID=2963
3. IUCN. (2015). *Camellia oleifera*. The IUCN Red List of Threatened Species. Retrieved December 20, 2021, from https://www.iucnredlist.org/species/62056210/62056214

Camellia oleifera C. Abel

Scientific name *Camellia oleifera* C. Abel

Chinese name 油茶

Family Theaceae Mirb.

Local distribution status Native species

Illustration number H. Y. Wong 009

FLOWERING PERIOD	COLOUR OF FLOWER	FRUITING PERIOD	COLOUR OF FRUIT
December to following January	White	September to October	Reddish brown

Camellia oleifera C. Abel is a shrub or small tree and can be found in forests. *Camellia oleifera* C. Abel is an economically important plant. The practice of using the seeds of this plant to extract edible oil can be traced back to China's Ming dynasty; the product is called the 'Olive Oil of the East'. *Camellia oleifera* C. Abel is widely cultivated in mainland China, and many cultivars have been derived. The genetic diversity of the wild populations can help to breed better cultivars.

Camellia oleifera C. Abel is regulated under Hong Kong's Forestry Regulations (Cap. 96A), and is listed as 'Least Concern' in the IUCN Red List of Threatened Species.

Name in *Flora of China*
Camellia oleifera C. Abel (Family: Theaceae)

Name in *Flora of Hong Kong*
Camellia oleifera Abel (Family: Theaceae)

4. Min, T. L., & Bartholomew, B. (2007). Theaceae. In Z. Y. Wu, P. H. Raven & D. Y. Hong (Eds.), *Flora of China* (Vol. 12, pp. 366–478). Science Press & Missouri Botanical Garden Press.
5. Missouri Botanical Garden. (2021). *Name-Camellia oleifera*. Tropicos. Retrieved December 20, 2021, from https://www.tropicos.org/name/31600017
6. Qin, S. Y., Rong, J., Zhang, W. J., & Chen, J. K. (2018). Cultivation history of *Camellia oleifera* and genetic resources in the Yangtze River Basin. *Biodiversity Science 26*(4), 384–395. https://doi.org/10.17520/biods.2017254

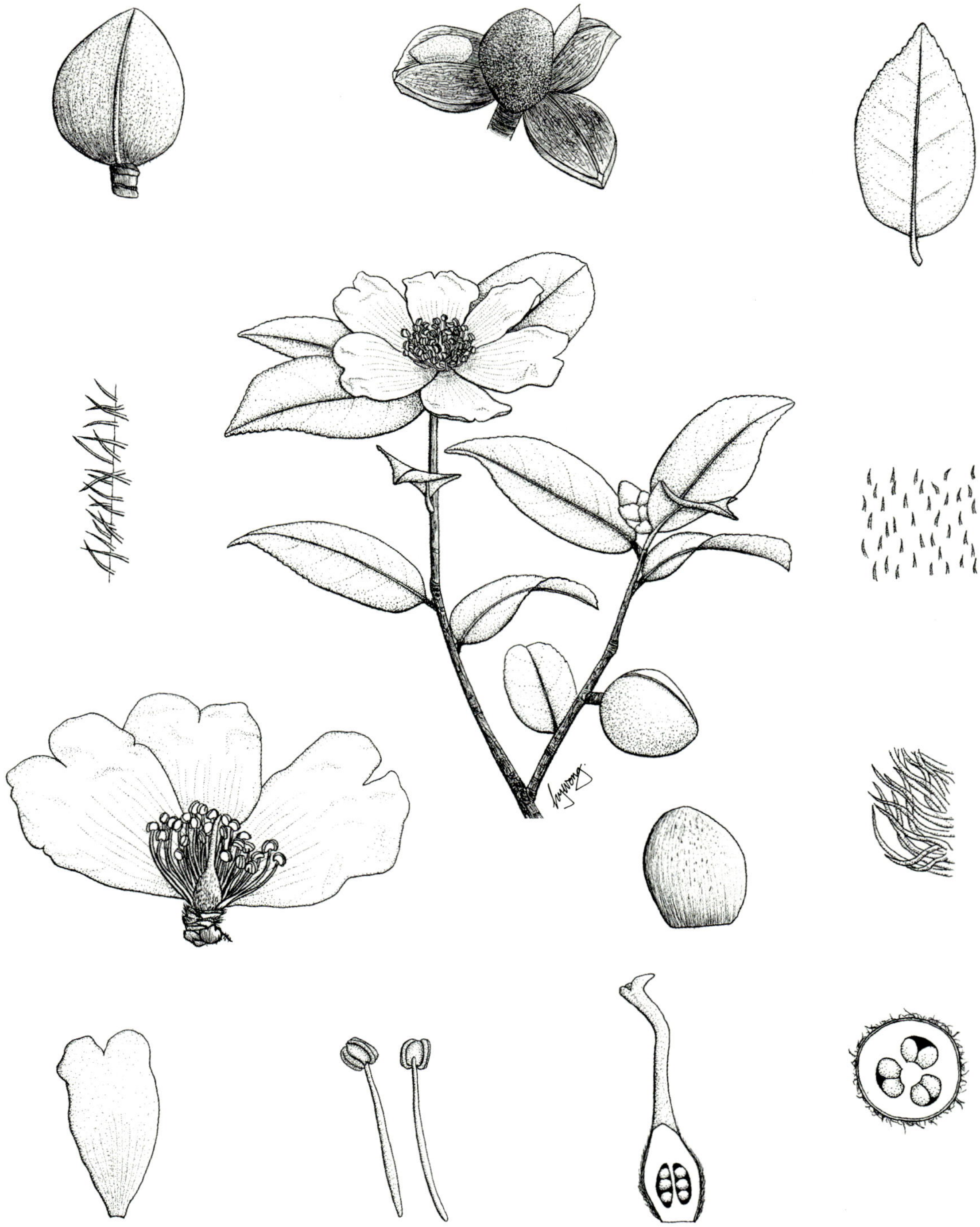

Camellia oleifera C. Abel 油茶

Illustration 009 by H. Y. Wong
Authentication by D. T. W. Lau

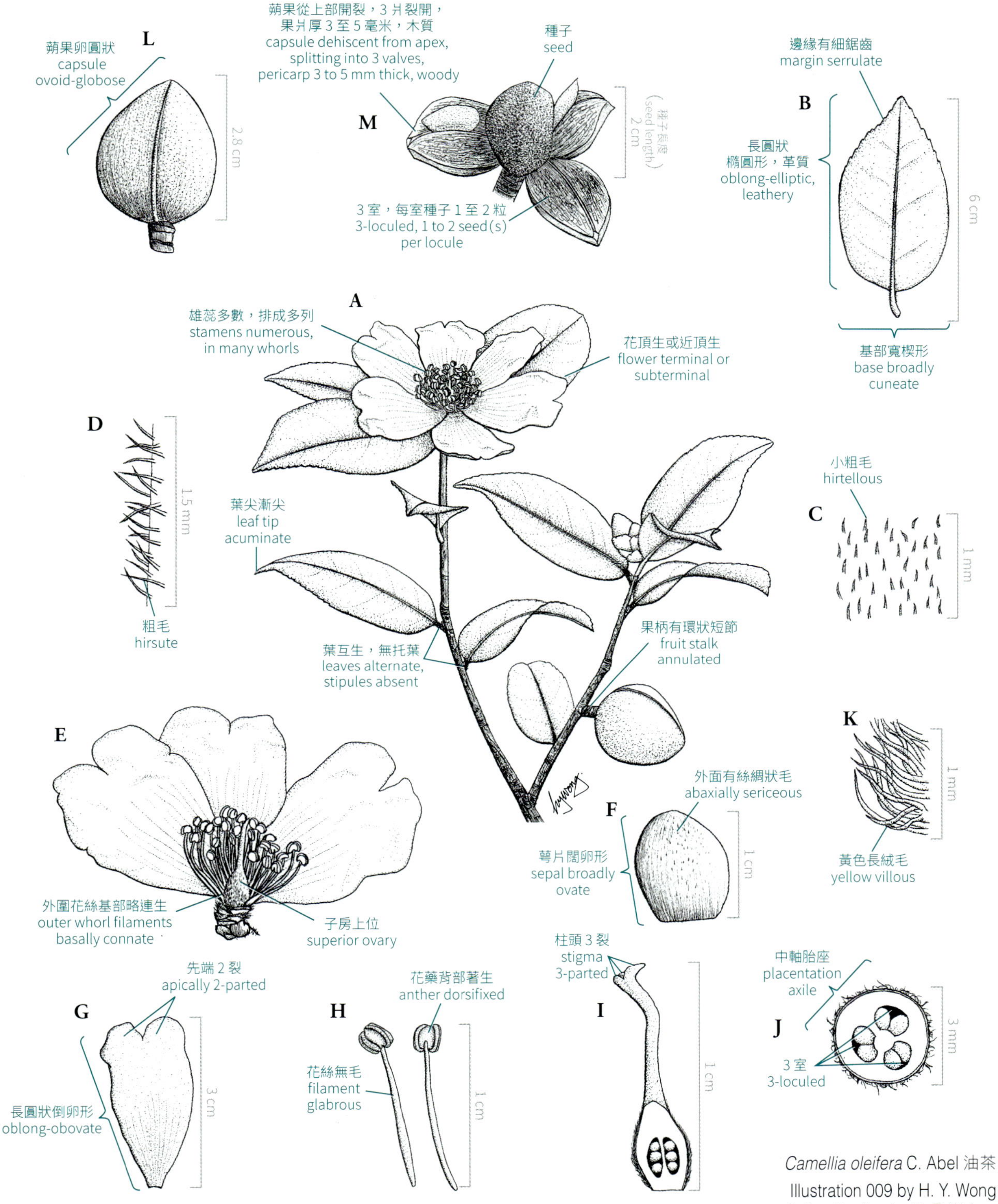

Camellia oleifera C. Abel 油茶

Illustration 009 by H. Y. Wong
Authentication by D. T. W. Lau

植物結構 PLANT STRUCTURES:
A. 花枝與果枝 Flowering and fruiting branch **B.** 葉底 Leaf, abaxial **C.** 葉柄上的被毛 Hairs on petiole **D.** 幼枝上的被毛 Hairs on young branch **E.** 花解剖圖，展示 3 片花瓣、多數雄蕊及雌蕊 Dissected flower, showing 3 petals, many stamens and gynoecium **F.** 萼片 Sepal **G.** 花瓣 Petal **H.** 雄蕊 Stamens **I.** 子房，縱切面 Ovary, in L. S. **J.** 子房，橫切面 Ovary, in C. S. **K.** 子房外壁的被毛 Hairs on ovary wall **L.** 未成熟果實 Immature fruit **M.** 成熟果實 Mature fruit

參考標本 REFERENCE SPECIMENS: Rare and endangered plants (D. T. W. Lau, K. W. Lam, S. W. Shek) 28B; Shiu Ying Hu 12546; Shiu Ying Hu 13806 (PE00629076); Shiu Ying Hu 11241

參考文獻 REFERENCES:

Chang, H. D., & Ren, S. X. (1998). Theaceae. In Delectis Florae Reipublicae Popularis Sinicae Agendae Academiae Sinicae Edita (Ed.), *Flora Reipublicae Popularis Sinicae* (Vol. 49(3), pp. 1–251). Science Press.
Deng, Y. F., & Xia, N. H. (2007). Theaceae. In Hong Kong Herbarium, AFCD & South China Botanical Garden, CAS (Eds.), *Flora of Hong Kong* (Vol. 1, pp. 178–193). Agriculture, Fisheries and Conservation Department, HKSAR Government.
Min, T. L., & Bartholomew, B. (2007). Theaceae. In Z. Y. Wu, P. H. Raven & D. Y. Hong (Eds.), *Flora of China* (Vol. 12, pp. 366–478). Science Press & Missouri Botanical Garden Press.
Stevens, P. F. (2021). Ericales. Angiosperm Phylogeny Website. Retrieved December 20, 2021, from http://www.mobot.org/MOBOT/research/APweb/welcome.html
Ye, C. X., & Shi, X. G. (2017). Theaceae. In Fairy Lake Botanical Garden, Shenzhen & Chinese Academy of Sciences (Eds.), *Flora of Shenzhen* (Vol. 1, pp. 600–625). China Forestry Publishing House.
Zhang, H. D. (1991). Theaceae. In South China Botanical Garden, the Chinese Academy of Sciences (Ed.), *Flora of Guangdong* (Vol. 2, pp. 123–177). Guangdong Science and Technology Press.

憑證標本
Voucher Specimen

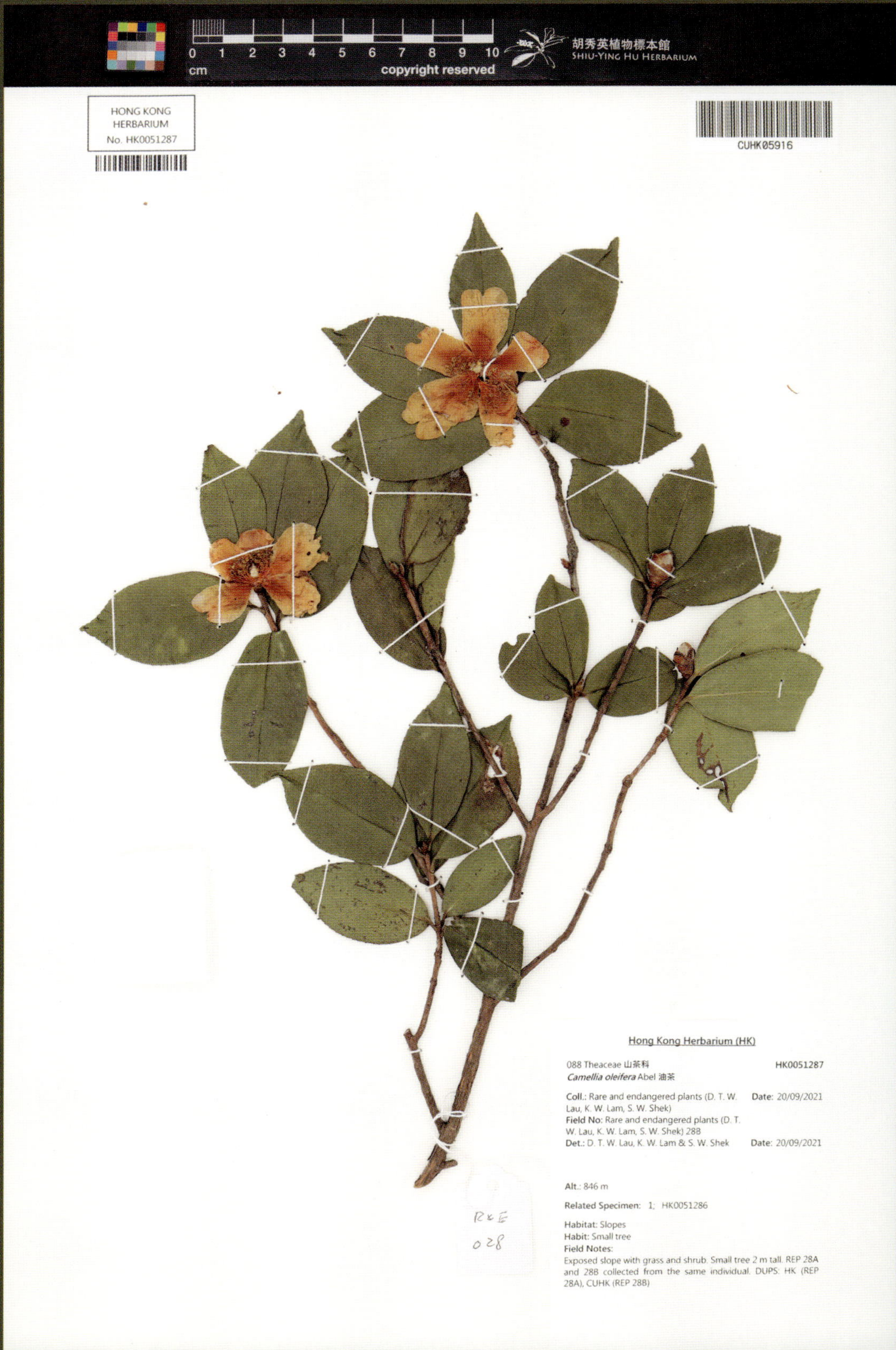

採集者編號 Collector no.
Rare and endangered plants (D. T. W. Lau, K. W. Lam, S. W. Shek) 28B

採集年份 Year of collection
2021

外觀 Overview

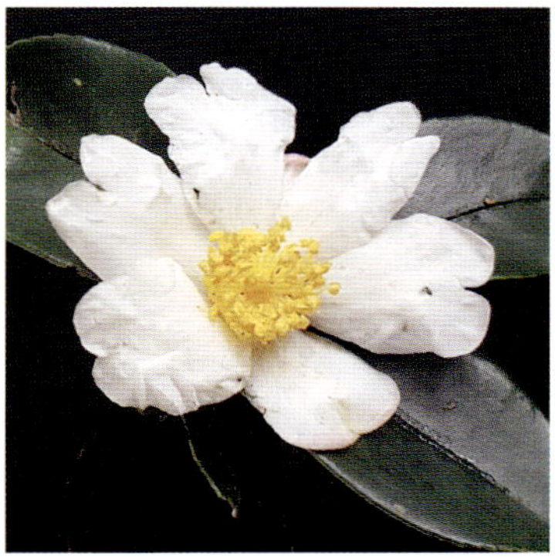

花枝 Flowering branch

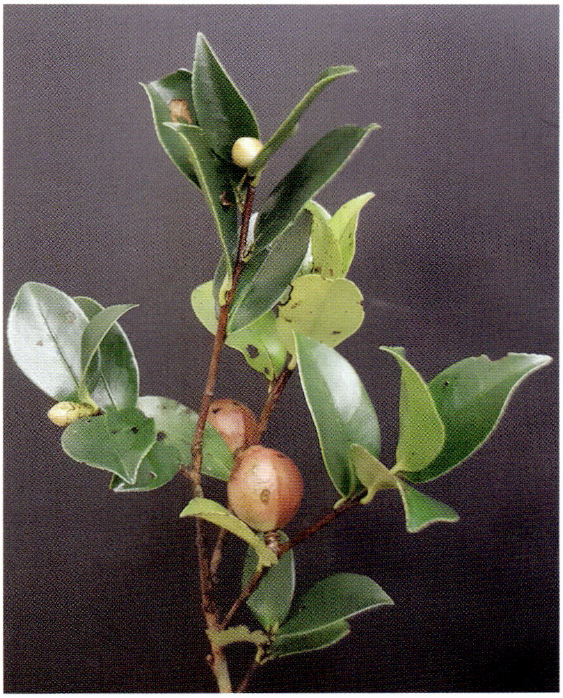

果枝 Fruiting branch

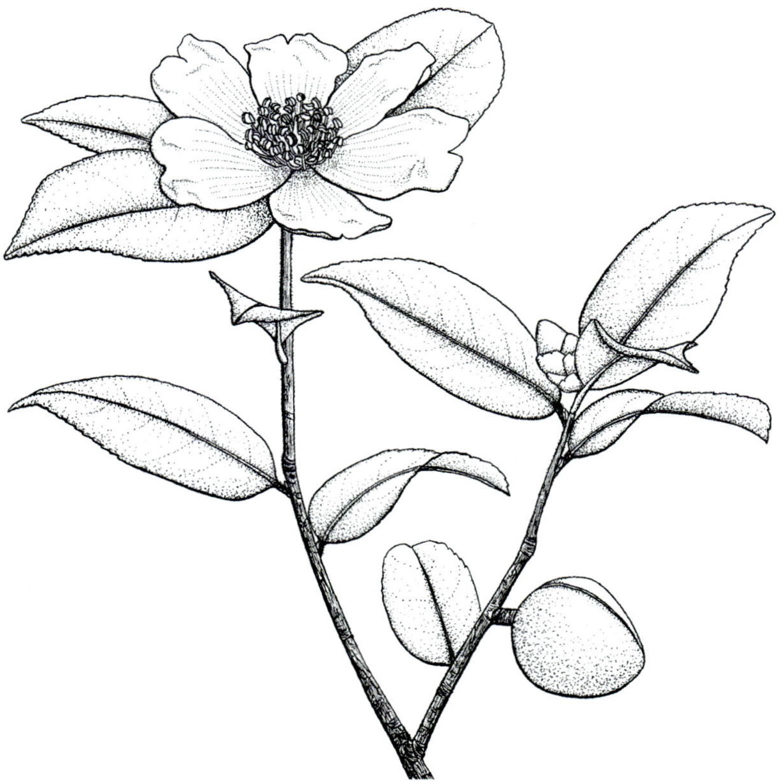

葉 Leaf

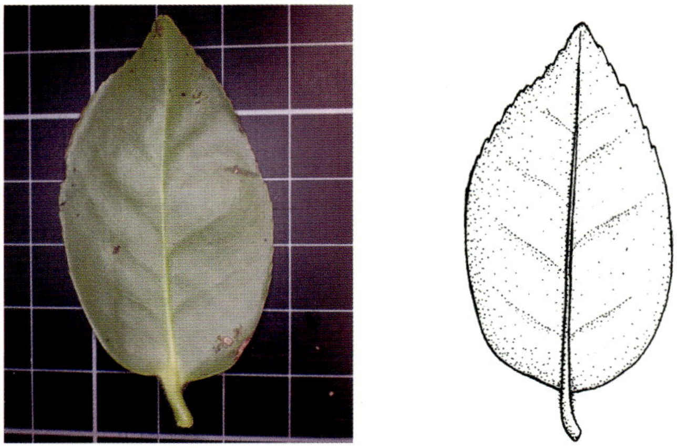

葉底 Leaf, abaxial

花 Flower

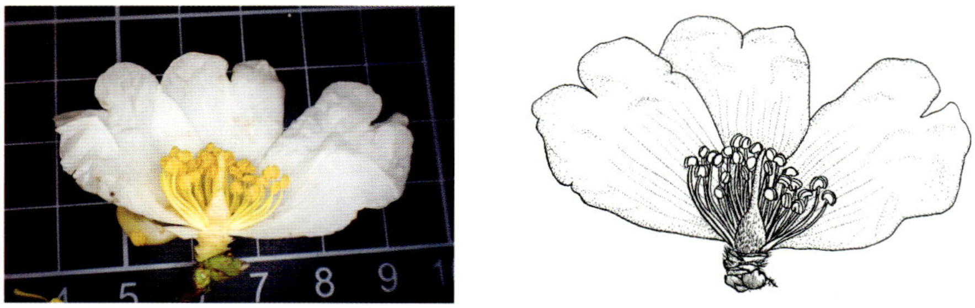

花解剖圖，展示 3 片花瓣、
多數雄蕊及雌蕊
Dissected flower, showing 3 petals,
many stamens and gynoecium

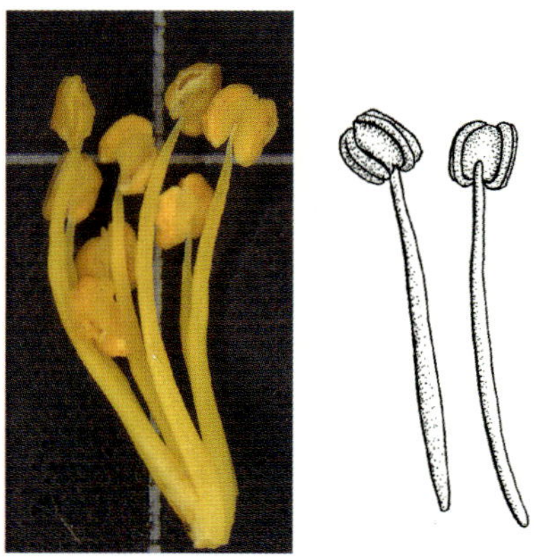

雄蕊 Stamens

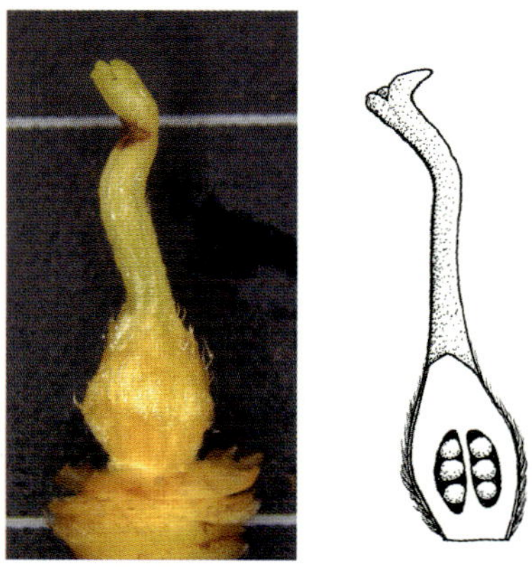

雌蕊 Gynoecium

花 Flower

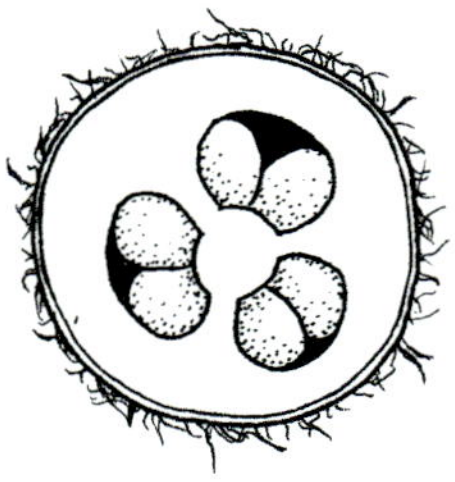
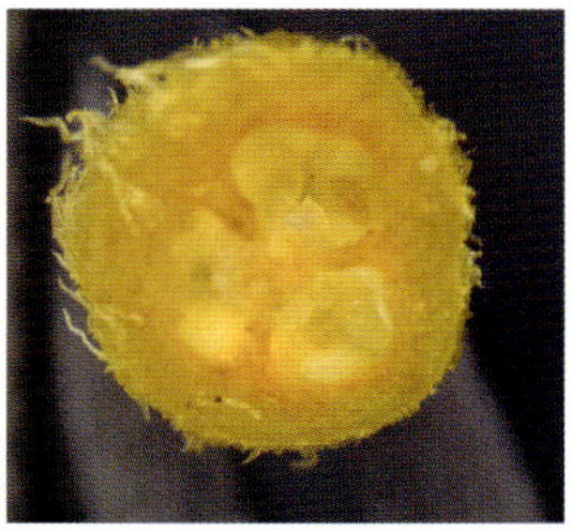

子房，橫切面
Ovary, in C. S.

果 Fruit

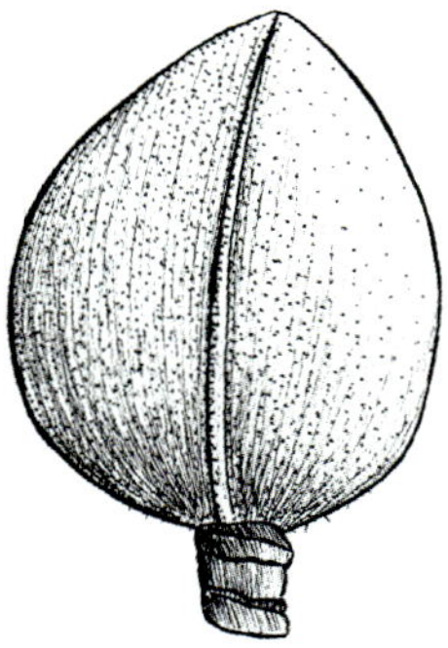
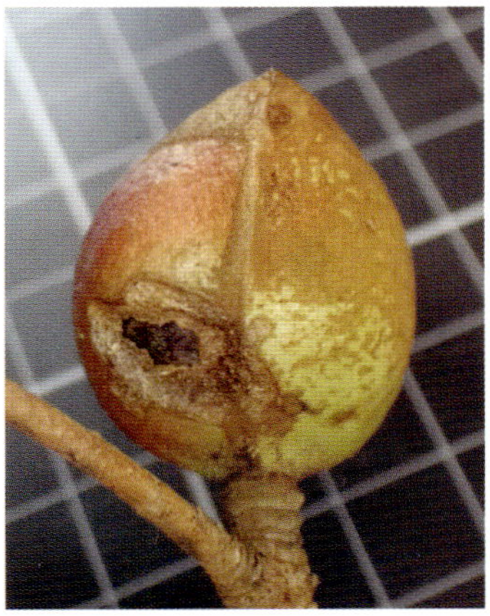

蒴果 Capsule

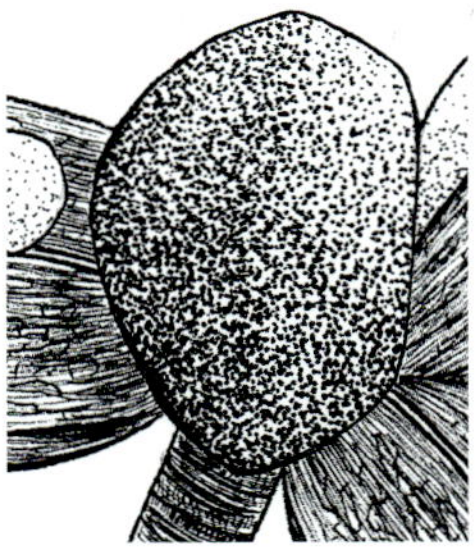
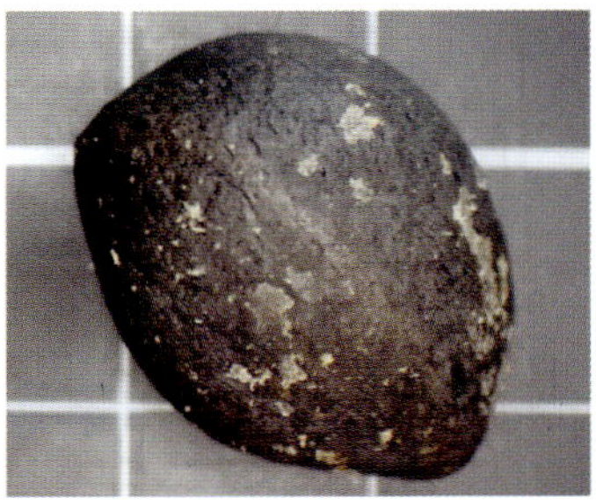

種子 Seed

Rhododendron simsii Planch.
紅杜鵑

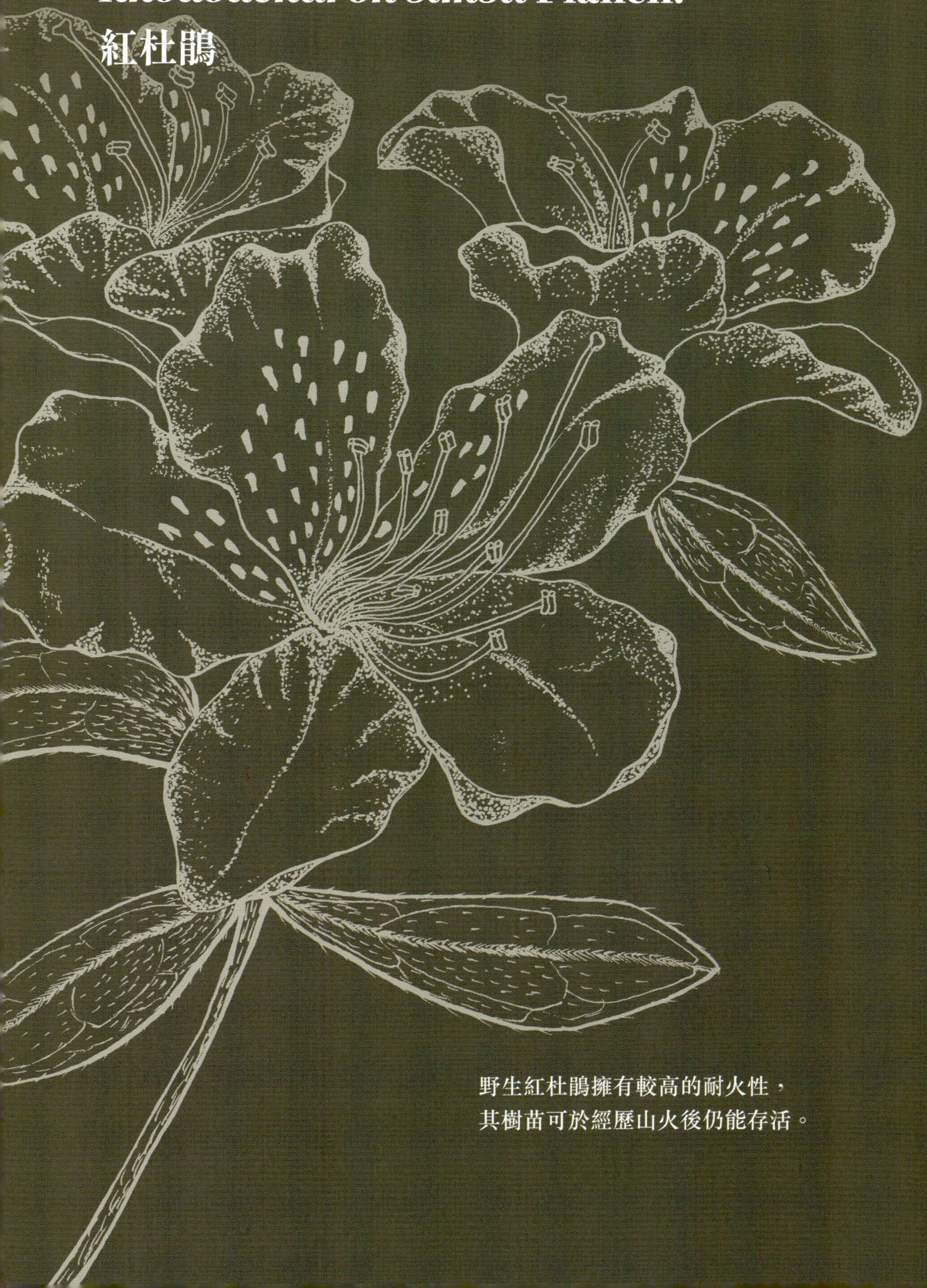

野生紅杜鵑擁有較高的耐火性，
其樹苗可於經歷山火後仍能存活。

胡秀英教授採集的標本
Specimen Collected by Professor Shiu-Ying Hu

採集者編號
Collector no.
S. Y. Hu & K. H. Yung 102

採集年份
Year of collection
1999

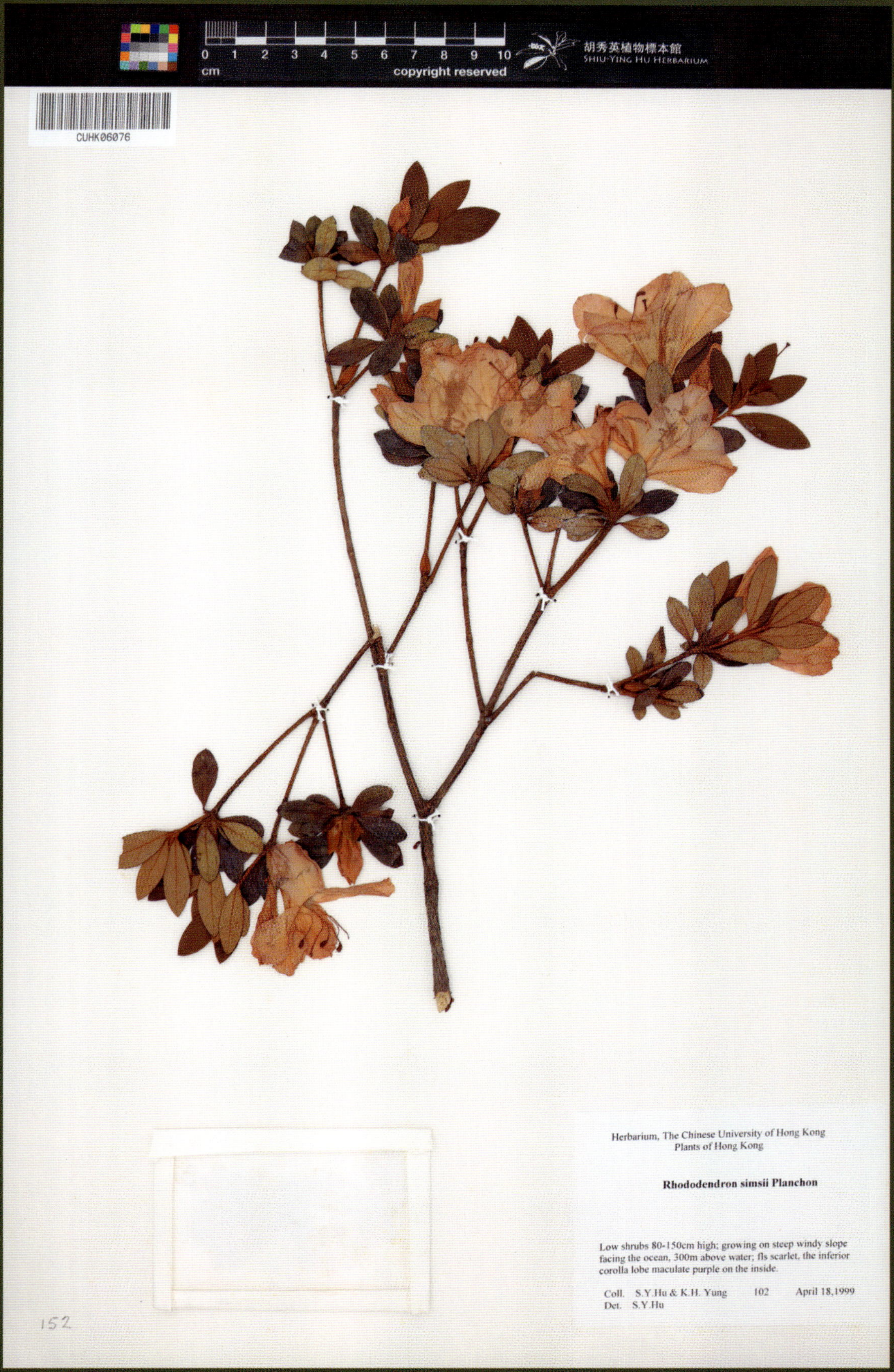
0 1 2 3 4 5 6 7 8 9 10
cm
copyright reserved
胡秀英植物標本館
SHIU-YING HU HERBARIUM
CUHK06076
Herbarium, The Chinese University of Hong Kong
Plants of Hong Kong
Rhododendron simsii Planchon
Low shrubs 80-150cm high; growing on steep windy slope facing the ocean, 300m above water; fls scarlet, the inferior corolla lobe maculate purple on the inside.
Coll. S.Y.Hu & K.H. Yung 102 April 18,1999
Det. S.Y.Hu
152

紅杜鵑

學名	*Rhododendron simsii* Planch.
中文名	紅杜鵑
科名	杜鵑花科
本地分布狀態	原生植物
繪圖編號	M. C. Li 004

花期	花色	果期	果色
12 至翌年 5 月	紅色	10 至 12 月	褐色

野生紅杜鵑廣泛分布於香港的次生灌木林。盛開的紅杜鵑不但為春季增添生氣，同時擁有較高的耐火性，其樹苗可於經歷山火後仍能存活。

紅杜鵑已被列入香港法例第 96A章《林務規例》規管。

《中國植物誌》(英文版)的採用名稱
Rhododendron simsii Planchon (科名：杜鵑花科)

《香港植物誌》的採用名稱
Rhododendron simsii Planch. (科名：杜鵑花科)

參考文獻 References

1. Fang, M. Y., Fang, R. Z., He, M. Y., Hu, L. Z., Yang, H. B., & Chamberlain, D. F. (2005). Ericaceae. In Z. Y. Wu, P. H. Raven & D. Y. Hong (Eds.), *Flora of China* (Vol. 14, pp. 260–455). Science Press & Missouri Botanical Garden Press.
2. Hong Kong Herbarium. (2022). *HK Herbarium-Rhododendron simsii*. HK Plant Database. Retrieved March 17, 2022, from https://herbarium.gov.hk/en/hk-plant-database/plant-detail/index.html?pType=species&oID=8099

Rhododendron simsii Planch.

Scientific name	*Rhododendron simsii* Planch.
Chinese name	紅杜鵑
Family	Ericaceae Juss.
Local distribution status	Native species
Illustration number	M. C. Li 004

FLOWERING PERIOD	COLOUR OF FLOWER	FRUITING PERIOD	COLOUR OF FRUIT
December to following May	Red	October to December	Brown

Rhododendron simsii Planch. is a native wild plant to Hong Kong, with diverse distribution in secondary shrublands. The blooming *Rhododendron simsii* Planch. not only enhances the vibrancy of springtime, but also, its seedlings can survive after hill fires due to its higher fire tolerance.

Rhododendron simsii Planch. is protected under Forestry Regulations (Cap. 96A).

Name in *Flora of China*
Rhododendron simsii Planchon (Family: Ericaceae)

Name in *Flora of Hong Kong*
Rhododendron simsii Planch. (Family: Ericaceae)

3. Missouri Botanical Garden. (2022). *Name-Rhododendron simsii.* Tropicos. Retrieved March 17, 2022, from https://www.tropicos.org/name/50054234
4. Ng, S. C., & Corlett, R. T. (2003). The ecology of six *Rhododendron* species (Ericaceae) with contrasting local abundance and distribution patterns in Hong Kong, China. *Plant Ecology 164*(2), 225–233.

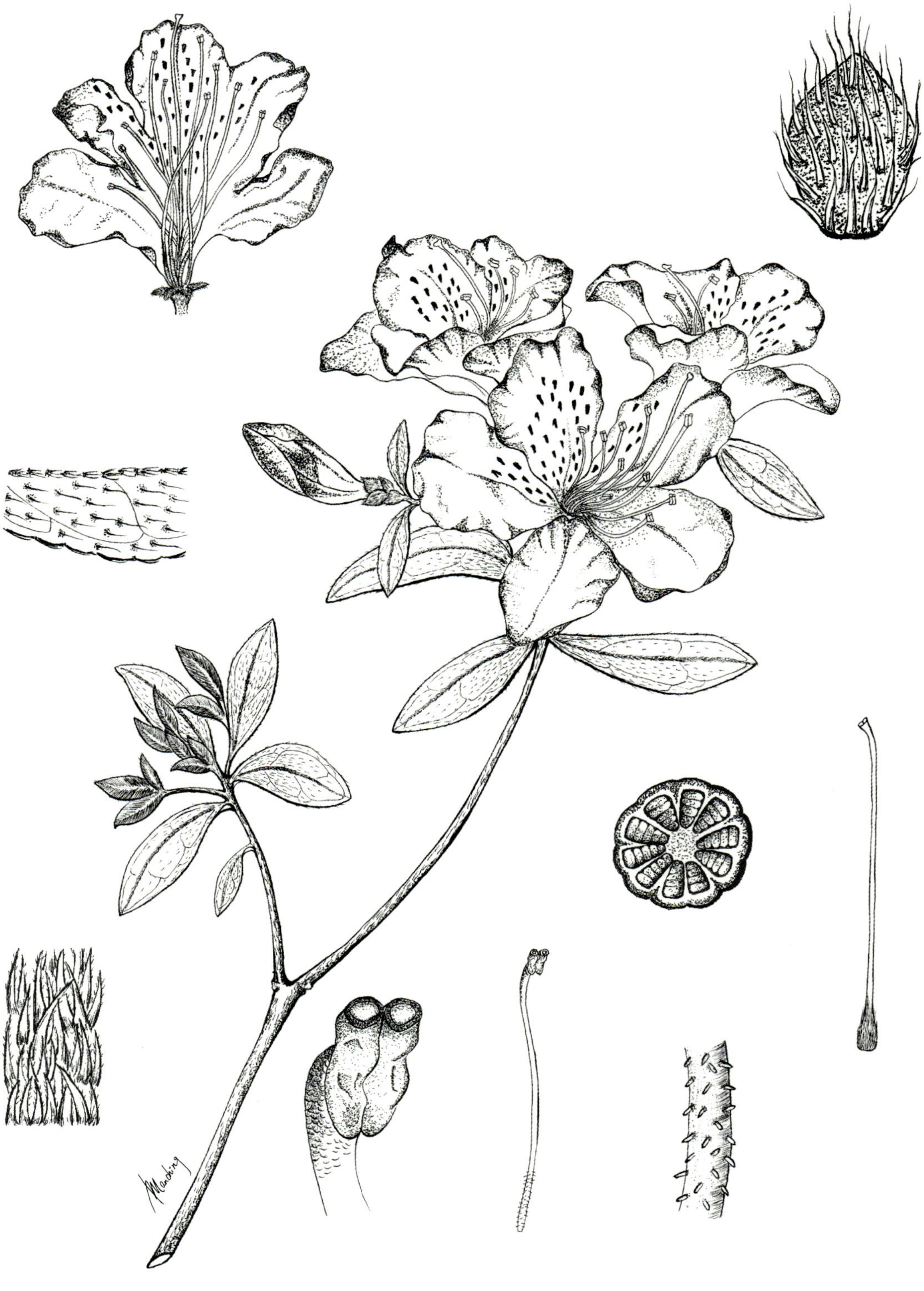

Rhododendron simsii Planch. 紅杜鵑

Illustration 004 by M. C. Li
Authenticated by D. T. W. Lau

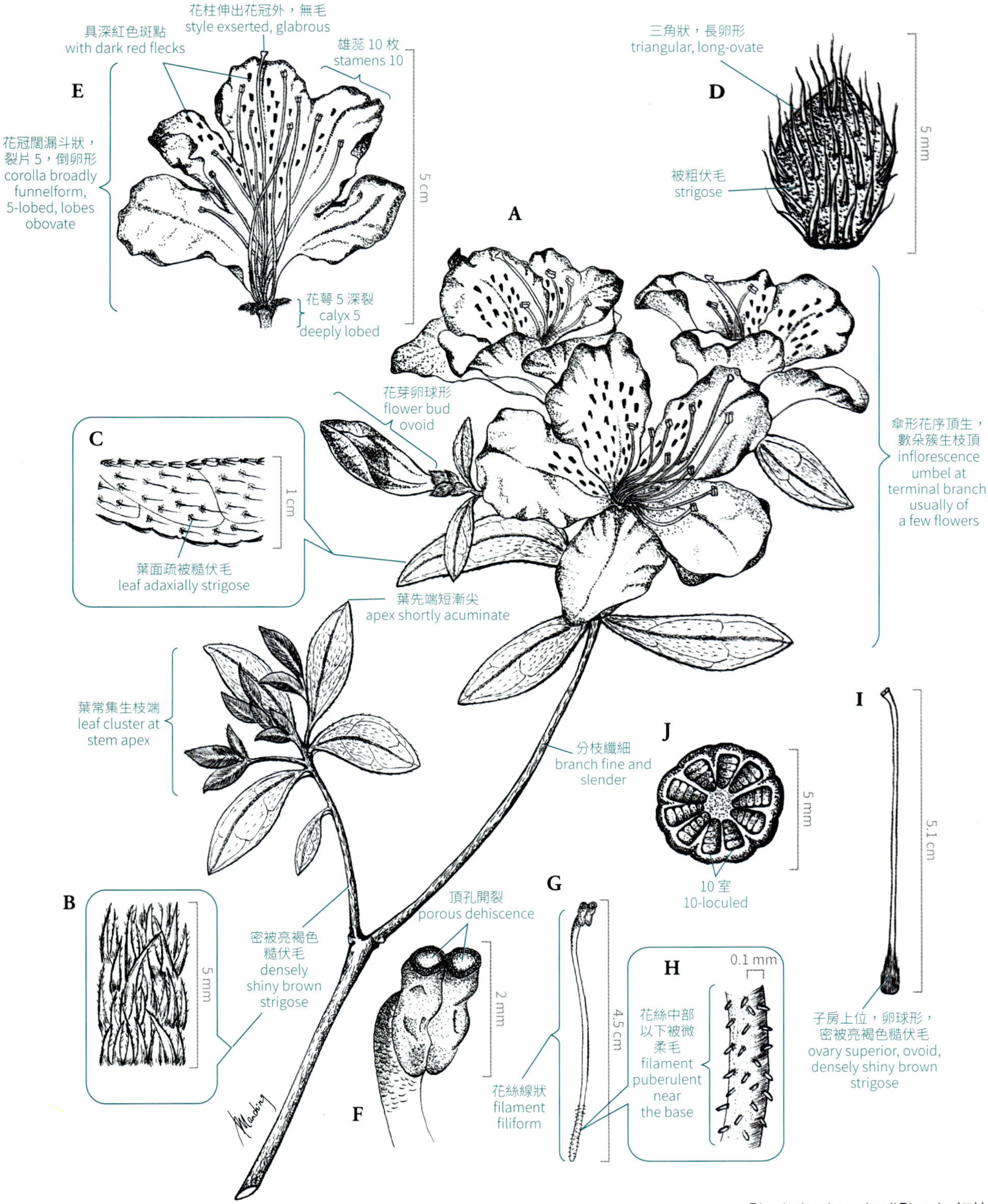

Rhododendron simsii Planch. 紅杜鵑

Illustration 004 by M. C. Li
Authenticated by D. T. W. Lau

植物結構 PLANT STRUCTURES:

A. 花枝 Flowering branch **B.** 小枝被糙伏毛局部放大 Close-up of strigose on branchlets **C.** 葉片被糙伏毛局部放大 Close-up of strigose on leaf blade **D.** 花萼裂片 Calyx lobe **E.** 花解剖圖 Dissected flower **F.** 花藥 Anther **G.** 雄蕊 Stamen **H.** 花絲中部以下被微柔毛局部放大 Close-up of hair on the lower part of the filament **I.** 雌蕊 Gynoecium **J.** 子房，橫切面 Ovary, in C. S.

參考標本 REFERENCE SPECIMENS: Rare and endangered plants (D. T. W. Lau, J. Y. Y. Lau, S. W. Shek, K. W. Lam) 6B; S. Y. Hu & K. H. Yung 102

參考文獻 REFERENCES:

Fang, M. Y., Fang, R. Z., He, M. Y., Hu, L. Z., Yang, H. B., & Chamberlain, D. F. (2005). Ericaceae. In Z. Y. Wu, P. H. Raven & D. Y. Hong (Eds.), *Flora of China* (Vol. 14, pp. 260–455). Science Press & Missouri Botanical Garden Press.

Hu, L. C., & Fang, M. Y. (1994). Ericaceae. In Delectis Florae Reipublicae Popularis Sinicae Agendae Academiae Sinicae Edita (Ed.), *Flora Reipublicae Popularis Sinicae* (Vol. 57(2), pp. 1–438). Science Press.

Ng, S. C. (2007). Ericaceae. In Hong Kong Herbarium, AFCD & South China Botanical Garden, CAS (Eds.), *Flora of Hong Kong* (Vol. 1, pp. 272–281). Agriculture, Fisheries and Conservation Department, HKSAR Government.

Stevens, P. F. (2022). Ericales. Angiosperm Phylogeny Website. Retrieved March 17, 2022, from http://www.mobot.org/MOBOT/research/APweb/welcome.html

Tang, G. D. (2010). Ericaceae. In Fairy Lake Botanical Garden, Shenzhen & Chinese Academy of Sciences (Eds.), *Flora of Shenzhen* (Vol. 2, pp. 126–136). China Forestry Publishing House.

憑證標本
Voucher Specimen

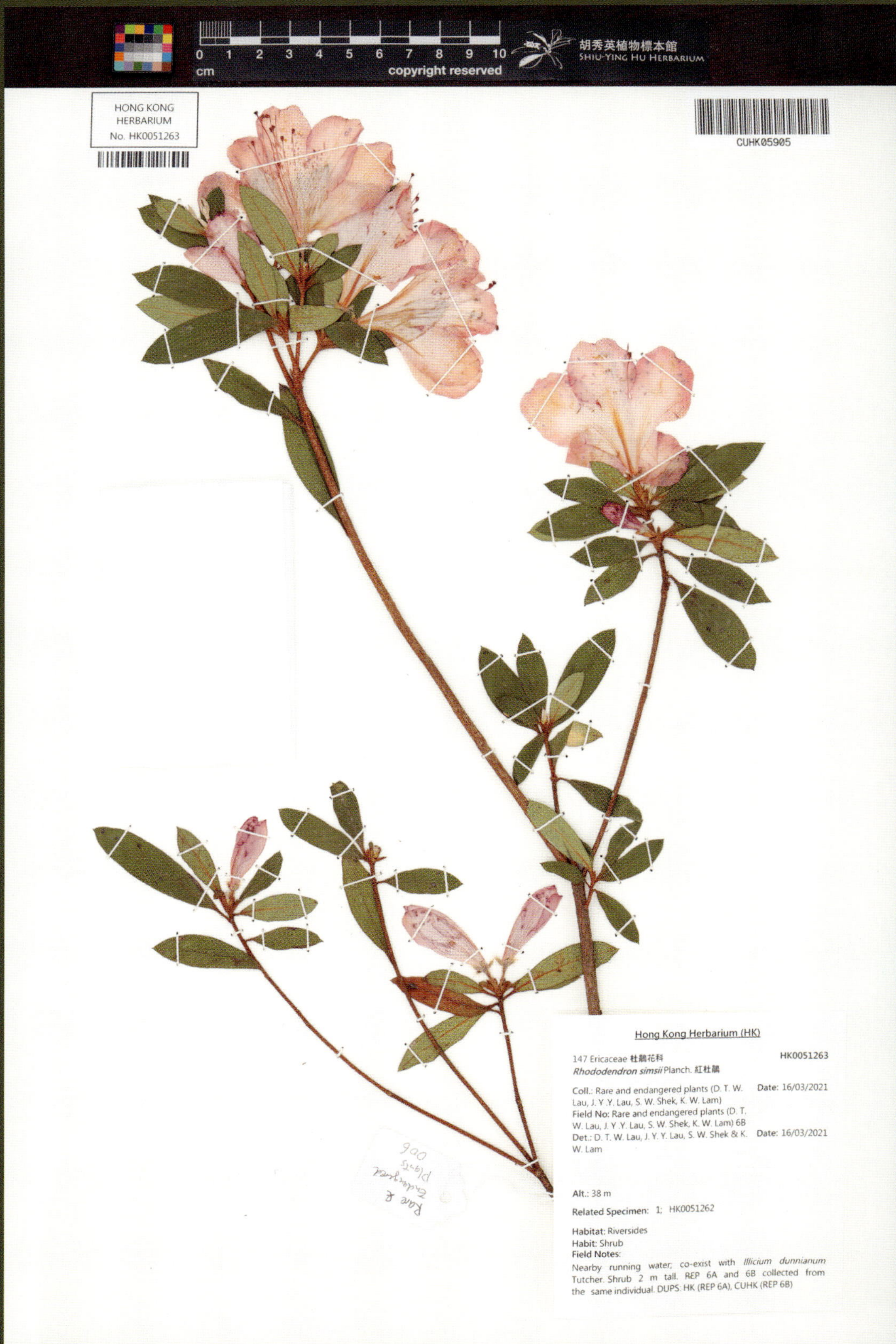

採集者編號 Collector no.
Rare and endangered plants (D. T. W. Lau, J. Y. Y. Lau, S. W. Shek, K. W. Lam) 6B

採集年份 Year of collection
2021

外觀 Overview

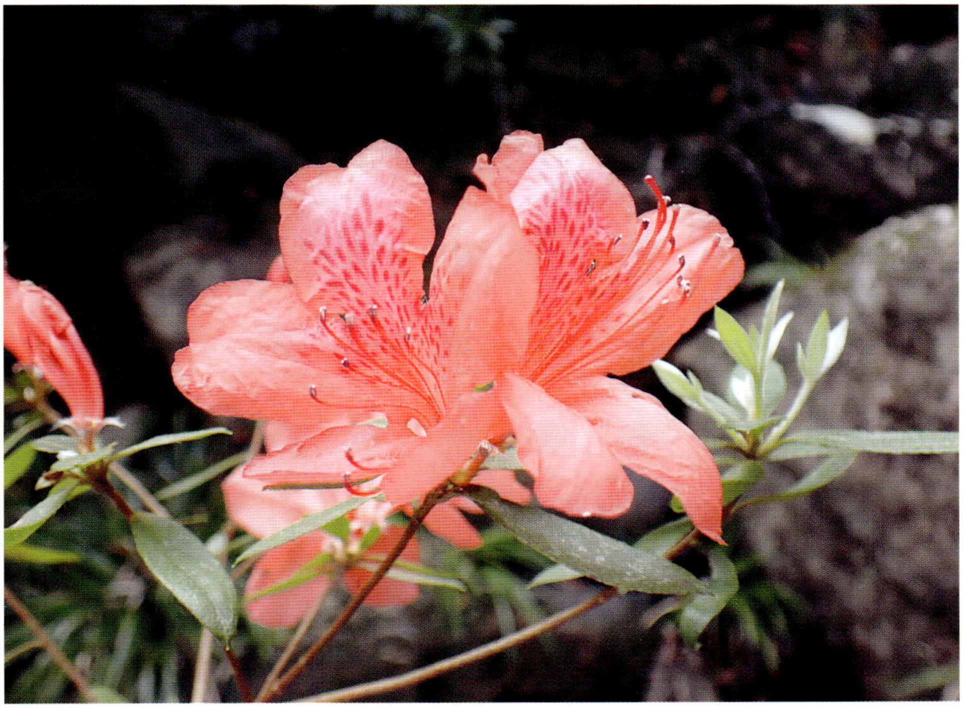

花枝 Flowering branch

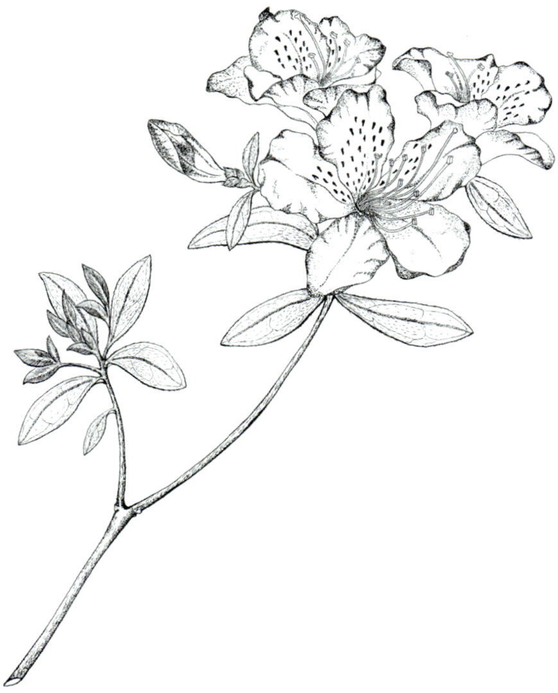

葉 Leaf

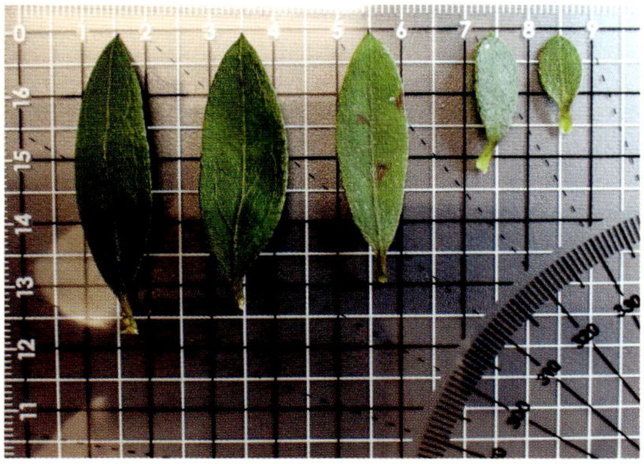

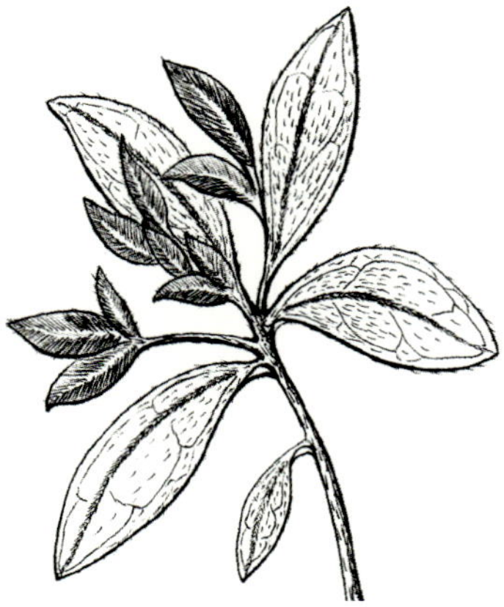

葉面 Leaves, adaxial

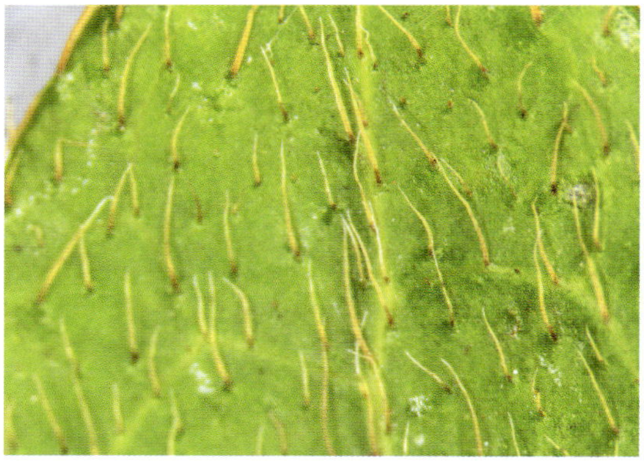

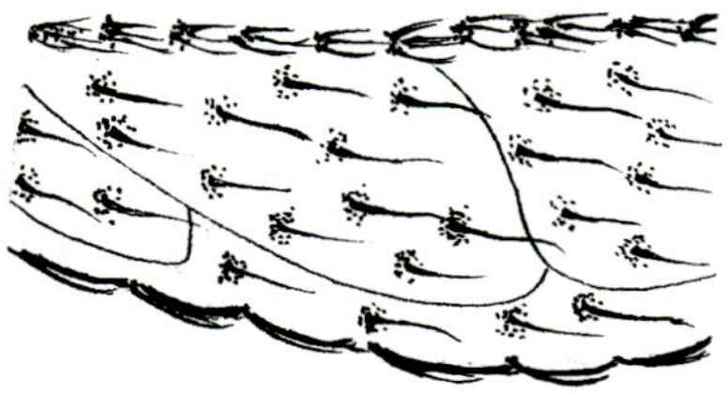

葉片被糙伏毛局部放大
Close-up of strigose on leaf blade

小枝 Branchlet

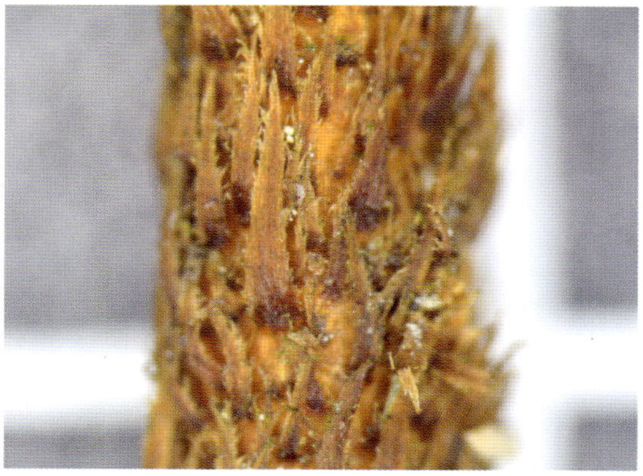

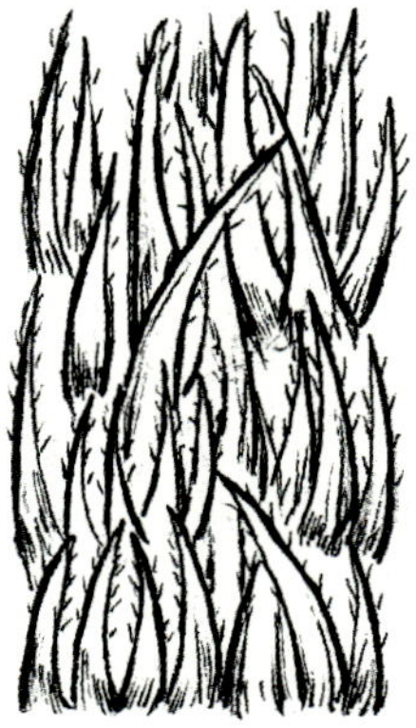

小枝被糙伏毛局部放大
Close-up of strigose on branchlets

花 Flower

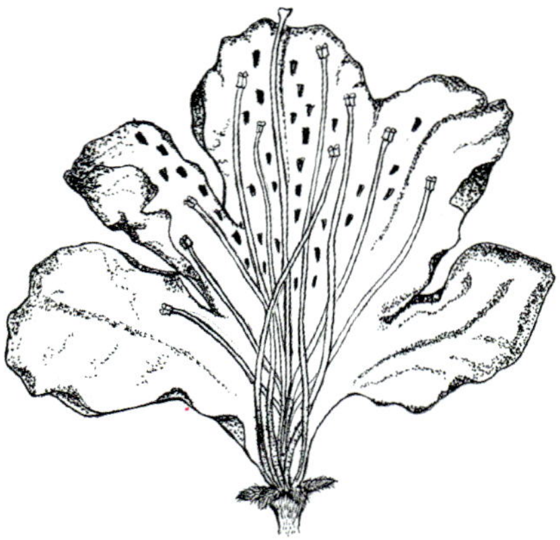
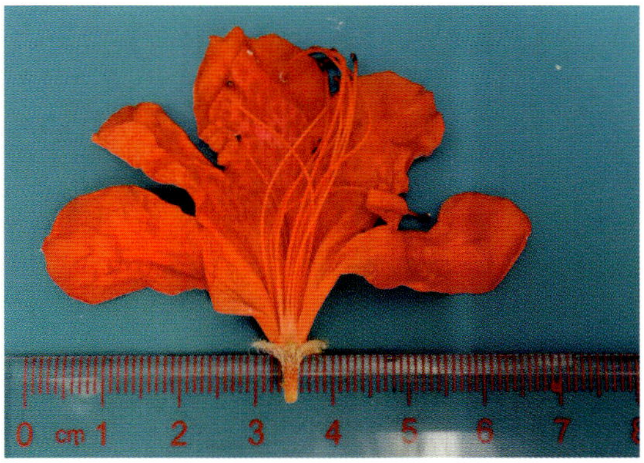

花解剖圖 Dissected flower

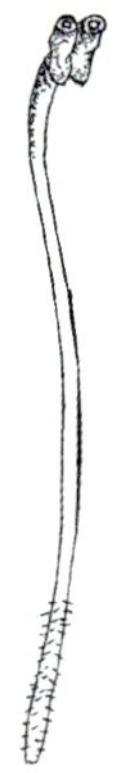
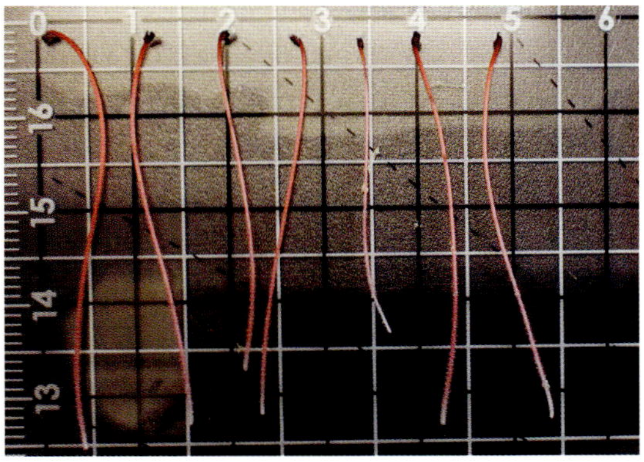

雄蕊 Stamens

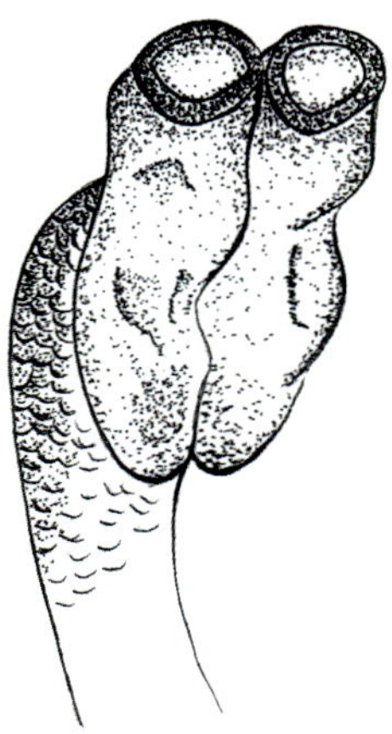
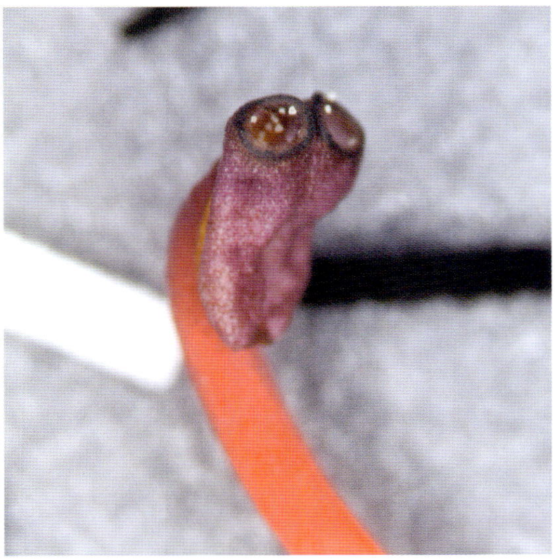

花藥 Anther

花 Flower

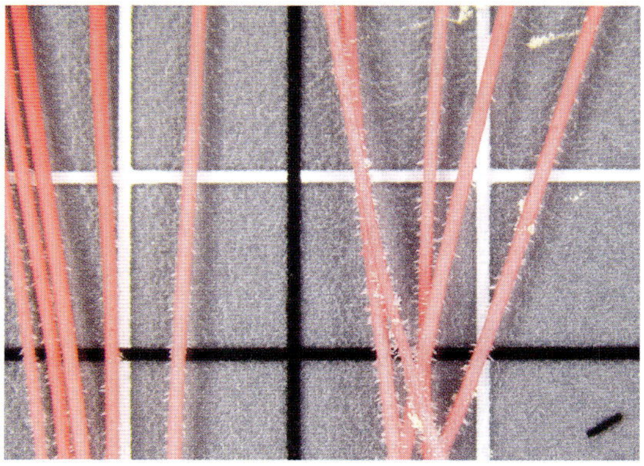

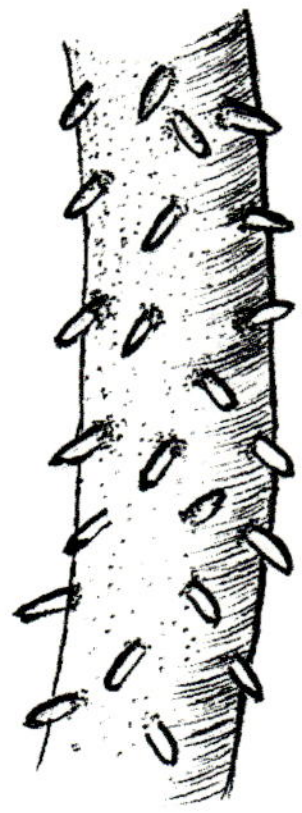

花絲中部以下被微柔毛局部放大
Close-up of hair on the lower part of the filament

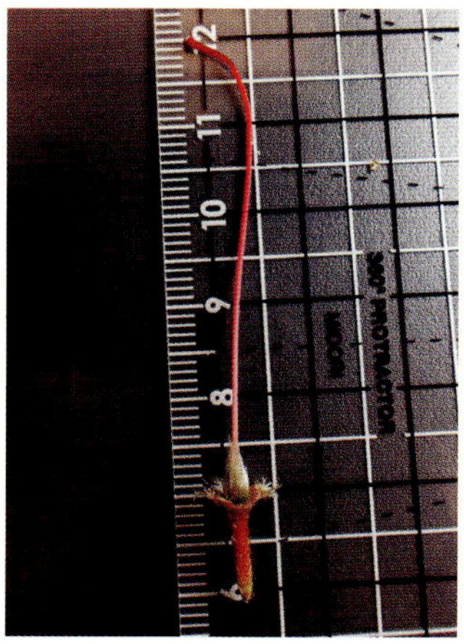

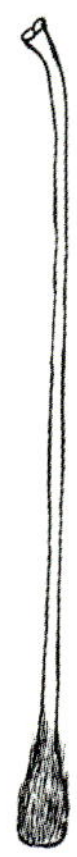

雌蕊 Gynoecium

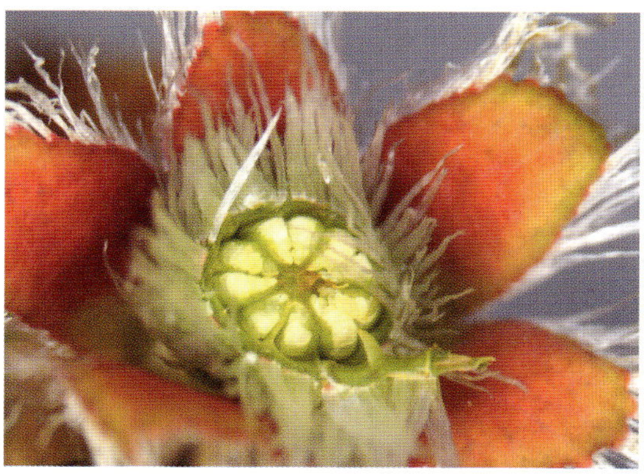

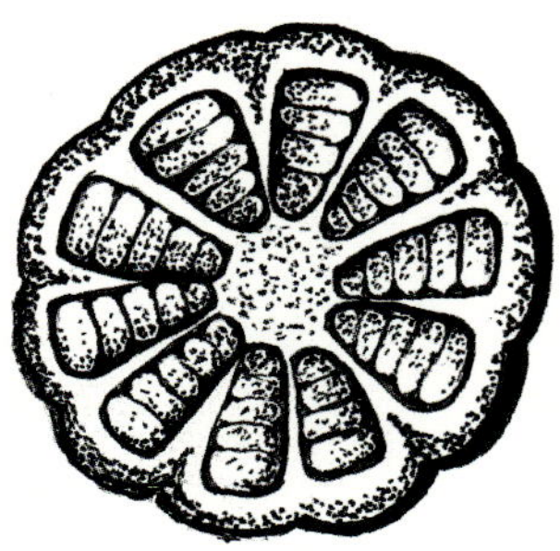

子房，橫切面 Ovary, in C. S.

Pavetta hongkongensis Bremek.
香港大沙葉

除了香港大沙葉的葉上擁有固氮的菌瘤，
很多同屬品種的葉上都有共生細菌。

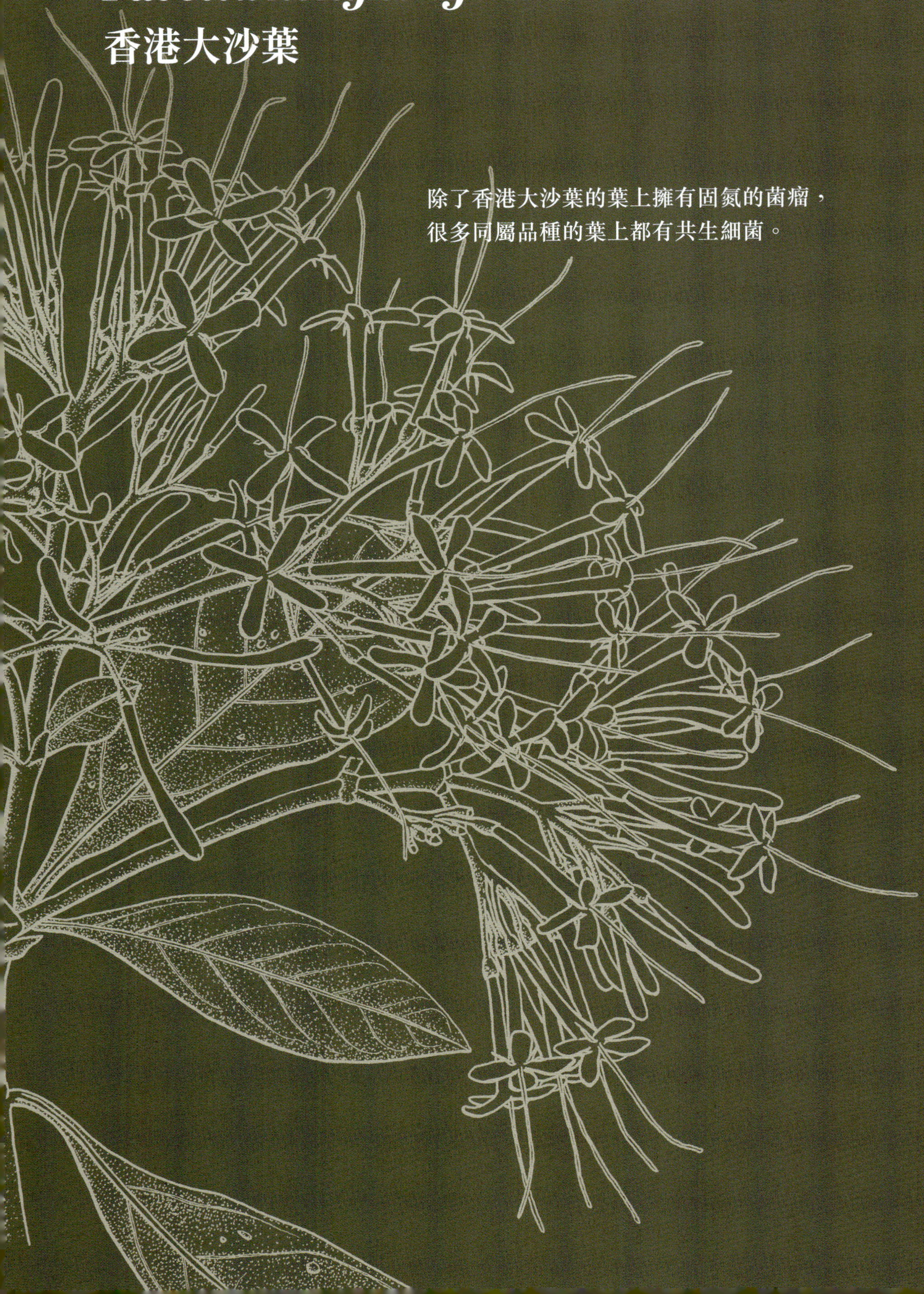

採集者編號
Collector no.
S. Y. Hu & P. P. H. But 23965

採集年份
Year of collection
2003

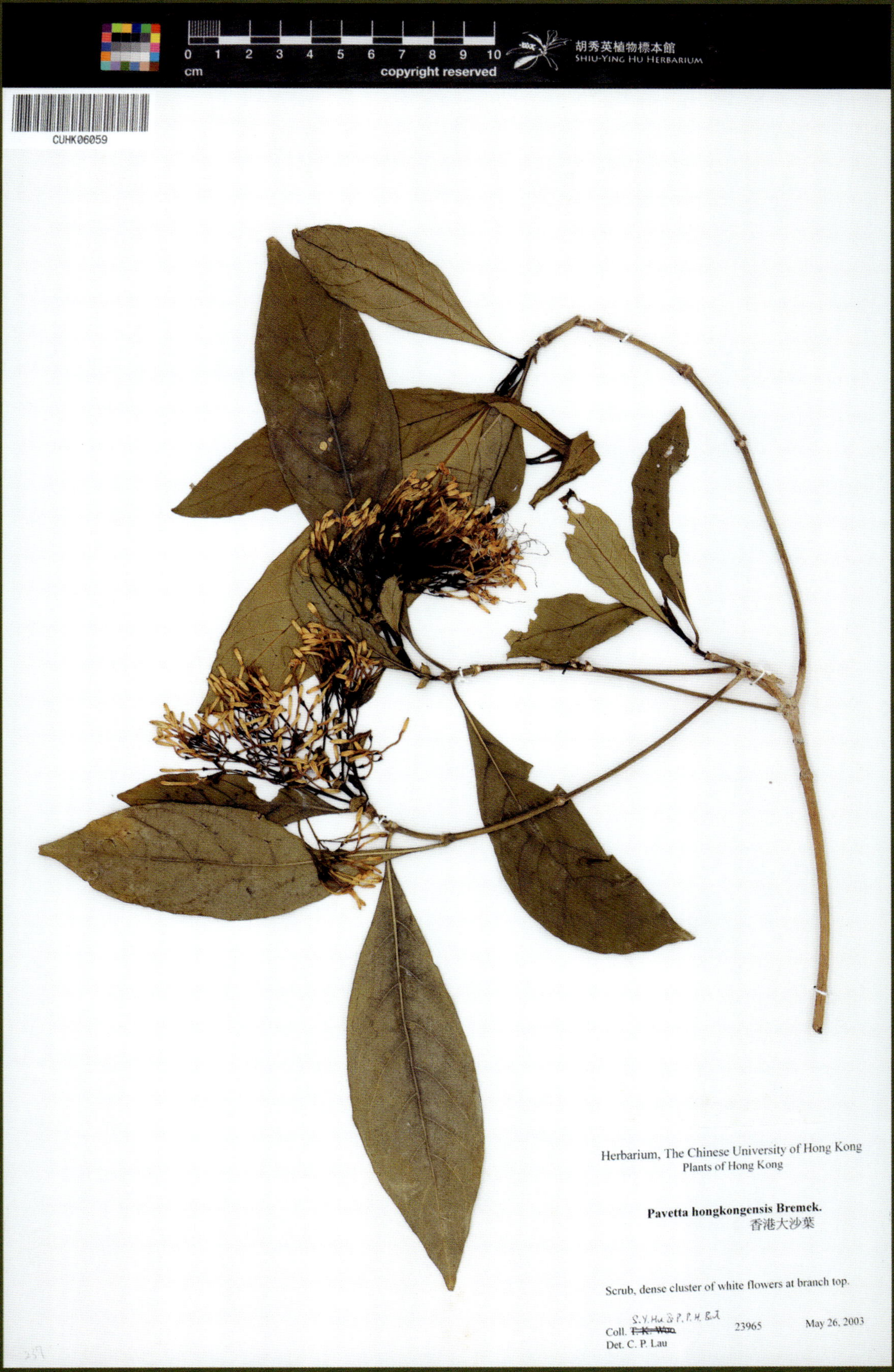

0 1 2 3 4 5 6 7 8 9 10
cm
copyright reserved
胡秀英植物標本館
SHIU-YING HU HERBARIUM
CUHK06059
Herbarium, The Chinese University of Hong Kong
Plants of Hong Kong
Pavetta hongkongensis Bremek.
香港大沙葉
Scrub, dense cluster of white flowers at branch top.
S. Y. Hu & P. P. H. But
Coll. ~~T. K. Woo~~
23965
May 26, 2003
Det. C. P. Lau

香港大沙葉

學名	*Pavetta hongkongensis* Bremek.
中文名	香港大沙葉
科名	茜草科
本地分布狀態	原生植物
繪圖編號	H. Y. Wong 005

花期	花色	果期	果色
3 至 10 月	白色	6 至 12 月	未有記錄

香港大沙葉為灌木或小喬木，生於灌木叢及森林。其模式標本由 J. G. Champion 於 1847 至 1850 年間在跑馬地採集。

除了香港大沙葉的葉上擁有固氮的菌瘤，很多同屬品種的葉上都有共生菌。有些學者提出這種共生關係為必需，而且會出現在植物的整個生長週期。有研究發現，共生菌亦存在於某些大沙葉屬植物的胚珠和種子內。

香港大沙葉在國際自然保護聯盟瀕危物種紅色名錄中被列為「無危」，受香港法例第 96A章《林務規例》規管。

《中國植物誌》(英文版) 的採用名稱

Pavetta hongkongensis Bremekamp (科名：茜草科)

《香港植物誌》的採用名稱

Pavetta hongkongensis Bremek. (科名：茜草科)

參考文獻 References

1. Chen, T., Zhu, H., Chen, J. R., Taylor, C. M., Ehrendorfer, F., Lantz, H., Funston, A. M., & Puff, C. (2011). Rubiaceae. In Z. Y. Wu, P. H. Raven & D. Y. Hong (Eds.), *Flora of China* (Vol. 19, pp. 57–368). Science Press & Missouri Botanical Garden Press.
2. Hong Kong Herbarium. (2021). *HK Herbarium-Pavetta hongkongensis.* HK Plant Database. Retrieved June 13, 2021, from https://www.herbarium.gov.hk/en/hk-plant-database/plant-detail/index.html?pType=species&oID=2491
3. IUCN. (2018). *Pavetta hongkongensis.* The IUCN Red List of Threatened Species. Retrieved June 18, 2021, from https://www.iucnredlist.org/species/147632811/147632838

Pavetta hongkongensis Bremek.

Scientific name	*Pavetta hongkongensis* Bremek.
Chinese name	香港大沙葉
Family	Rubiaceae Juss.
Local distribution status	Native species
Illustration number	H. Y. Wong 005

FLOWERING PERIOD	COLOUR OF FLOWER	FRUITING PERIOD	COLOUR OF FRUIT
March to October	White	June to December	Not Recorded

Pavetta hongkongensis Bremek. is a shrub or small tree, growing in thickets and forests. Its type specimen was collected by J. G. Champion in Happy Valley between 1847 and 1850.

Apart from *Pavetta hongkongensis* Bremek., which has nitrogen-fixing bacterial nodules on its leaves, many other species in the same genus have symbiotic bacteria on their leaves as well. Some scholars hypothesized that these symbiotic relationships are necessary and happen throughout the entire life cycle of the plants. A study found that symbiotic bacteria also exist in the ovules and seeds of certain *Pavetta* species.

Pavetta hongkongensis Bremek. is listed as 'Least Concern' in the IUCN Red List of Threatened Species. It is regulated under Hong Kong's Forestry Regulations (Cap. 96A).

Name in *Flora of China*
Pavetta hongkongensis Bremekamp (Family: Rubiaceae)

Name in *Flora of Hong Kong*
Pavetta hongkongensis Bremek. (Family: Rubiaceae)

4. Lemaire, B., Van Oevelen, S., De Block, P., Verstraete, B., Smets, E., Prinsen, E., & Dessein, S. (2012). Identification of the bacterial endosymbionts in leaf nodules of *Pavetta* (Rubiaceae). *International Journal of Systematic and Evolutionary Microbiology 62* (1), 202–209. https://doi.org/10.1099/ijs.0.028019-0
5. Missouri Botanical Garden. (2021). *Name-Pavetta hongkongensis*. Tropicos. Retrieved June 18, 2021, from https://www.tropicos.org/name/27913234

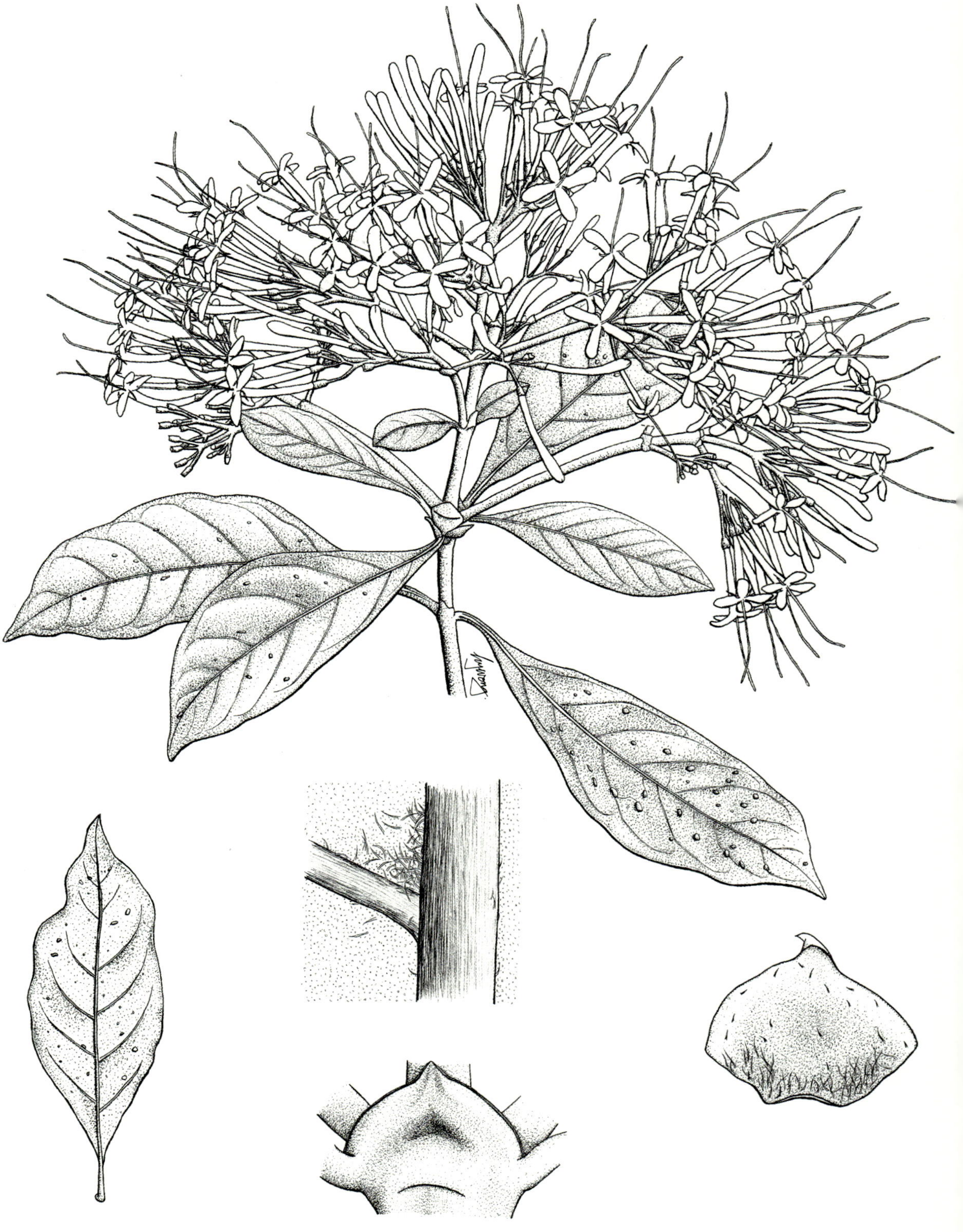

Pavetta hongkongensis Bremek. 香港大沙葉
Illustration 005 by H. Y. Wong
Authentication by D. T. W. Lau

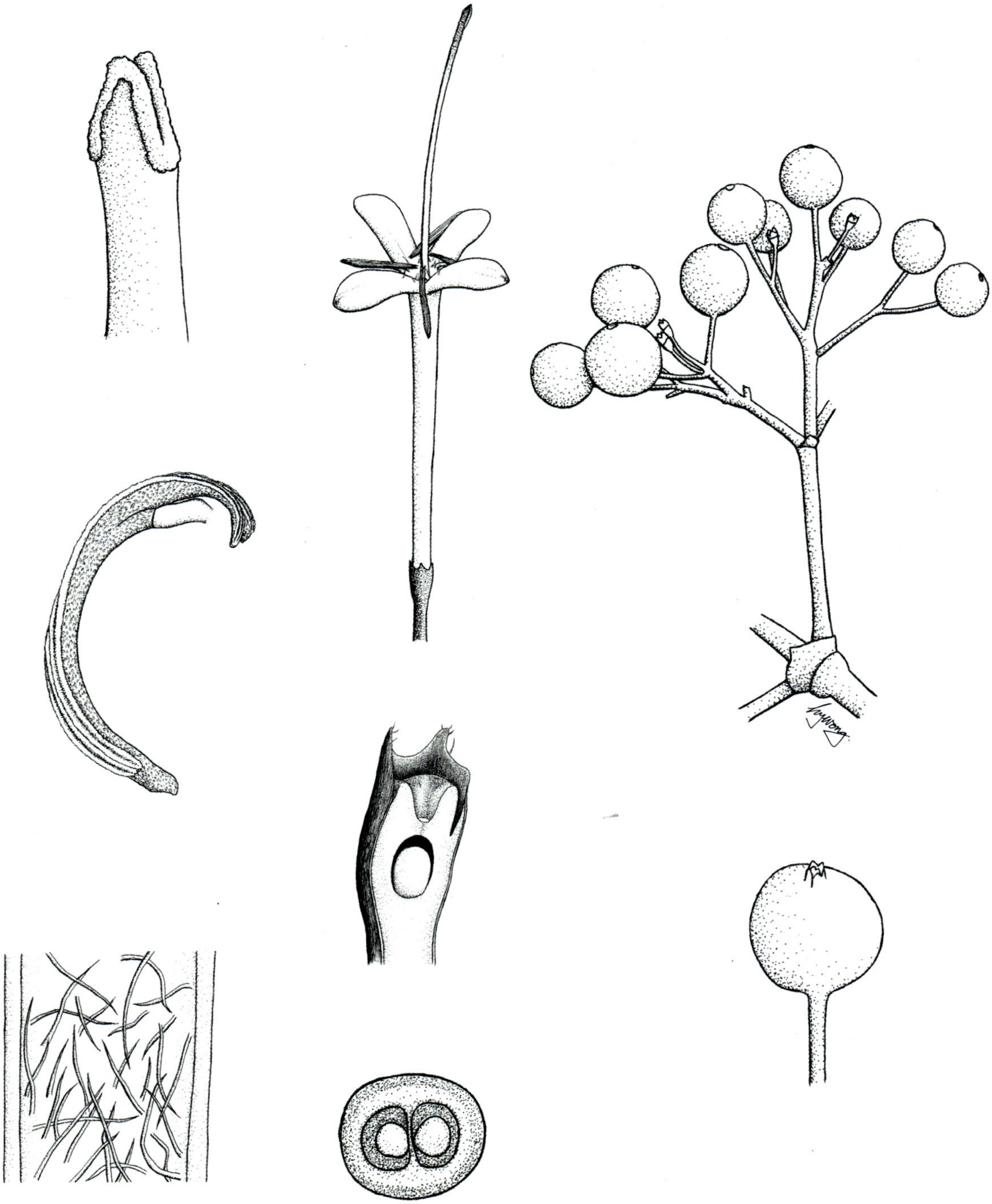

Pavetta hongkongensis Bremek. 香港大沙葉
Illustration 005 by H. Y. Wong
Authentication by D. T. W. Lau

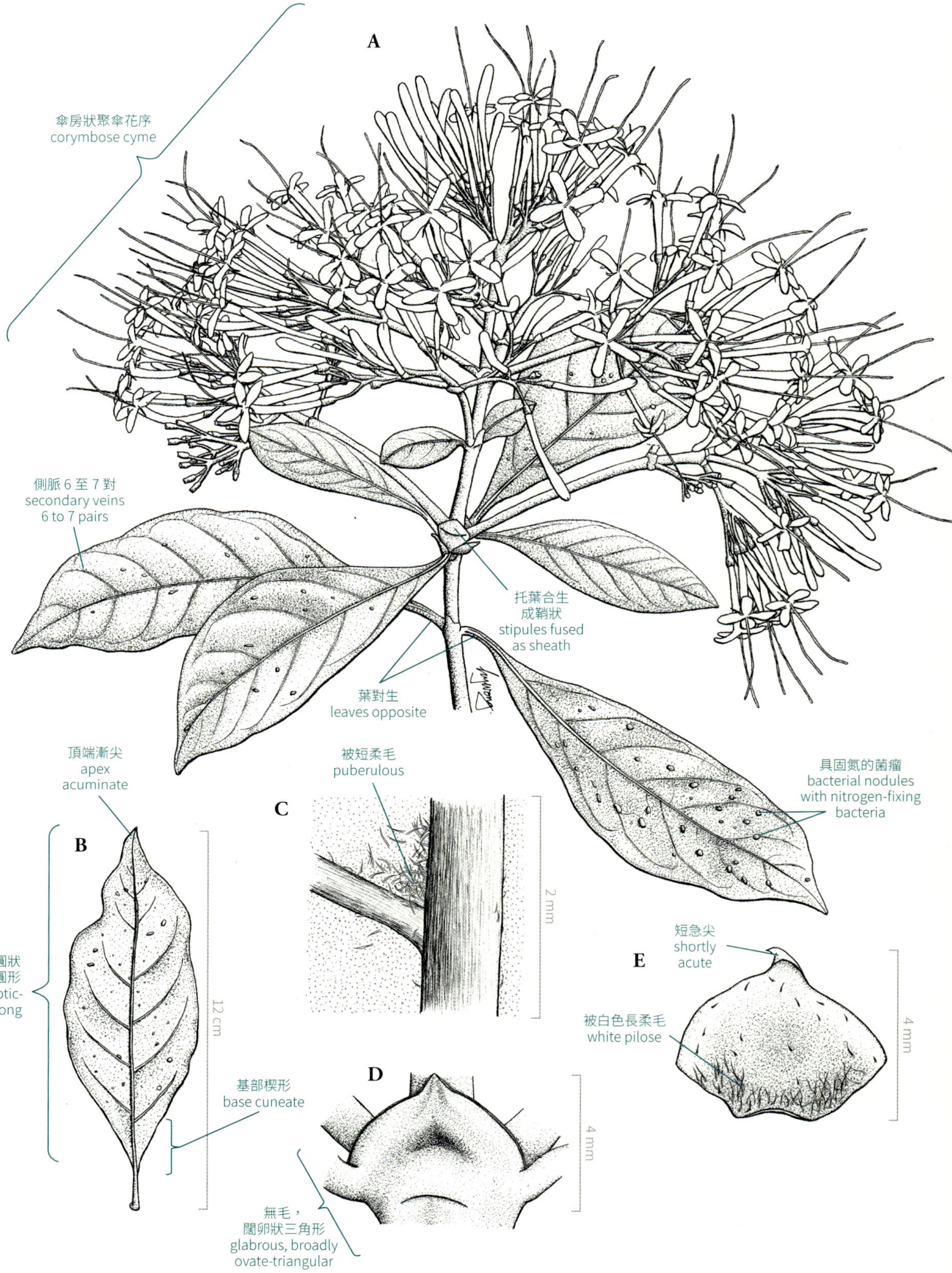

Pavetta hongkongensis Bremek. 香港大沙葉

Illustration 005 by H. Y. Wong
Authentication by D. T. W. Lau

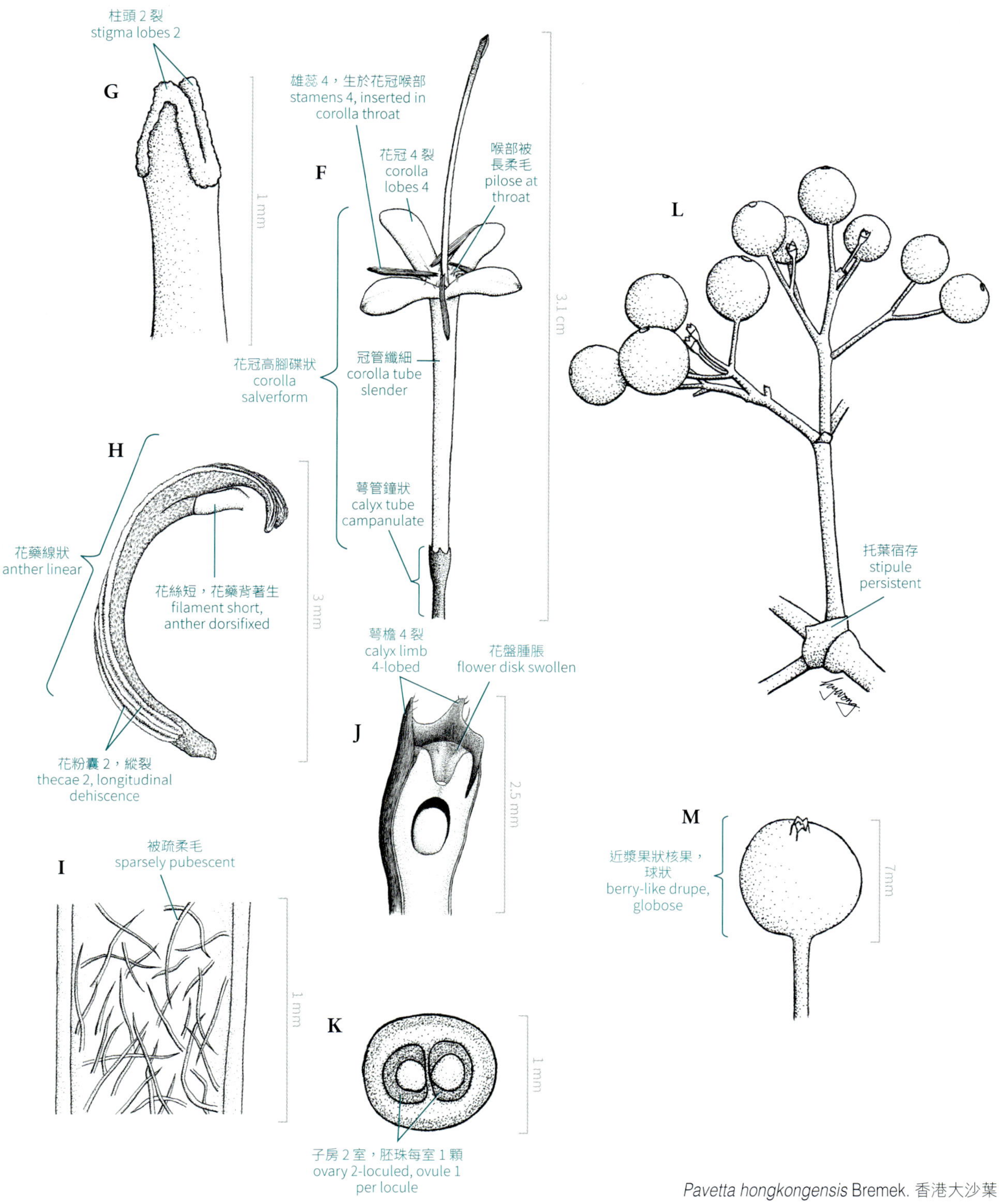

Pavetta hongkongensis Bremek. 香港大沙葉

Illustration 005 by H. Y. Wong
Authentication by D. T. W. Lau

植物結構 PLANT STRUCTURES:
A. 花枝 Flowering branch **B.** 葉底 Leaf, abaxial **C.** 在葉底脈腋上的被毛 Hairs on axillary part of midvein of abaxial leaf **D.** 托葉，背軸面 Stipule, abaxial **E.** 托葉，向軸面 Stipule, adaxial **F.** 花 Flower **G.** 柱頭 Stigma **H.** 雄蕊 Stamen **I.** 花冠管基部內壁的被毛 Hairs on the base of inner corolla tube **J.** 子房，縱切面 Ovary, in L. S. **K.** 子房，橫切面 Ovary, in C. S. **L.** 果序 Infructescence **M.** 果實 Fruit

參考標本 REFERENCE SPECIMENS: Rare and endangered plants (J. Y. Y. Lau, D. T. W. Lau, K. W. Lam, S. W. Shek) 12B; H.Y. Wong 004 (CUSLSH3026); H. Y. Wong 008 (CUSLSH3102); S. Y. Hu & P. P. H. But 23965

參考文獻 REFERENCES:
Chen, T. (2012). Rubiaceae. In Fairy Lake Botanical Garden, Shenzhen & Chinese Academy of Sciences (Eds.), *Flora of Shenzhen* (Vol. 3, pp. 457–523). China Forestry Publishing House.
Chen, T., Zhu, H., Chen, J. R., Taylor, C. M., Ehrendorfer, F., Lantz, H., Funston, A. M., & Puff, C. (2011). Rubiaceae. In Z. Y. Wu, P. H. Raven & D. Y. Hong (Eds.), *Flora of China*, (Vol. 19, pp. 57–368). Science Press & Missouri Botanical Garden Press.
Gao, Y. Z., Chen, W. Q., & Ruan, Y. Z. (2005). Rubiaceae. In South China Botanical Garden, the Chinese Academy of Sciences (Ed.), *Flora of Guangdong* (Vol. 6, pp. 108–248). Guangdong Science and Technology Press.
Hu, Q. M. (2009). Rubiaceae. In Hong Kong Herbarium, AFCD & South China Botanical Garden, CAS (Eds.), *Flora of Hong Kong* (Vol. 3, pp. 203–240). Agriculture, Fisheries and Conservation Department, HKSAR Government.
Lo, H. S., Ko, W. C., Chen, W. C., & Ruan, Y. Z. (1999). Rubiaceae. In Delectis Florae Reipublicae Popularis Sinicae Agendae Academiae Sinicae Edita (Ed.), *Flora Reipublicae Popularis Sinicae* (Vol. 71(2), pp. 2–338). Science Press.
Stevens, P. F. (2021). Gentianales. Angiosperm Phylogeny Website. Retrieved June 15, 2021, from http://www.mobot.org/MOBOT/research/APweb/welcome.html

憑證標本
Voucher Specimen

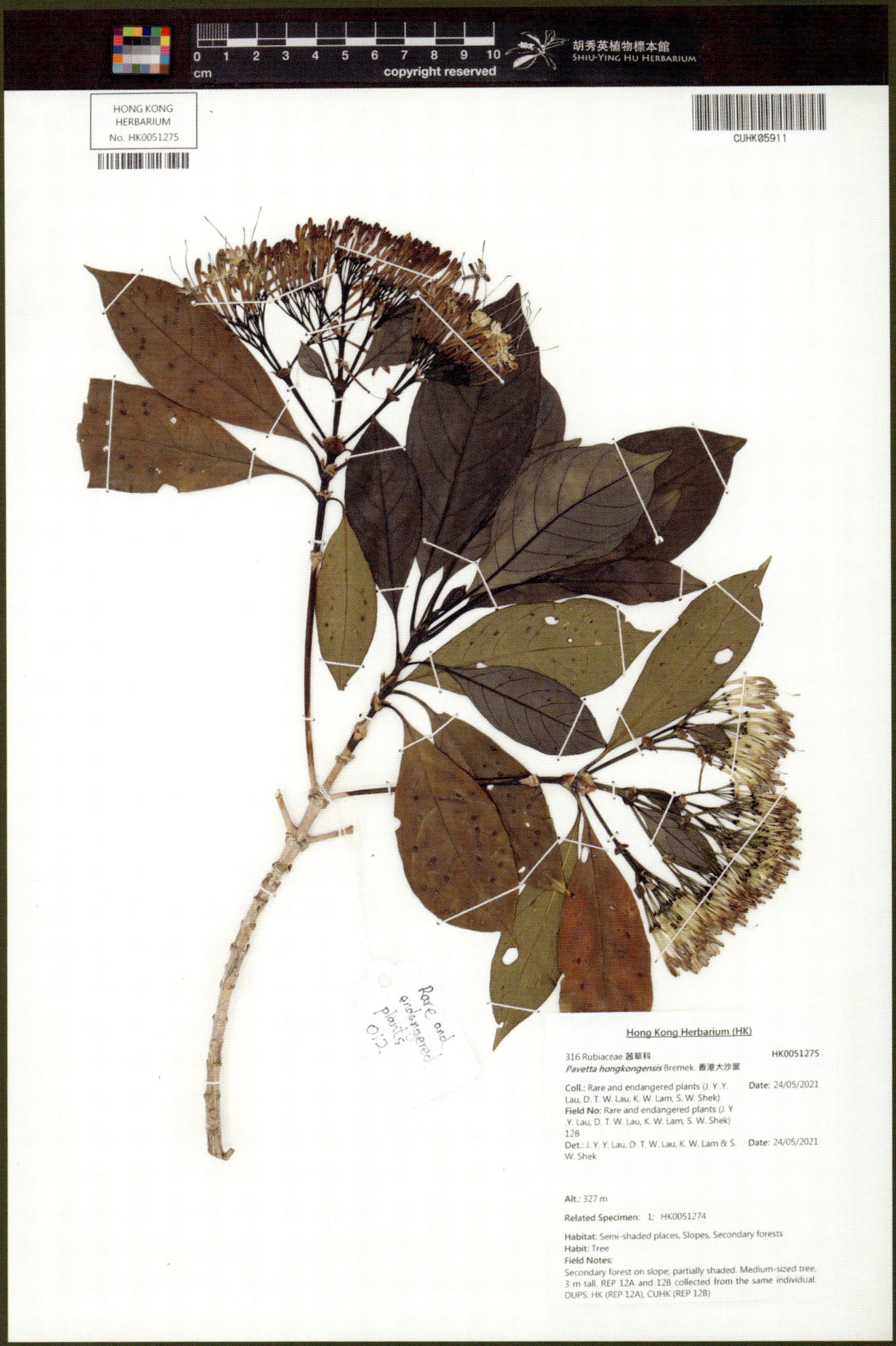

採集者編號 Collector no.
Rare and endangered plants (J. Y. Y. Lau, D. T. W. Lau, K. W. Lam, S. W. Shek) 12B

採集年份 Year of collection
2021

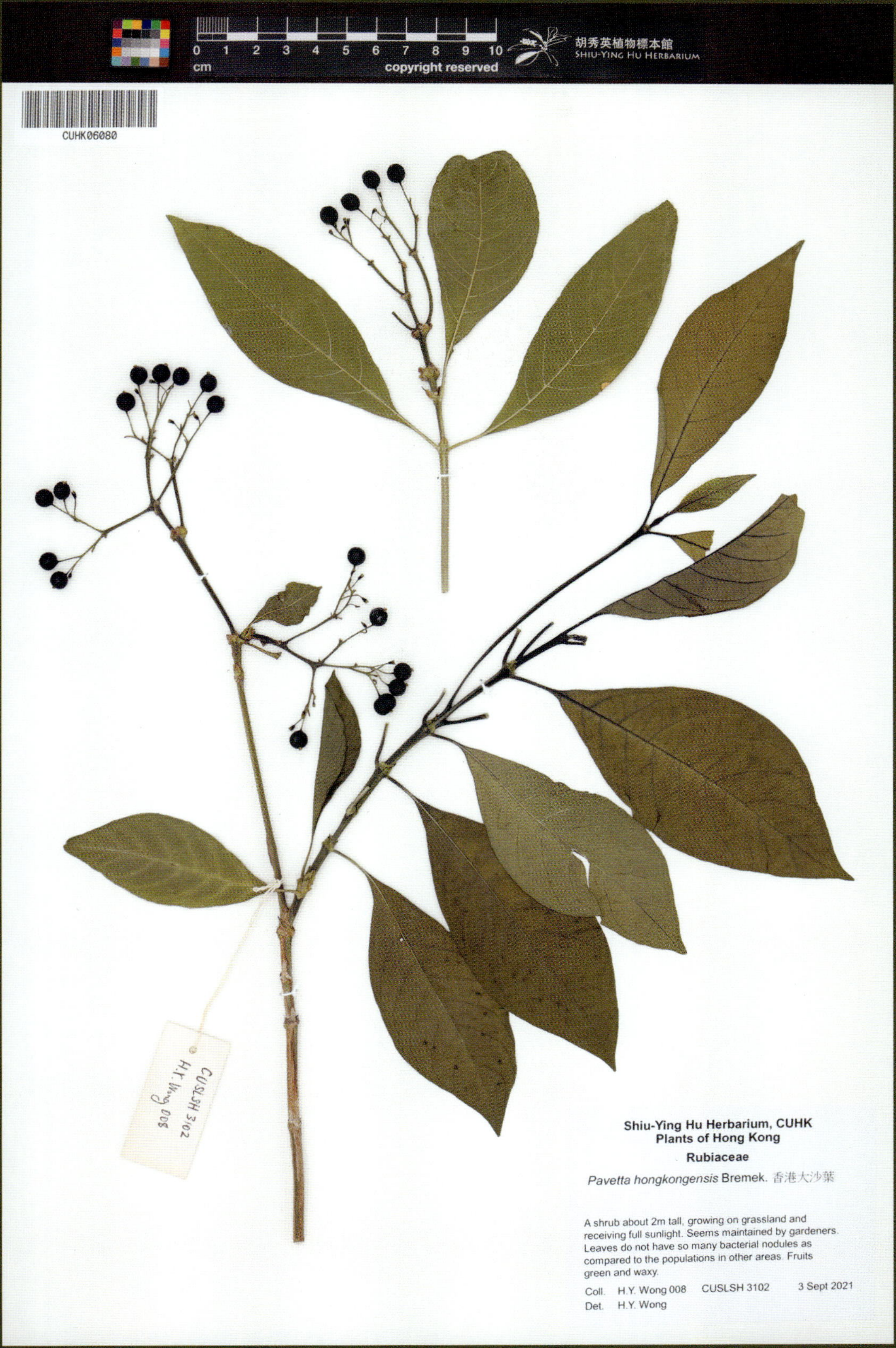

採集者編號 Collector no.
H. Y. Wong 008

採集年份 Year of collection
2021

外觀 Overview

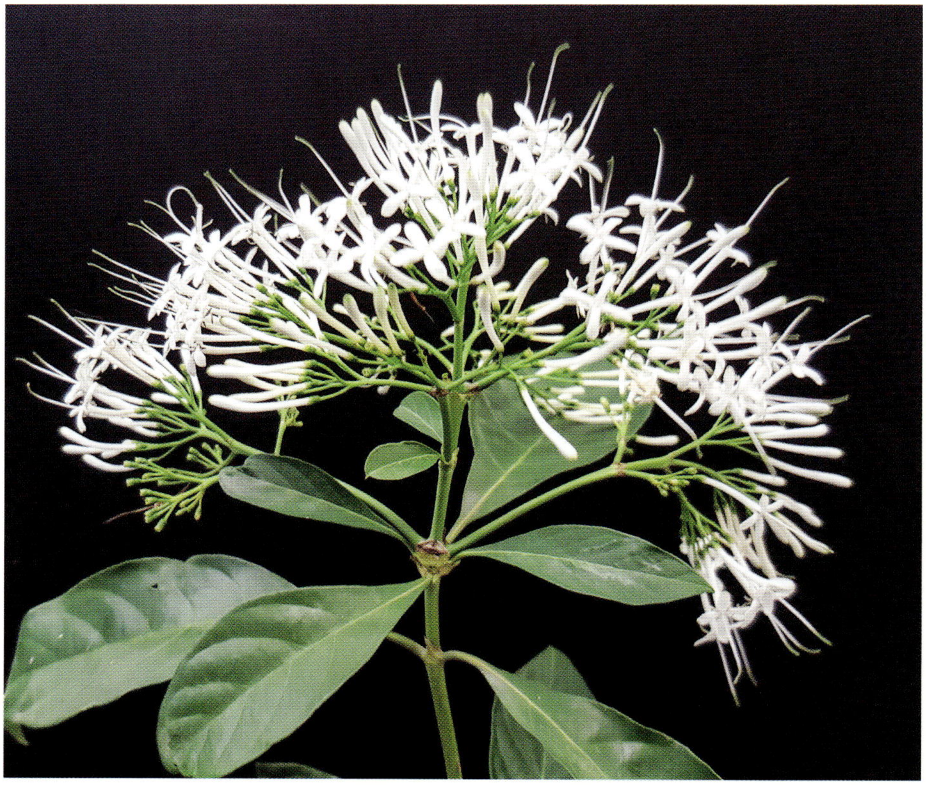

花枝 Flowering branch

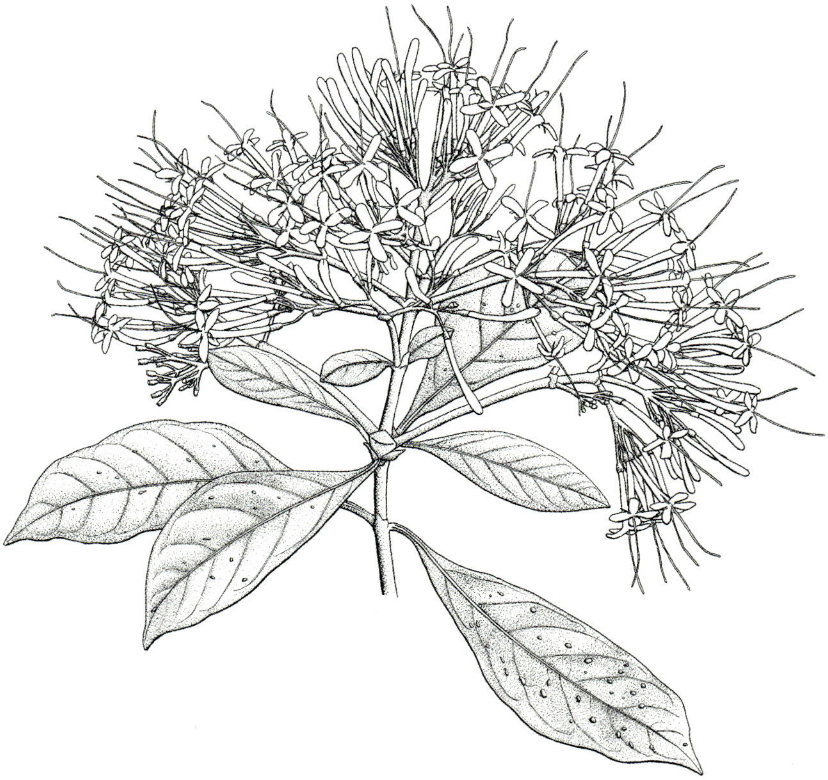

葉 Leaf

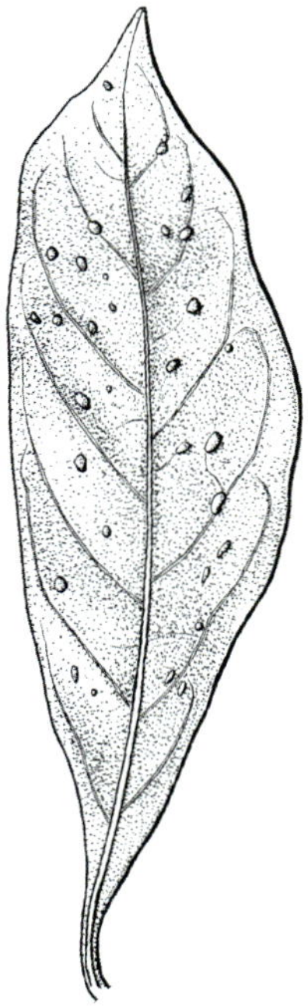

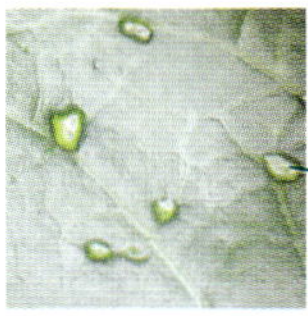

固氮的菌瘤
Bacterial nodules with nitrogen-fixing bacteria

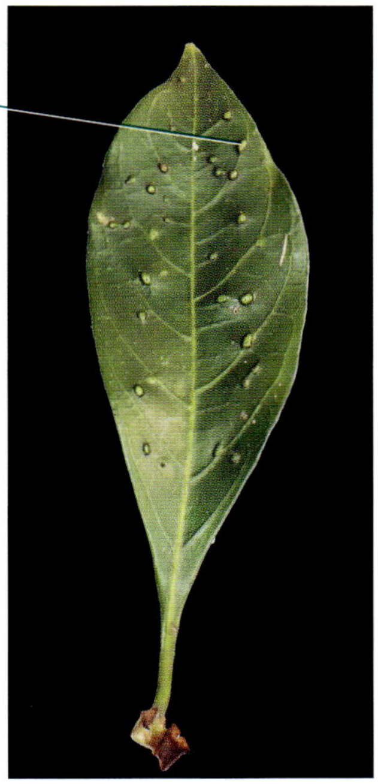

葉面 Leaf, adaxial

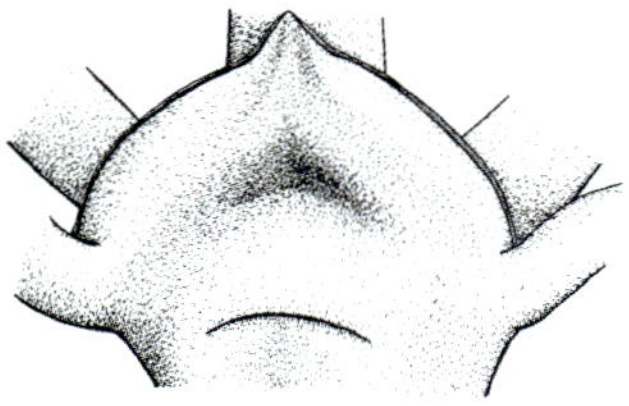

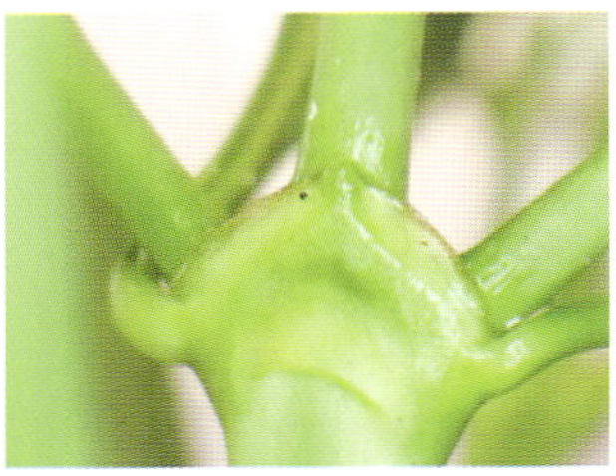

托葉 Stipule

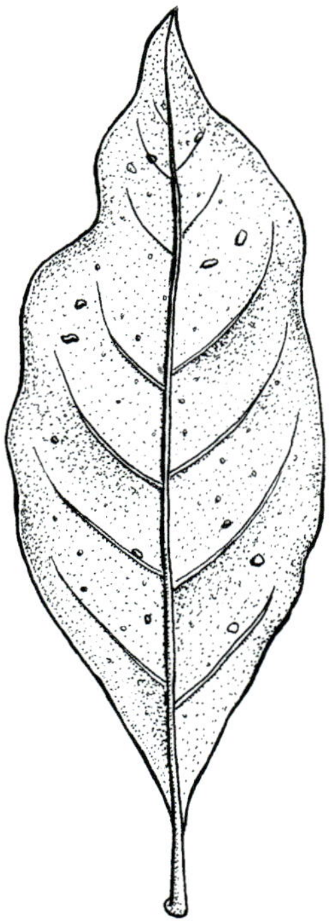

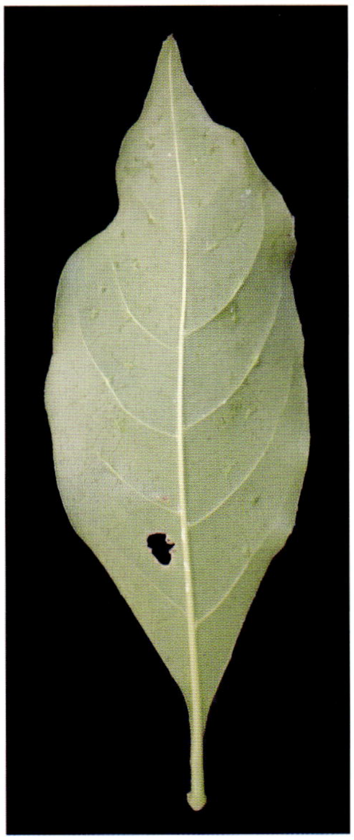

葉底 Leaf, abaxial

花 Flower

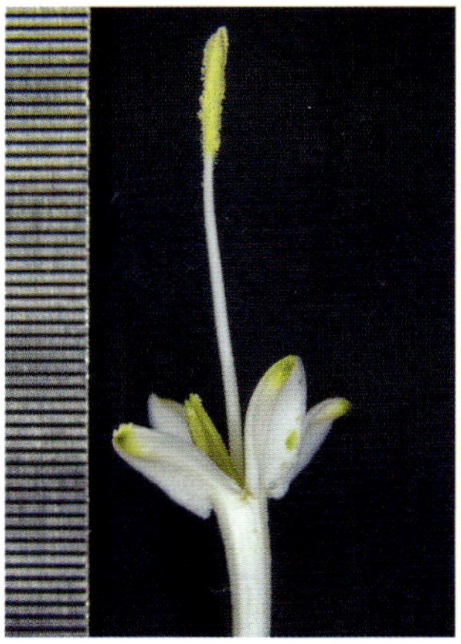
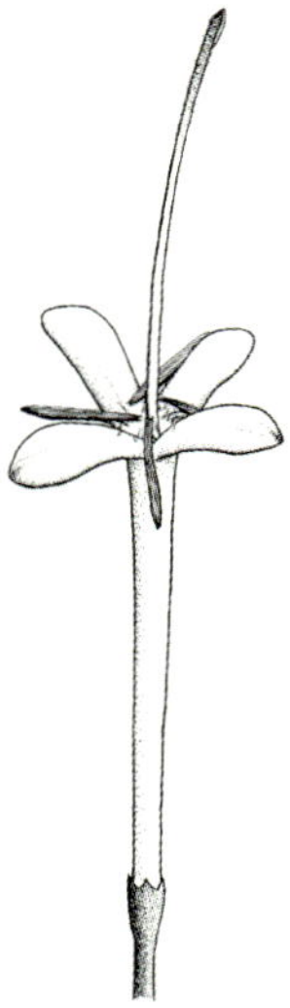

花
Flower

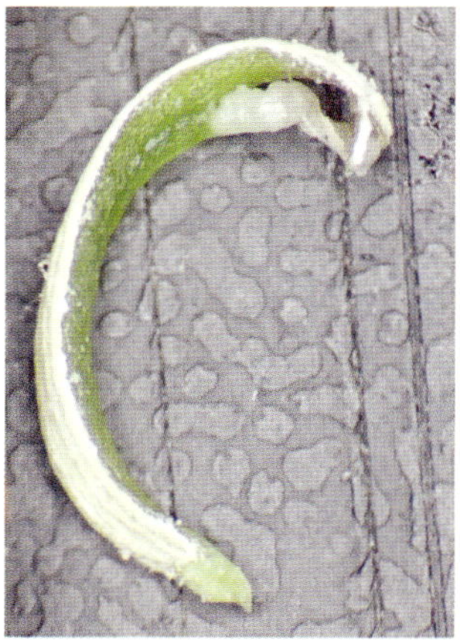
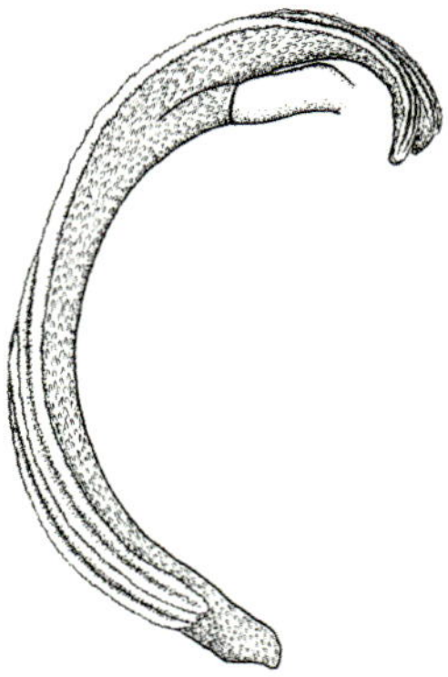

雄蕊 Stamen

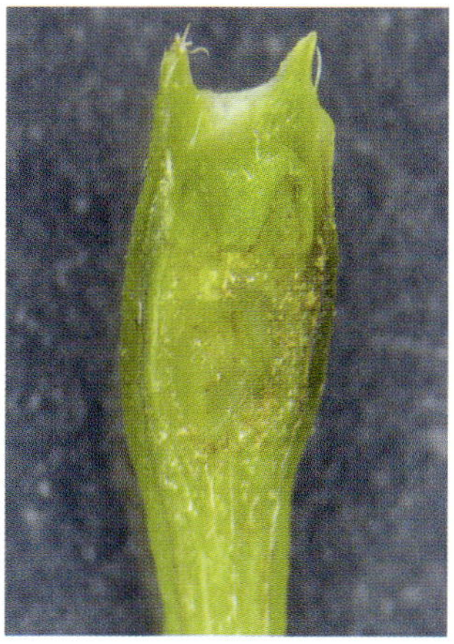
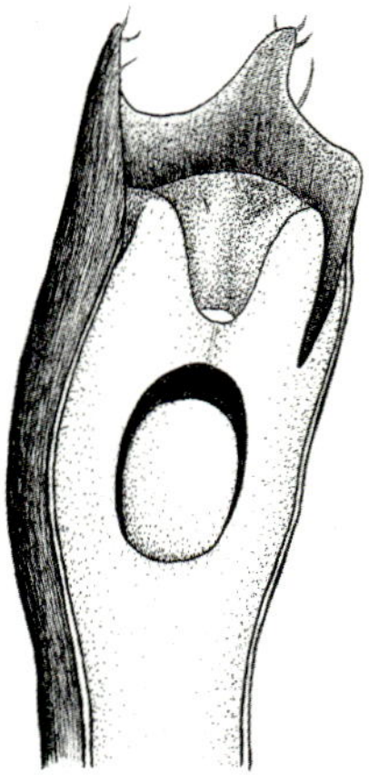

子房，縱切面
Ovary, in L. S.

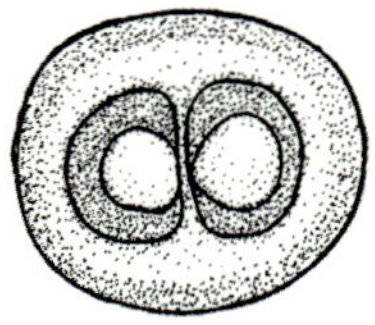
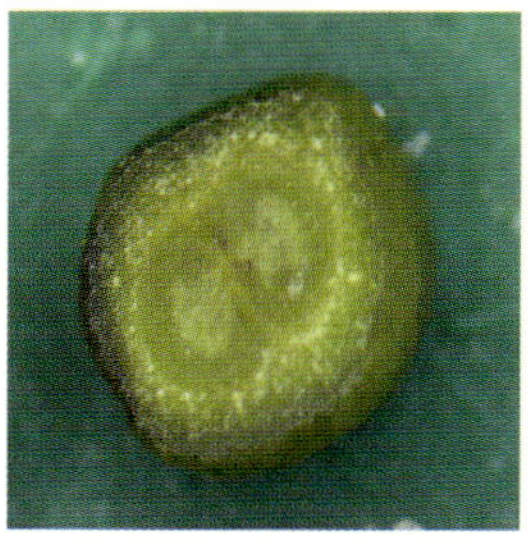

子房，橫切面
Ovary, in C. S.

果 Fruit

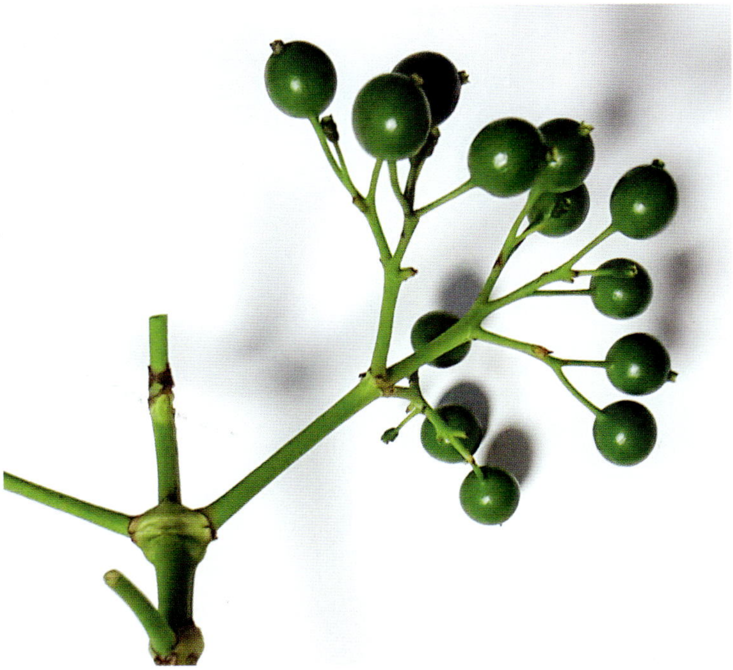

果序 Infructescence

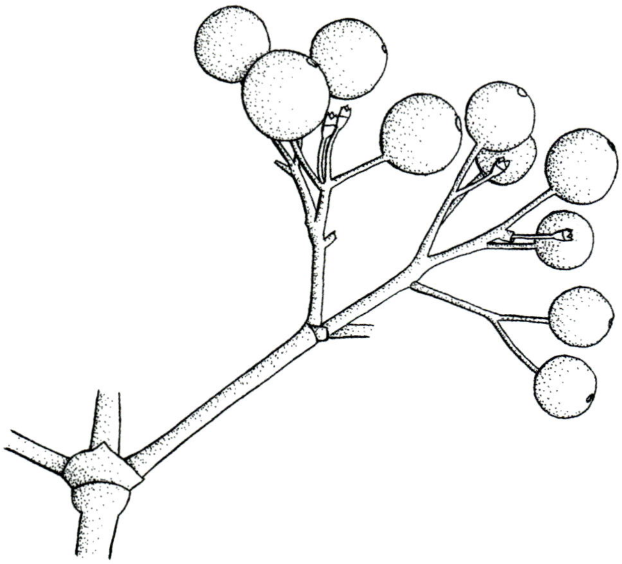

Oroxylum indicum (L.) Benth. ex Kurz
木蝴蝶

木蝴蝶在東南亞多個國家均作為傳統藥物，其化學提取物具有抗菌、抗炎、保肝等作用。

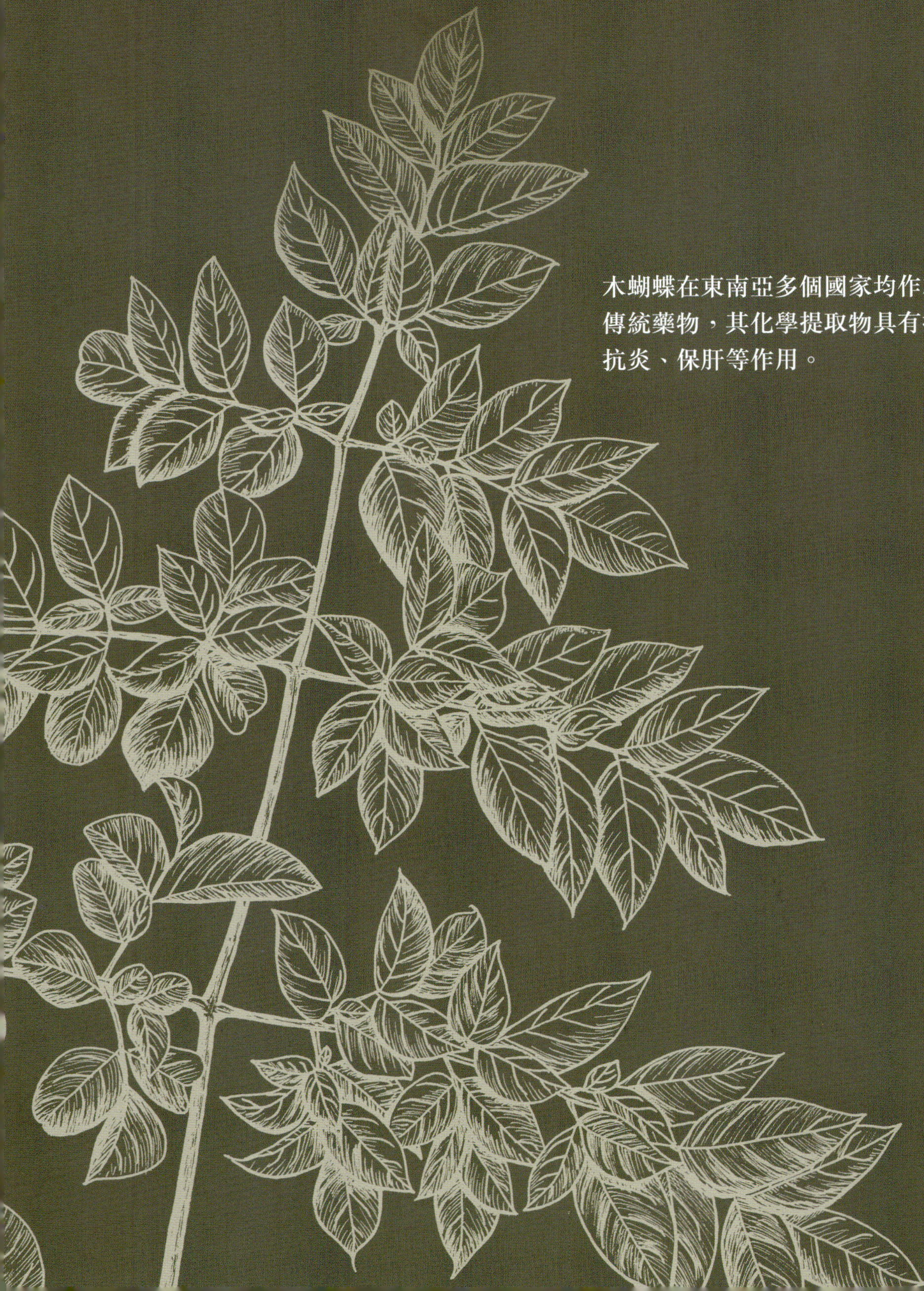

採集者編號
Collector no.
Shiu Ying Hu 12390

採集年份
Year of collection
1972

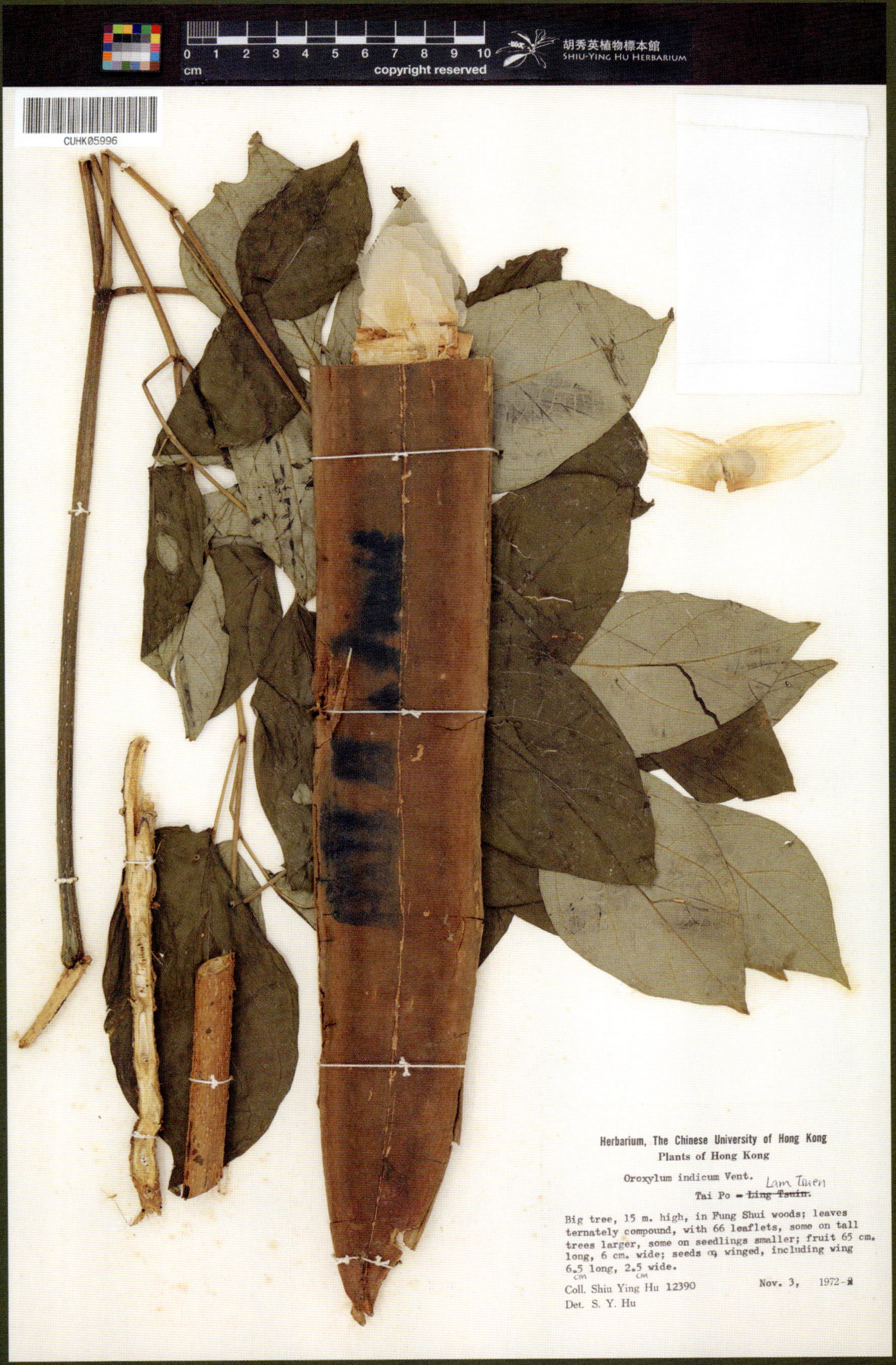
0 1 2 3 4 5 6 7 8 9 10
cm
copyright reserved
胡秀英植物標本館
SHIU-YING HU HERBARIUM
CUHK05996
Herbarium, The Chinese University of Hong Kong
Plants of Hong Kong
Oroxylum indicum Vent.
Lam Tsuen
Tai Po - ~~Ling Tsuin.~~
Big tree, 15 m. high, in Fung Shui woods; leaves ternately compound, with 66 leaflets, some on tall trees larger, some on seedlings smaller; fruit 65 cm. long, 6 cm. wide; seeds ∞ winged, including wing 6.5 long, 2.5 wide.
cm cm
Coll. Shiu Ying Hu 12390
Nov. 3, 1972
Det. S. Y. Hu

木蝴蝶

學名	*Oroxylum indicum* (L.) Benth. ex Kurz
中文名	木蝴蝶
科名	紫葳科
本地分布狀態	原生植物
繪圖編號	H. Y. Wong 001

花期	花色	果期	果色
9 至 11 月	紫紅色	11 至翌年 1 月	褐色

木蝴蝶為 6 至 10 米高的喬木，常見於村邊風水林。花於晚上綻放。

木蝴蝶在東南亞多個國家均作為傳統藥物，其化學提取物具有抗菌、抗炎、保肝等作用。從中醫角度，木蝴蝶種子能治支氣管炎及咳嗽等問題。在泰國，木蝴蝶的花及幼果則被當作蔬菜食用。除了藥用及食用，木蝴蝶因其獨特的外觀，亦被種植作觀賞用途。

《中國植物誌》(英文版）的採用名稱

Oroxylum indicum (Linnaeus) Bentham ex Kurz（科名：紫葳科）

《香港植物誌》的採用名稱

Oroxylum indicum Vent.（科名：紫葳科）

參考文獻 References

1. Dinda, B., SilSarma, I., Dinda, M., & Rudrapaul, P. (2015). *Oroxylum indicum* (L.) Kurz, an important Asian traditional medicine: from traditional uses to scientific data for its commercial exploitation. *Journal of Ethnopharmacology 161*, 255–278. https://doi.org/10.1016/j.jep.2014.12.027
2. Harminder, V. S., & Chaudhary, A. K. (2011). A review on the taxonomy, ethnobotany, chemistry and pharmacology of *Oroxylum indicum* Vent. *Indian Journal of Pharmaceutical Sciences 73*(5), 483–490. https://doi.org/10.4103/0250-474X.98981
3. Hong Kong Herbarium. (2021). *HK Herbarium-Oroxylum indicum*. HK Plant Database. Retrieved August 25, 2021, from https://www.herbarium.gov.hk/en/hk-plant-database/plant-detail/index.html?pType=species&oID=4514

Oroxylum indicum (L.) Benth. ex Kurz

Scientific name	*Oroxylum indicum* (L.) Benth. ex Kurz
Chinese name	木蝴蝶
Family	Bignoniaceae Juss.
Local distribution status	Native species
Illustration number	H. Y. Wong 001

FLOWERING PERIOD	COLOUR OF FLOWER	FRUITING PERIOD	COLOUR OF FRUIT
September to November	Purplish red	November to following January	Brown

Oroxylum indicum (L.) Benth. ex Kurz is a 6 to 10 m tall tree, and is usually found in Fung Shui Woods near villages. Flowers open at night.

In many Southeast Asian countries, *Oroxylum indicum* (L.) Benth. ex Kurz is a traditional medicine. Its chemical extracts are anti-bacterial, anti-inflammatory, a tonic for the liver, and more. In traditional Chinese medicine, its seeds can treat bronchitis and coughing. In Thailand, the flowers and young fruits are consumed as vegetables. Apart from medicine and food, *Oroxylum indicum* (L.) Benth. ex Kurz is also planted for ornamental purposes due to its unique appearance.

Name in *Flora of China*
Oroxylum indicum (Linnaeus) Bentham ex Kurz (Family: Bignoniaceae)

Name in *Flora of Hong Kong*
Oroxylum indicum Vent. (Family: Bignoniaceae)

4. Missouri Botanical Garden. (2021). *Name-Oroxylum indicum*. Tropicos. Retrieved August 25, 2021, from https://www.tropicos.org/name/3701303
5. Zhang, Z. Y., & Santisuk, T. (1998). Bignoniaceae. In Z. Y. Wu, P. H. Raven & D. Y. Hong (Eds.), *Flora of China* (Vol. 18, pp. 213–225). Science Press & Missouri Botanical Garden Press.

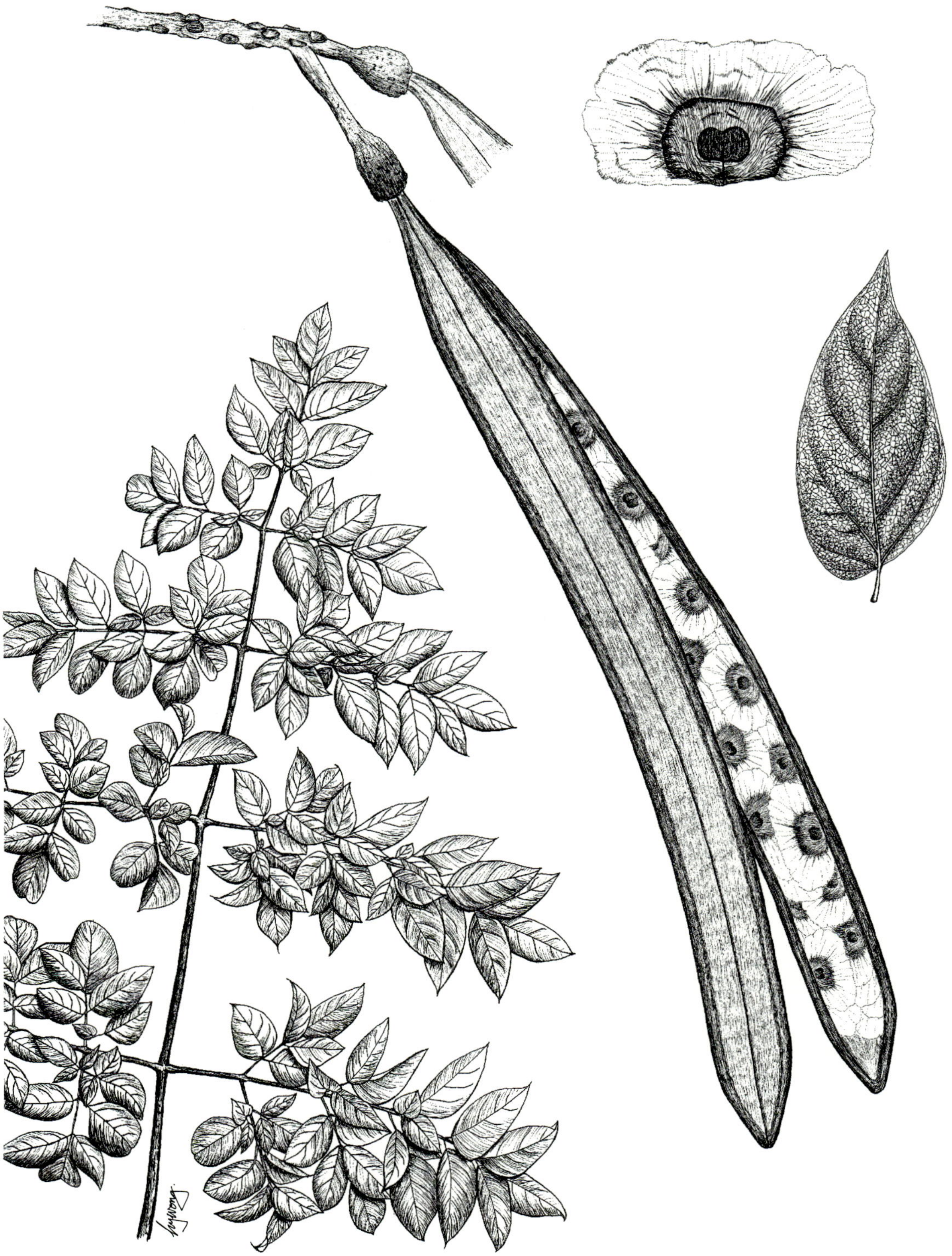

Oroxylum indicum (L.) Benth. ex Kurz 木蝴蝶

Illustration 001 by H. Y. Wong
Authentication by D. T. W. Lau

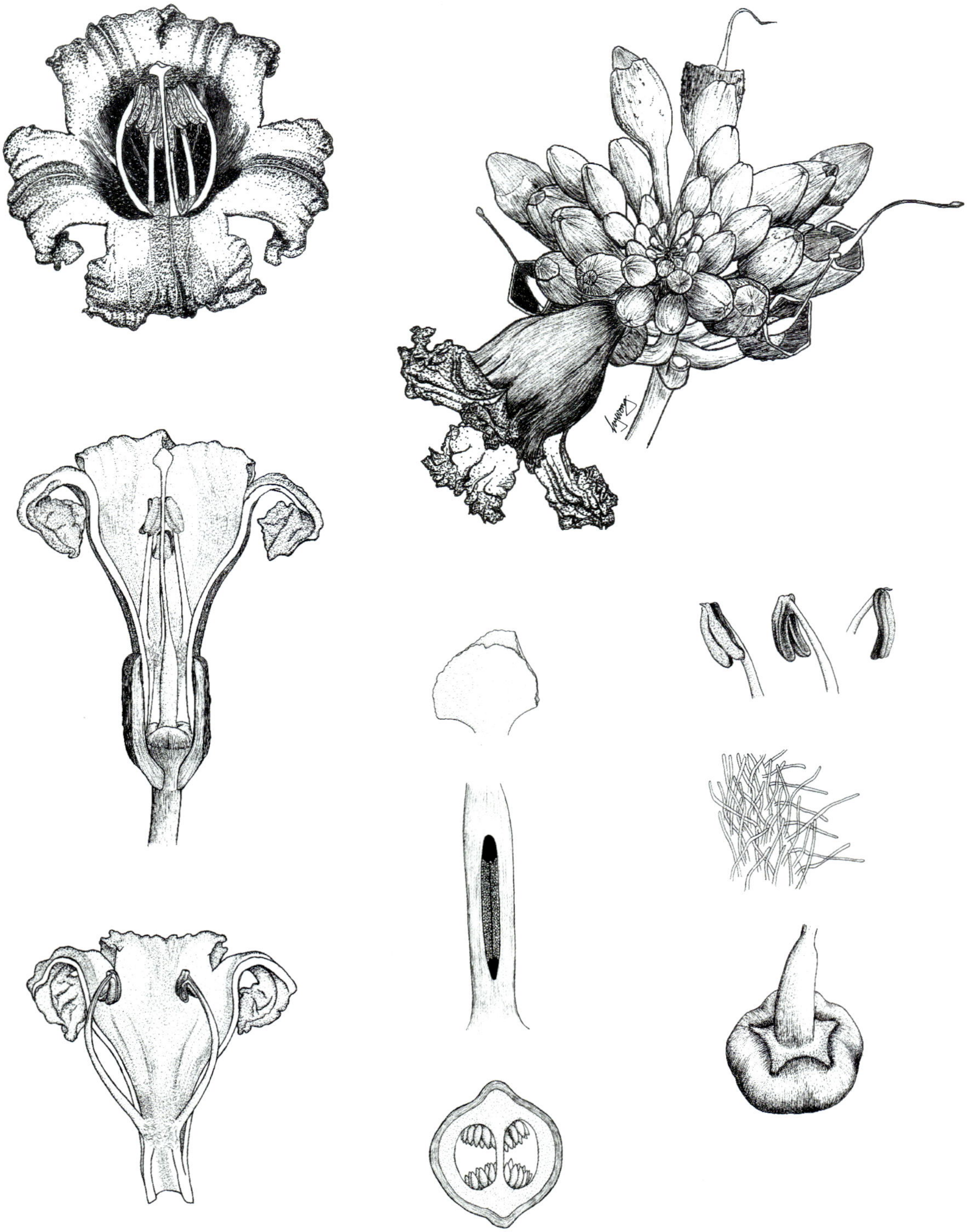

Oroxylum indicum (L.) Benth. ex Kurz 木蝴蝶
Illustration 001 by H. Y. Wong
Authentication by D. T. W. Lau

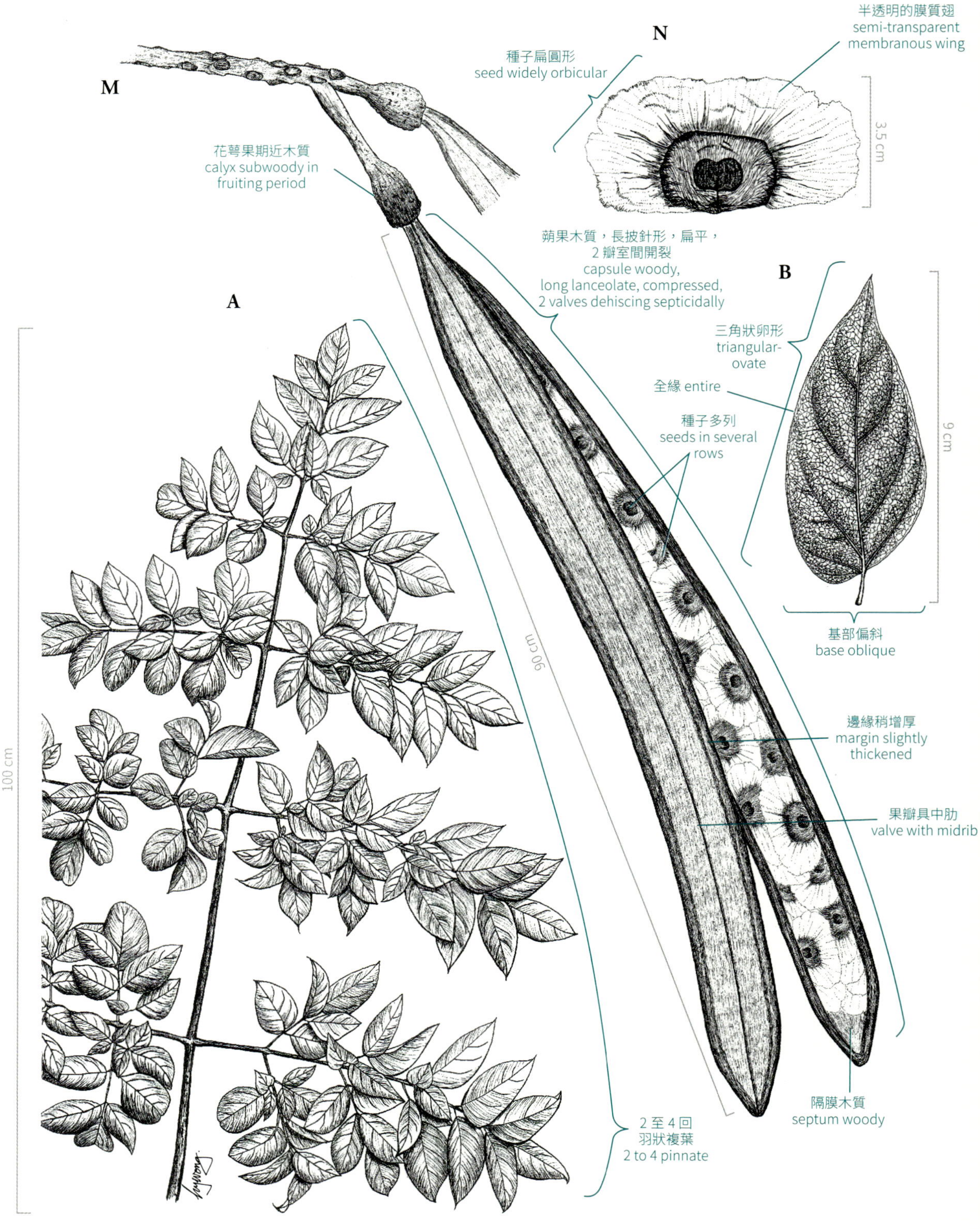

Oroxylum indicum (L.) Benth. ex Kurz 木蝴蝶

Illustration 001 by H. Y. Wong

Authentication by D. T. W. Lau

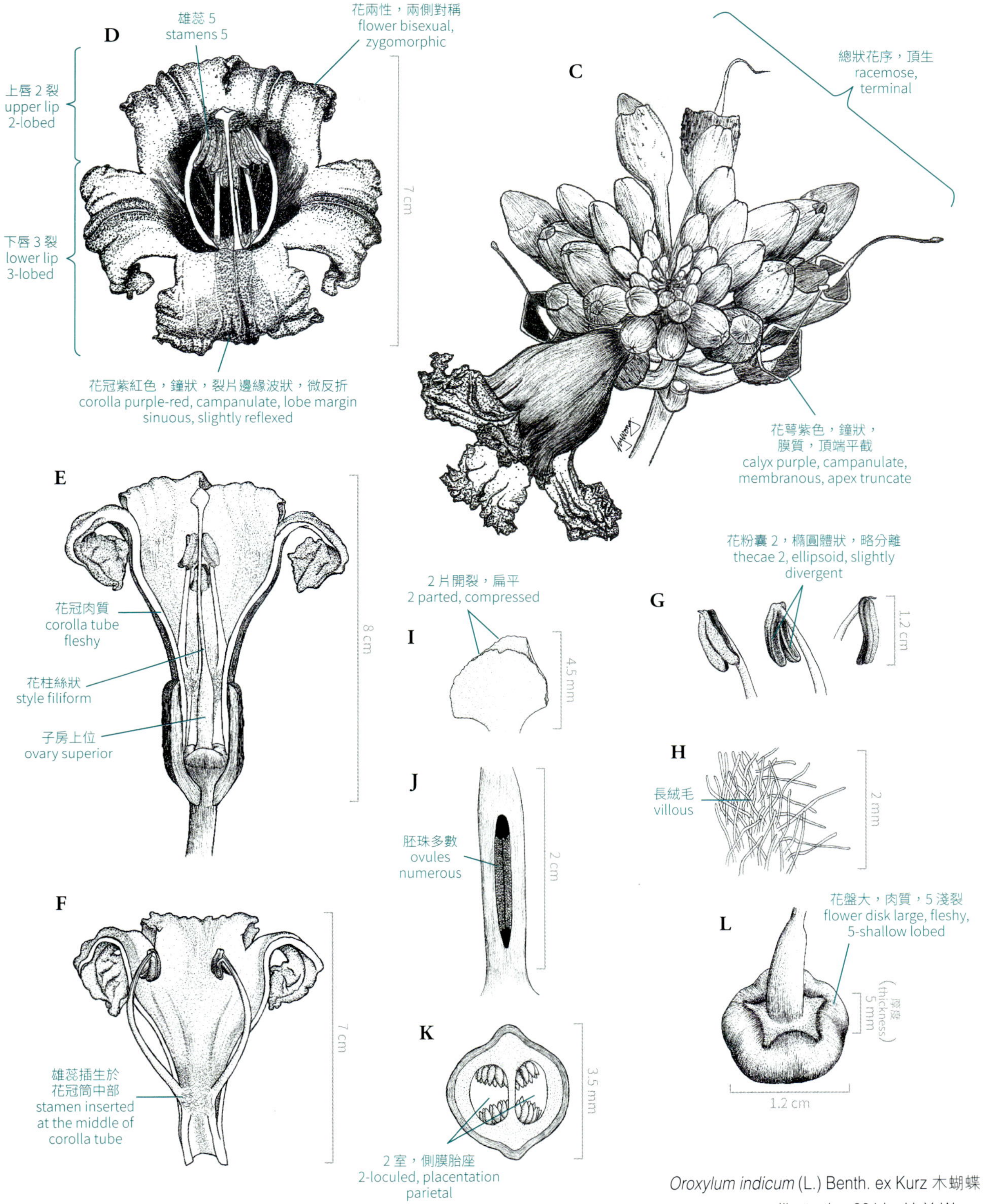

Oroxylum indicum (L.) Benth. ex Kurz 木蝴蝶

Illustration 001 by H. Y. Wong
Authentication by D. T. W. Lau

植物結構 PLANT STRUCTURES:
A. 葉面 Leaf, adaxial **B.** 小葉葉面 Leaflet, adaxial **C.** 花序 Inflorescence **D.** 花，正面觀 Flower, frontal view **E.** 花解剖圖，展示上唇內部 Dissected flower, showing the inner surface of upper lip **F.** 花解剖圖，展示下唇內部 Dissected flower, showing the inner surface of lower lip **G.** 花藥 Anthers **H.** 花絲基部的被毛 Hairs on filament base **I.** 柱頭 Stigma **J.** 子房，縱切面 Ovary, in L. S. **K.** 子房，橫切面 Ovary, in C. S. **L.** 子房與花盤 Ovary and flower disk **M.** 果實 Fruit **N.** 種子 Seed

參考標本 REFERENCE SPECIMENS: H. Y. Wong 006 (CUSLSH3056); 鄧良 3580 (IBSC0563238); Shiu Ying Hu 12390

參考文獻 REFERENCES:

Li, P. Q. (2012). Bignoniaceae. In Fairy Lake Botanical Garden, Shenzhen & Chinese Academy of Sciences (Eds.), *Flora of Shenzhen* (Vol. 3, pp. 423–436). China Forestry Publishing House.
Stevens, P. F. (2021). Lamiales. Angiosperm Phylogeny Website. Retrieved August 25, 2021, from http://www.mobot.org/MOBOT/research/APweb/welcome.html
Tao, D. D., & Yin, W. Q. (1990). Bignoniaceae. In Delectis Florae Reipublicae Popularis Sinicae Agendae Academiae Sinicae Edita (Ed.), *Flora Reipublicae Popularis Sinicae* (Vol. 69, pp. 1–62). Science Press.
Xia, N. H. (2009). Bignoniaceae. In Hong Kong Herbarium, AFCD & South China Botanical Garden, CAS (Eds.), *Flora of Hong Kong* (Vol. 3, pp. 186–191). Agriculture, Fisheries and Conservation Department, HKSAR Government.
Zhang, Z. Y., & Santisuk, T. (1998). Bignoniaceae. In Z. Y. Wu, P. H. Raven & D. Y. Hong (Eds.), *Flora of China* (Vol. 18, pp. 213–225). Science Press & Missouri Botanical Garden Press.

憑證標本
Voucher Specimen

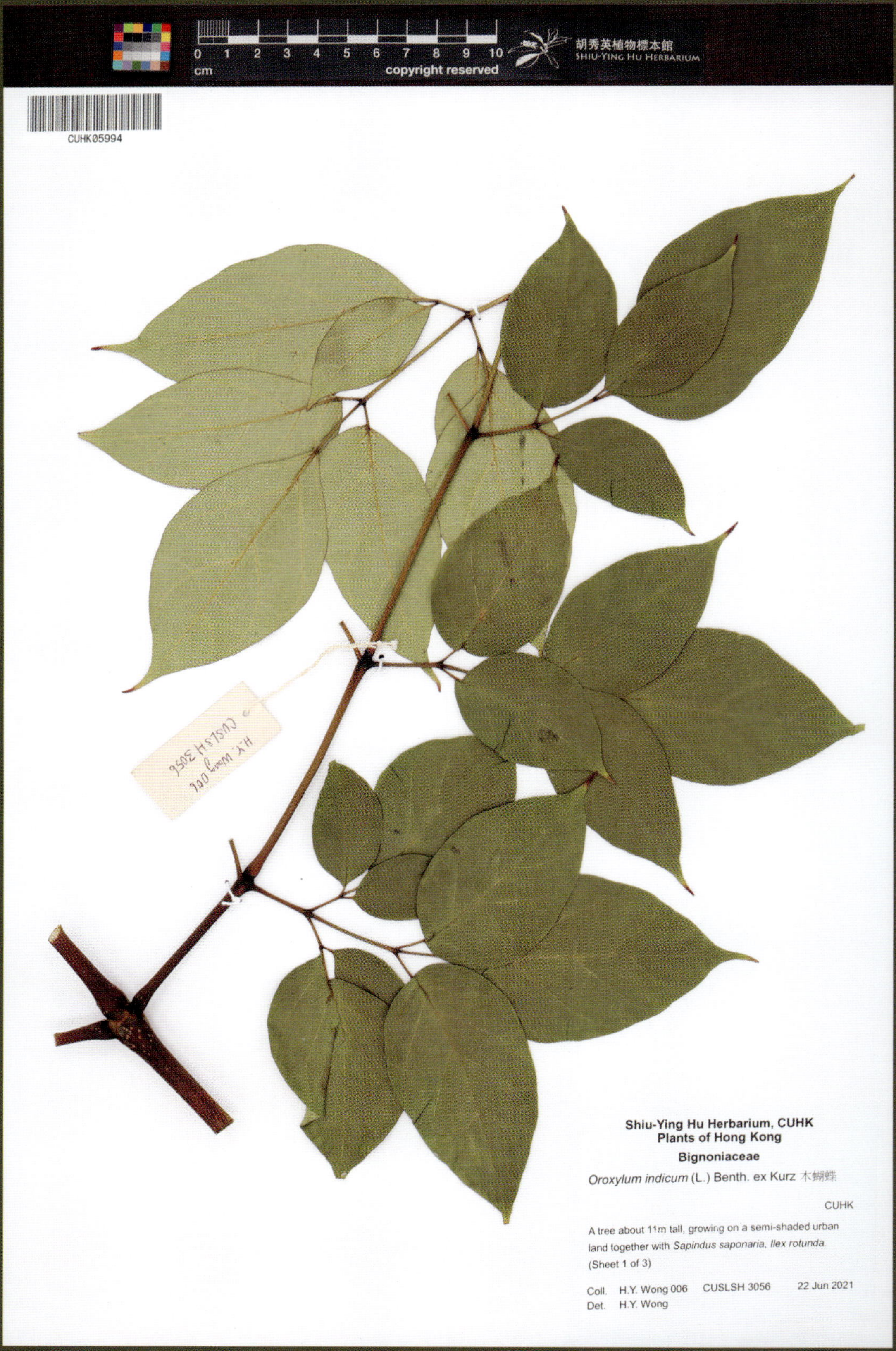

採集者編號 Collector no.
H. Y. Wong 006

採集年份 Year of collection
2021

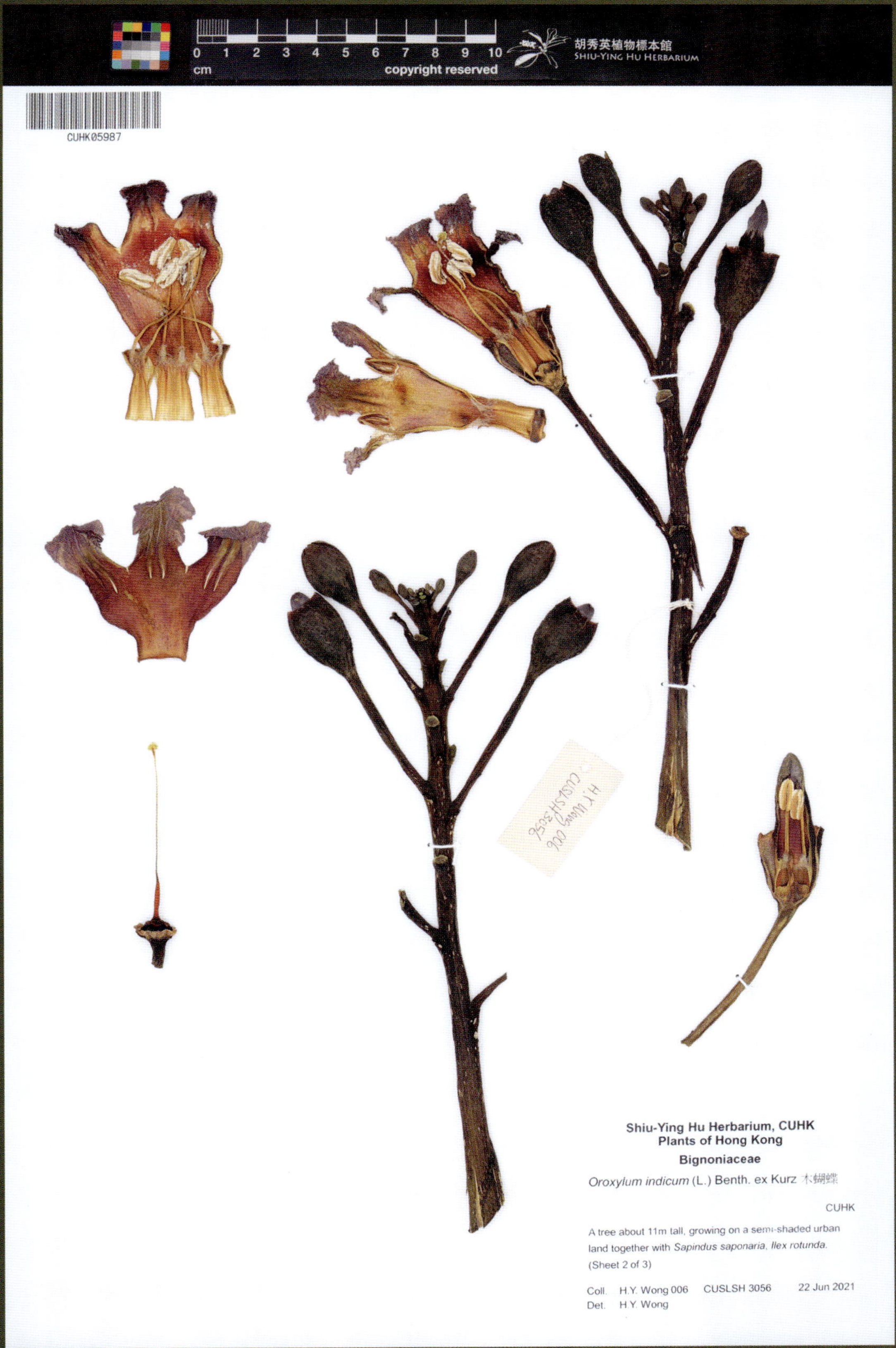

採集者編號 Collector no.
H. Y. Wong 006

採集年份 Year of collection
2021

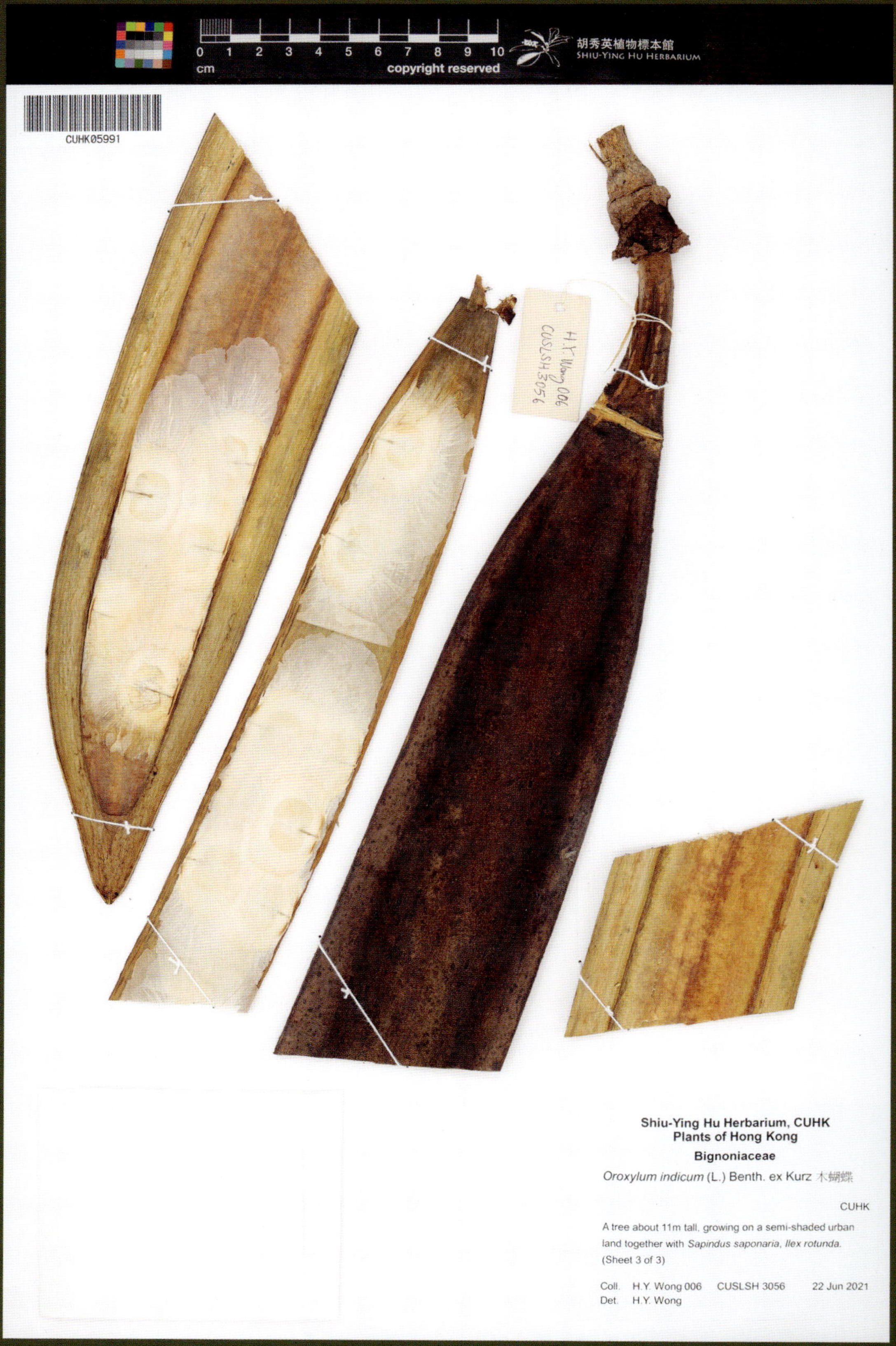

採集者編號 Collector no.
H. Y. Wong 006

採集年份 Year of collection
2021

葉 Leaf

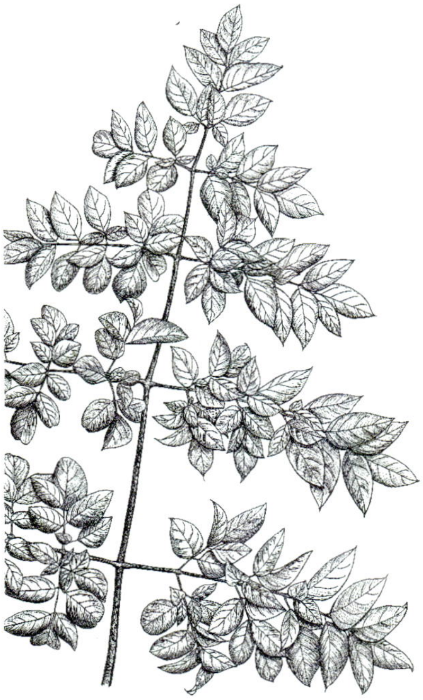

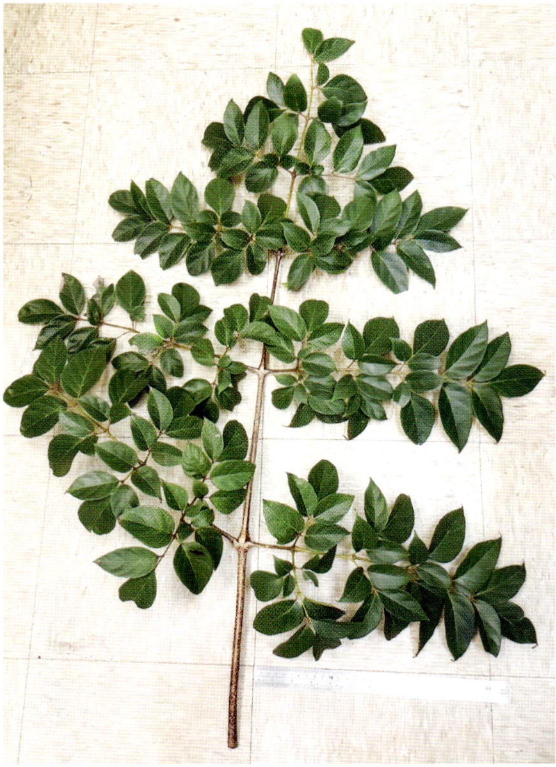

葉面 Leaf, adaxial

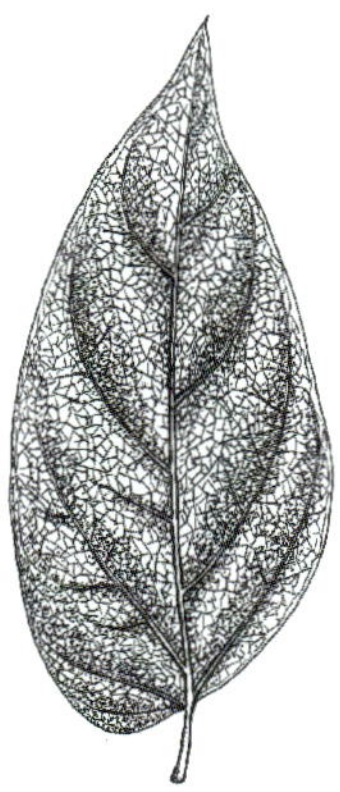

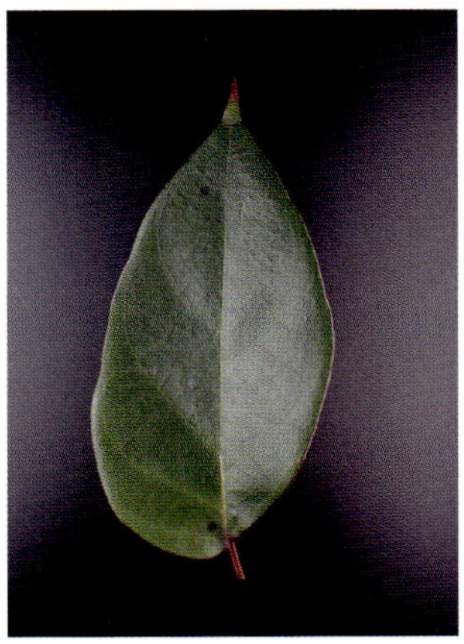

小葉葉面 Leaflet, adaxial

花 Flower

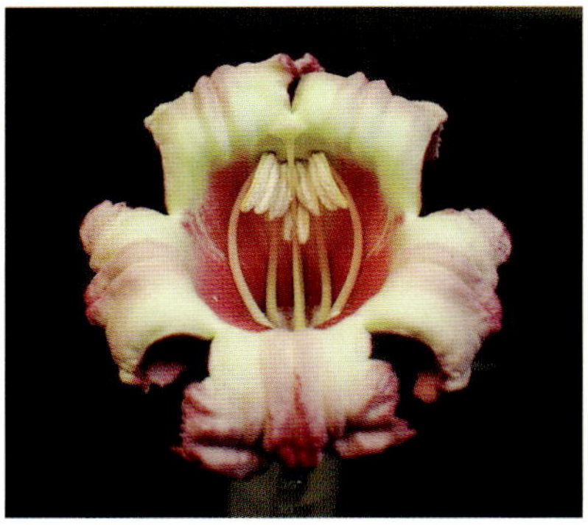
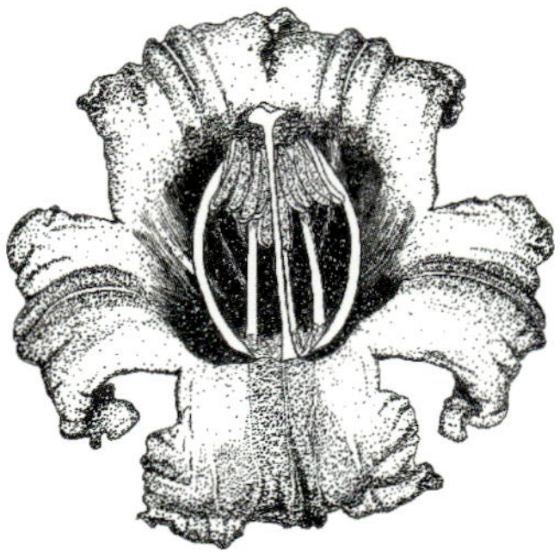

花，正面觀 Flower, frontal view

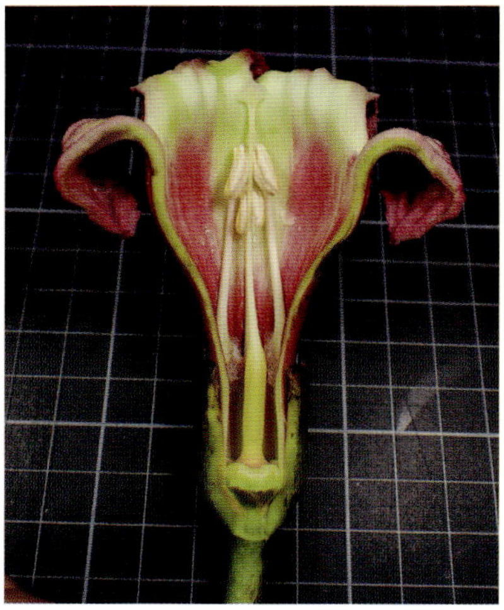
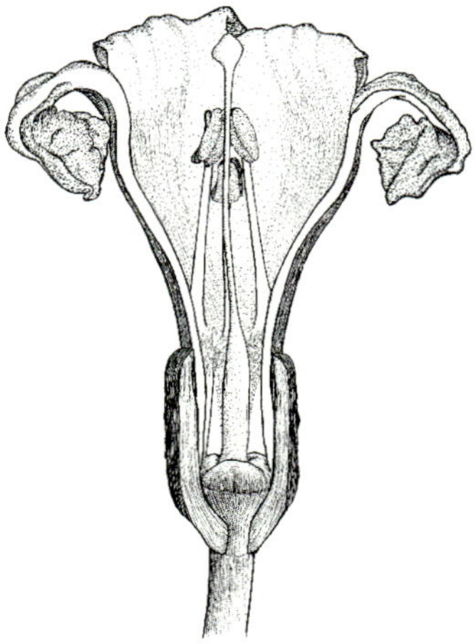

花解剖圖，展示上唇內部
Dissected flower, showing the
inner surface of upper lip

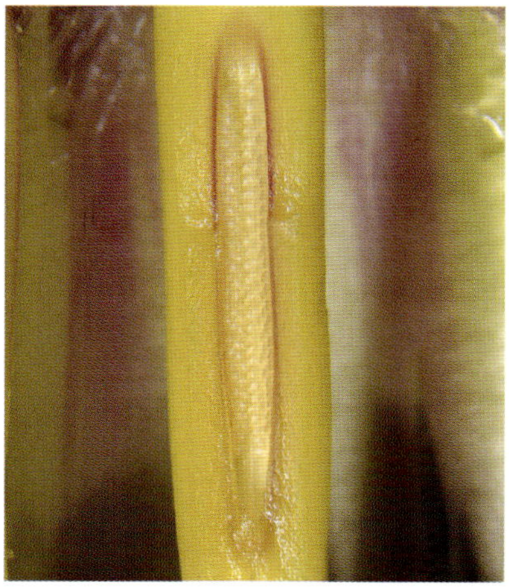
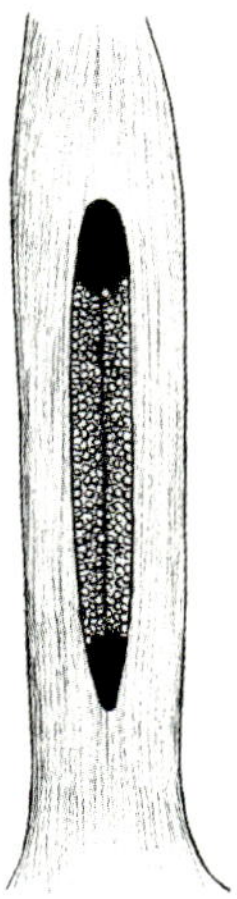

子房，縱切面 Ovary, in L. S.

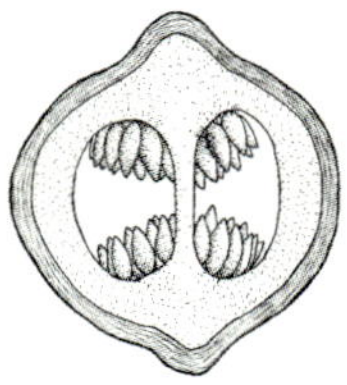
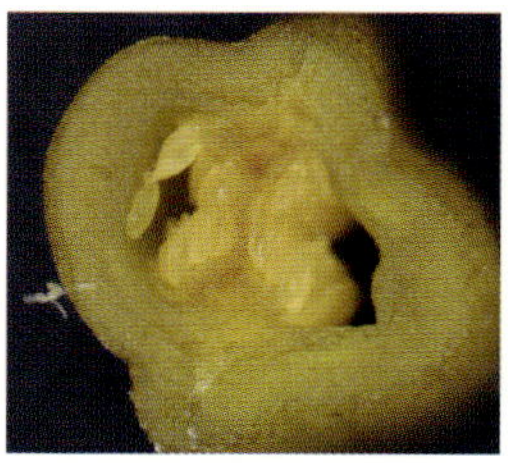

子房，橫切面
Ovary, in C. S.

花 Flower

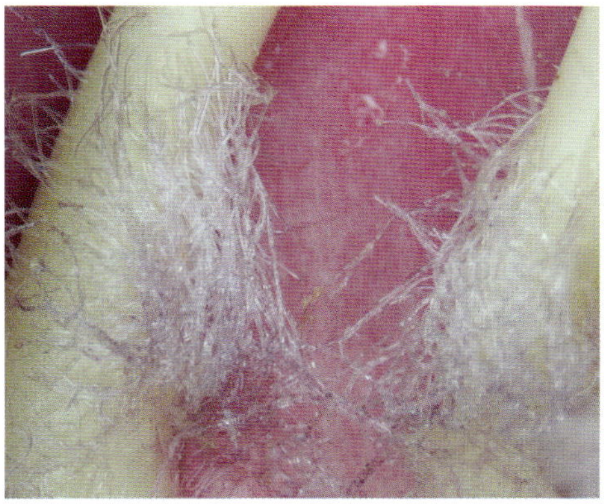

花絲基部的被毛 Hairs on filament base

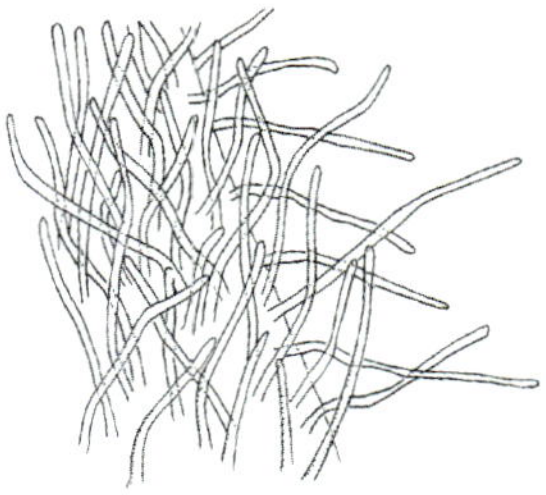

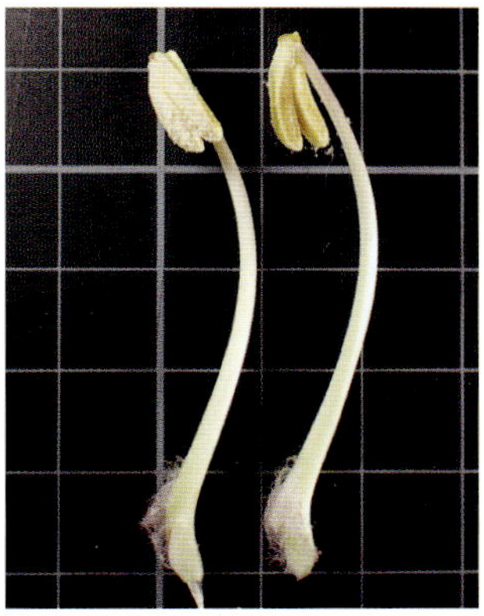

雄蕊 Stamens

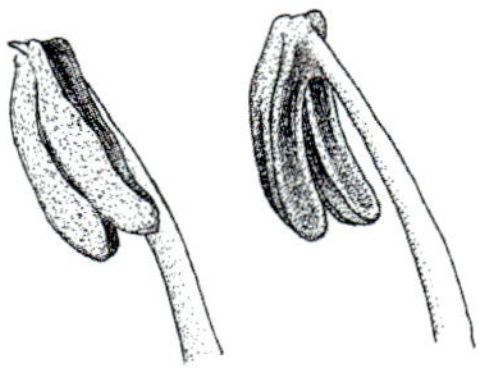

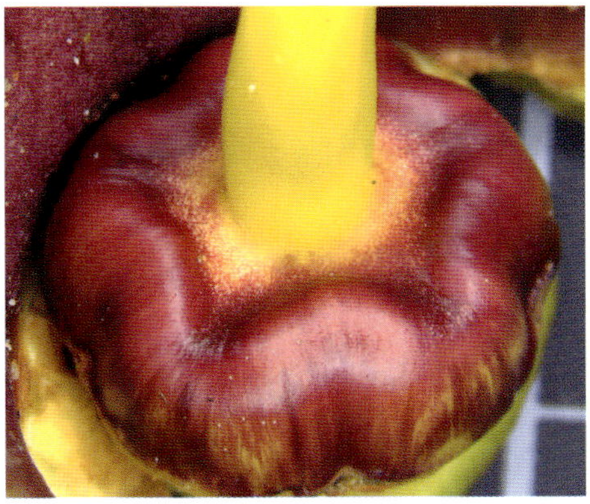

花盤 Flower disk

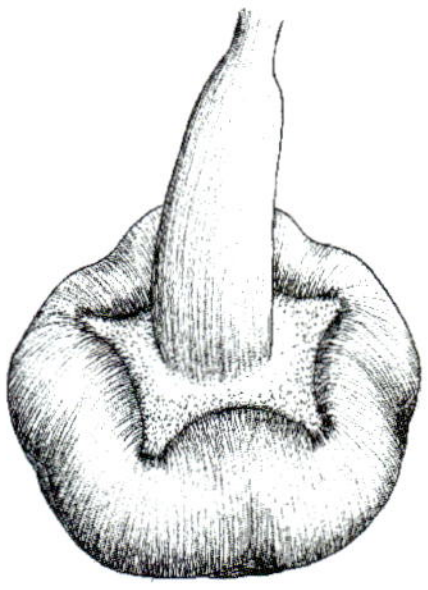

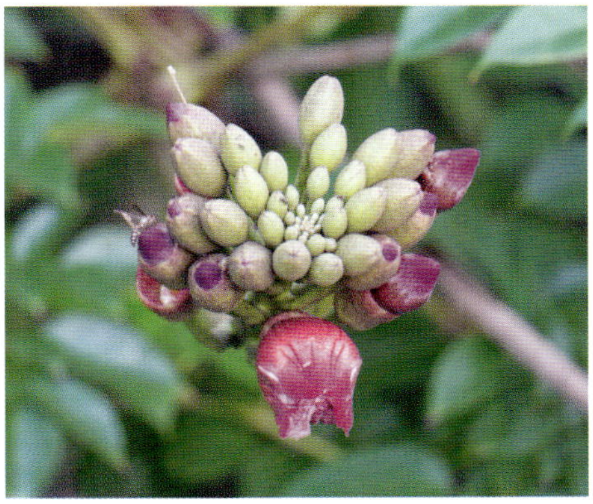

花序 Inflorescence

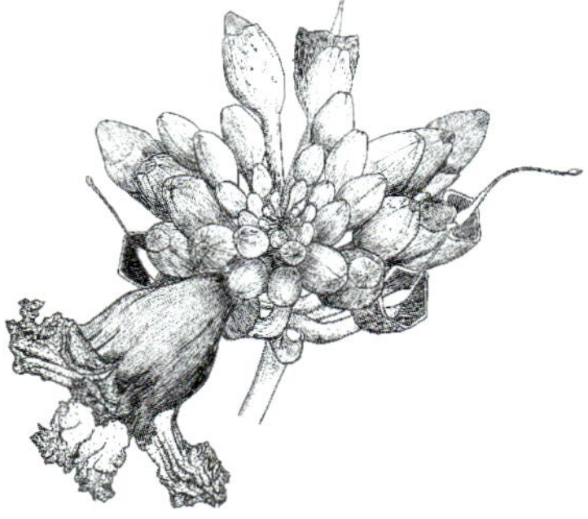

果 Fruit

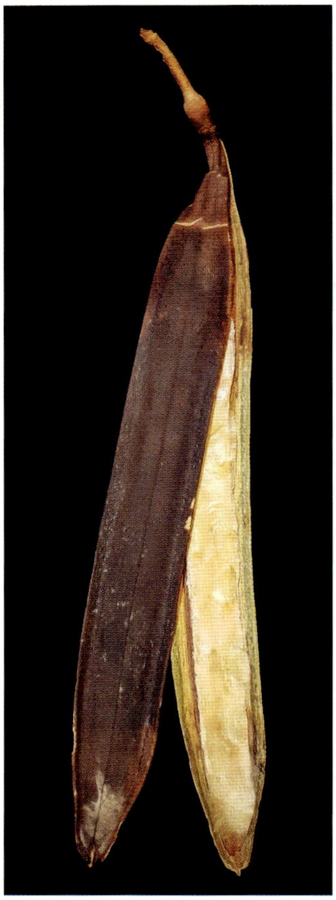

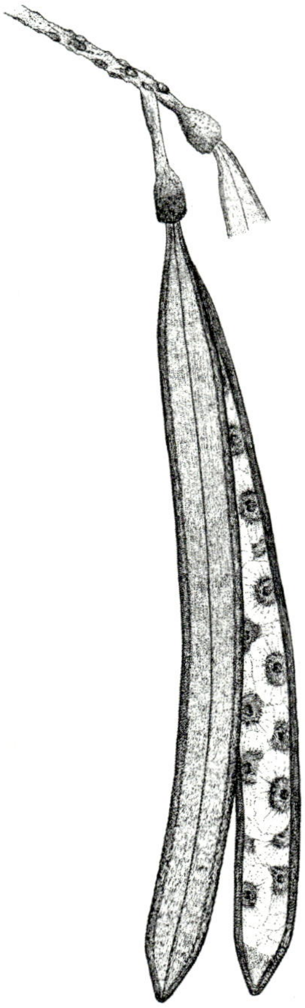

蒴果 Capsule

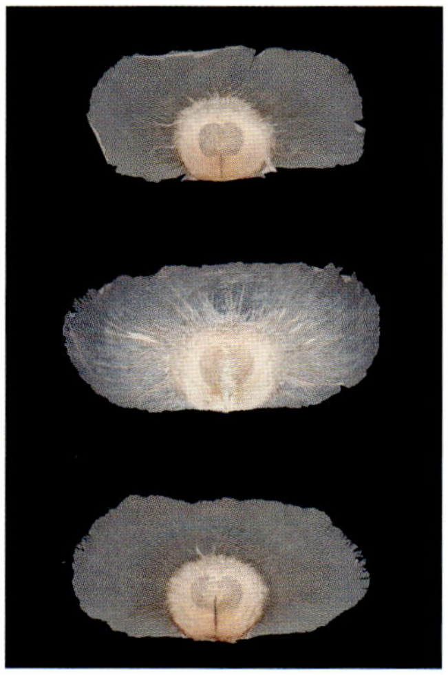

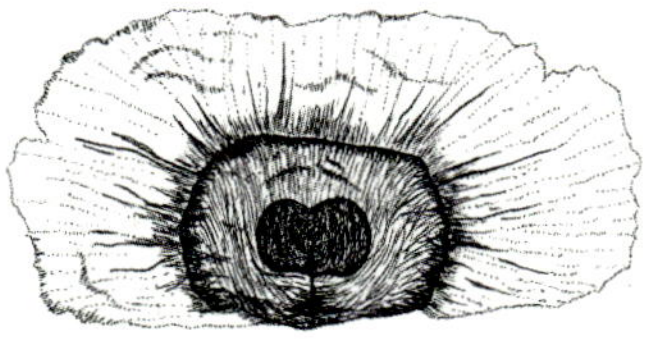

種子 Seeds

Ilex rotunda var. *microcarpa* (Lindl. ex Paxton) S. Y. Hu
小果鐵冬青

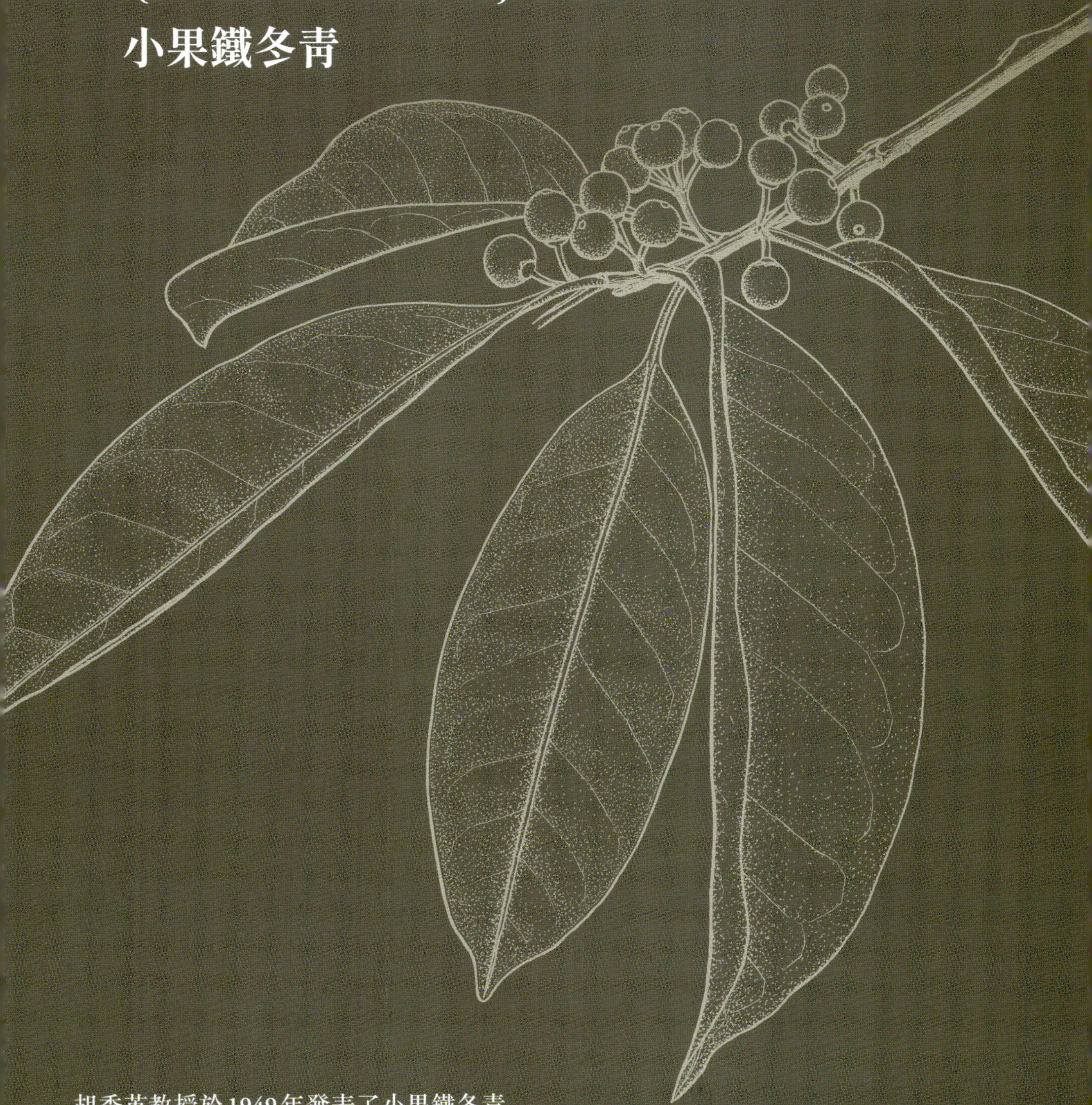

胡秀英教授於1949年發表了小果鐵冬青為鐵冬青（*Ilex rotunda* Thunb.）的變種。

胡秀英教授採集的標本
Specimen Collected by Professor Shiu-Ying Hu

採集者編號
Collector no.
Shiu Ying Hu & Kung Hing Yung 315

採集年份
Year of collection
2001

CUHK00429

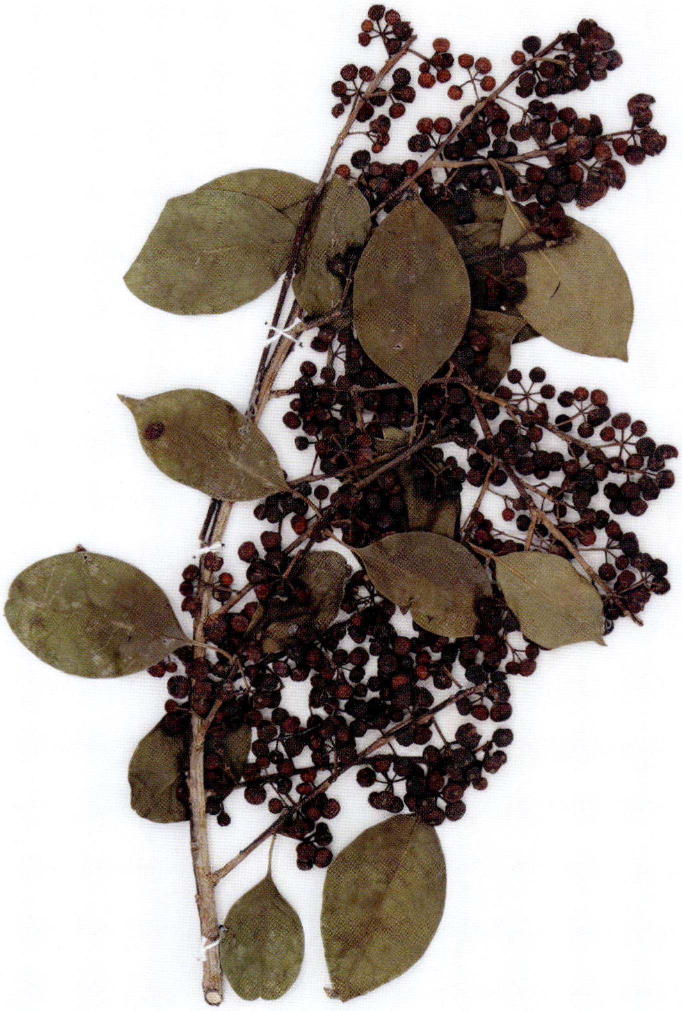

Herbarium, The Chinese University of Hong Kong
Plants of Hong Kong

Ilex rotunda (Thunb.) var. microcarpa (Lindl. ex Paxt.) S.Y. Hu

Garden of Gloria Barretto, Tai-po, N.T.

Trees 15 m high, trunk 20 cm in diameter, bark smooth, light green, used in Chinese medicine locally, fruits scarlet, pomiform, 3 mm long, 3-5 mm across, fruiting heavily, a beautiful specimen in any gardens.

Coll. Shiu Ying Hu & Kung Hing Yung   315   Jan 13, 2001
Det. Shiu Ying Hu

小果鐵冬青

學名	*Ilex rotunda* var. *microcarpa* (Lindl. ex Paxton) S. Y. Hu
中文名	小果鐵冬青
科名	冬青科
本地分布狀態	原生植物
繪圖編號	H. Y. Wong 007

花期	花色	果期	果色
3 至 5 月	白色	12 至翌年 2 月	紅色

小果鐵冬青是常綠喬木，長於次生林。胡秀英教授於 1949 年發表了小果鐵冬青為鐵冬青（*Ilex rotunda* Thunb.）的變種。兩者的分別在於，前者的果實體積較後者細小，並且花梗被微柔毛。

鐵冬青及其變種，除了能為雀鳥提供冬季食物，亦具觀賞價值，因此可作綠化用途。

《中國植物誌》（英文版）的採用名稱

Ilex rotunda Thunberg（科名：冬青科）

《香港植物誌》的採用名稱

Ilex rotunda Thunb. var. *microcarpa* (Lindl. ex Paxton) S. Y. Hu（科名：冬青科）

參考文獻 References

1. Chen, S. K., Ma, H. Y., Feng, Y. X., Barriera, G., & Loizeau, P. A. (2008). Aquifoliaceae. In Z. Y. Wu, P. H. Raven & D. Y. Hong (Eds.), *Flora of China* (Vol. 11, pp. 359–438). Science Press & Missouri Botanical Garden Press.
2. Development Bureau. (2021). *Small fruited Holly, Chinese Holly*. Community Outreach-Tree Labels with QR Codes. Retrieved Dec 3, 2021, from https://www.greening.gov.hk/tc/communityoutreach/qrcode tree labels/index_id_78.html
3. Hong Kong Herbarium. (2021). *HK Herbarium-Ilex rotunda* var. *microcarpa*. HK Plant Database. Retrieved August 10, 2021, from https://www.herbarium.gov.hk/en/hk-plant-database/plant-detail/index.html?pType=species&oID=11442

Ilex rotunda var. *microcarpa* (Lindl. ex Paxton) S. Y. Hu

Scientific name	*Ilex rotunda* var. *microcarpa* (Lindl. ex Paxton) S. Y. Hu
Chinese name	小果鐵冬青
Family	Aquifoliaceae Bercht. & J. Presl
Local distribution status	Native species
Illustration number	H. Y. Wong 007

FLOWERING PERIOD	COLOUR OF FLOWER	FRUITING PERIOD	COLOUR OF FRUIT
March to May	White	December to following February	Red

Ilex rotunda var. *microcarpa* (Lindl. ex Paxton) S. Y. Hu is an evergreen tree and grows in secondary forests. In 1949, Professor Shiu-Ying Hu published *Ilex rotunda* var. *microcarpa* (Lindl. ex Paxton) S. Y. Hu as a variety of *Ilex rotunda* Thunb. The differences between them are that the former has smaller fruits than the later, and its pedicels are puberulent.

Ilex rotunda Thunb. and its variety not only provide food for birds in winter but also have aesthetic value, therefore they are used for greening.

Name in *Flora of China*
Ilex rotunda Thunberg (Family: Aquifoliaceae)

Name in *Flora of Hong Kong*
Ilex rotunda Thunb. var. *microcarpa* (Lindl. ex Paxton) S. Y. Hu
(Family: Aquifoliaceae)

4. Hu, S. Y. (1949). The genus *Ilex* in China. *Journal of the Arnold Arboretum 30*, 233–344.
5. Missouri Botanical Garden. (2021). *Name-Ilex rotunda* var. *microcarpa*. Tropicos. Retrieved July 3, 2021, from https://www.tropicos.org/name/50000942
6. Zhang, S. H., Ling, L. T., Wang S. R., & Zhang, Q. L. (2016). Elite individual standard and selection of *Ilex rotunda* in landscape. *Fujian Linye 6* (6), 45–48. https://doi.org/10.3969/j.issn.1003-4382.2016.06.024

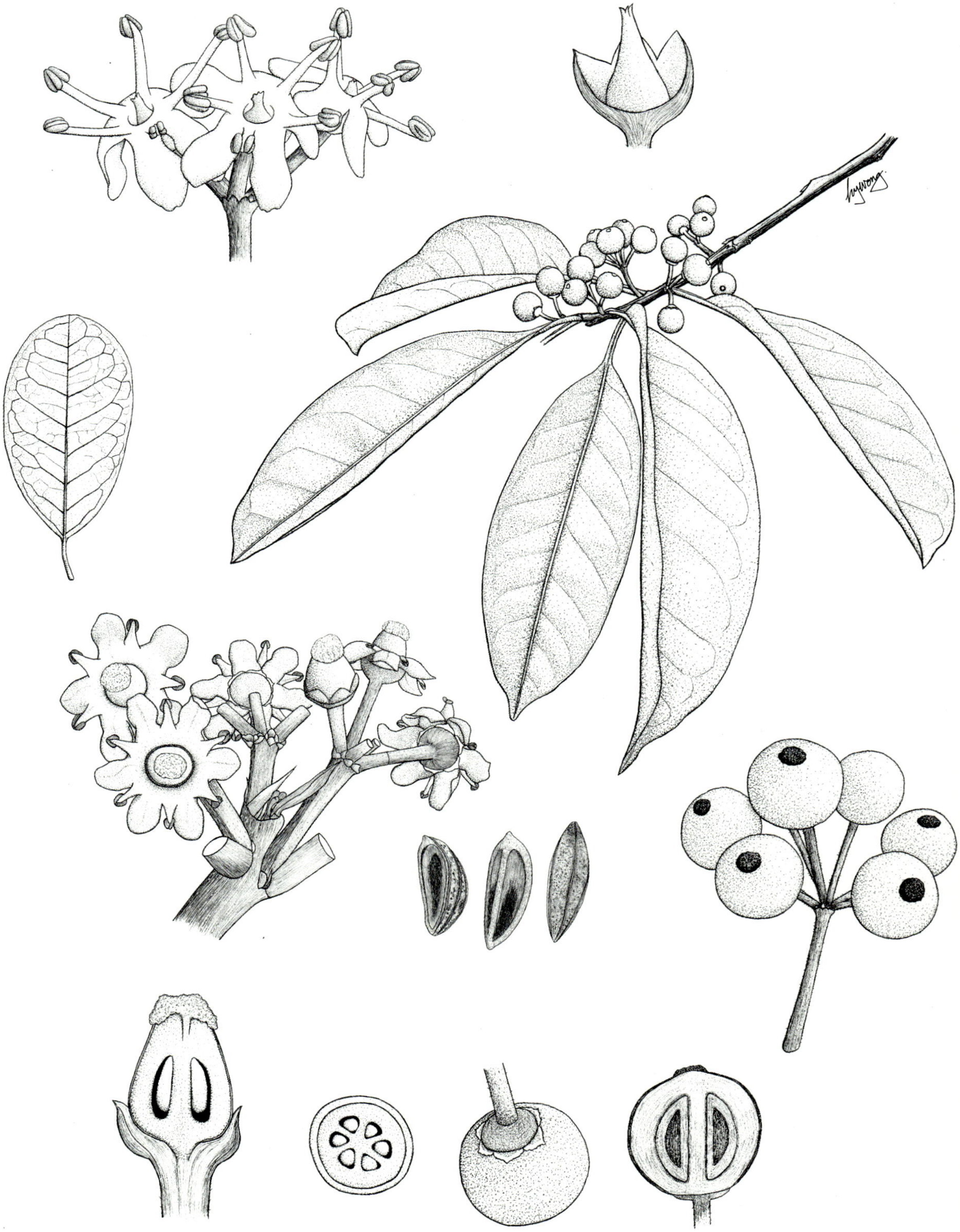

Ilex rotunda var. *microcarpa*
(Lindl. ex Paxton) S. Y. Hu 小果鐵冬青
Illustration 007 by H. Y. Wong
Authentication by D. T. W. Lau

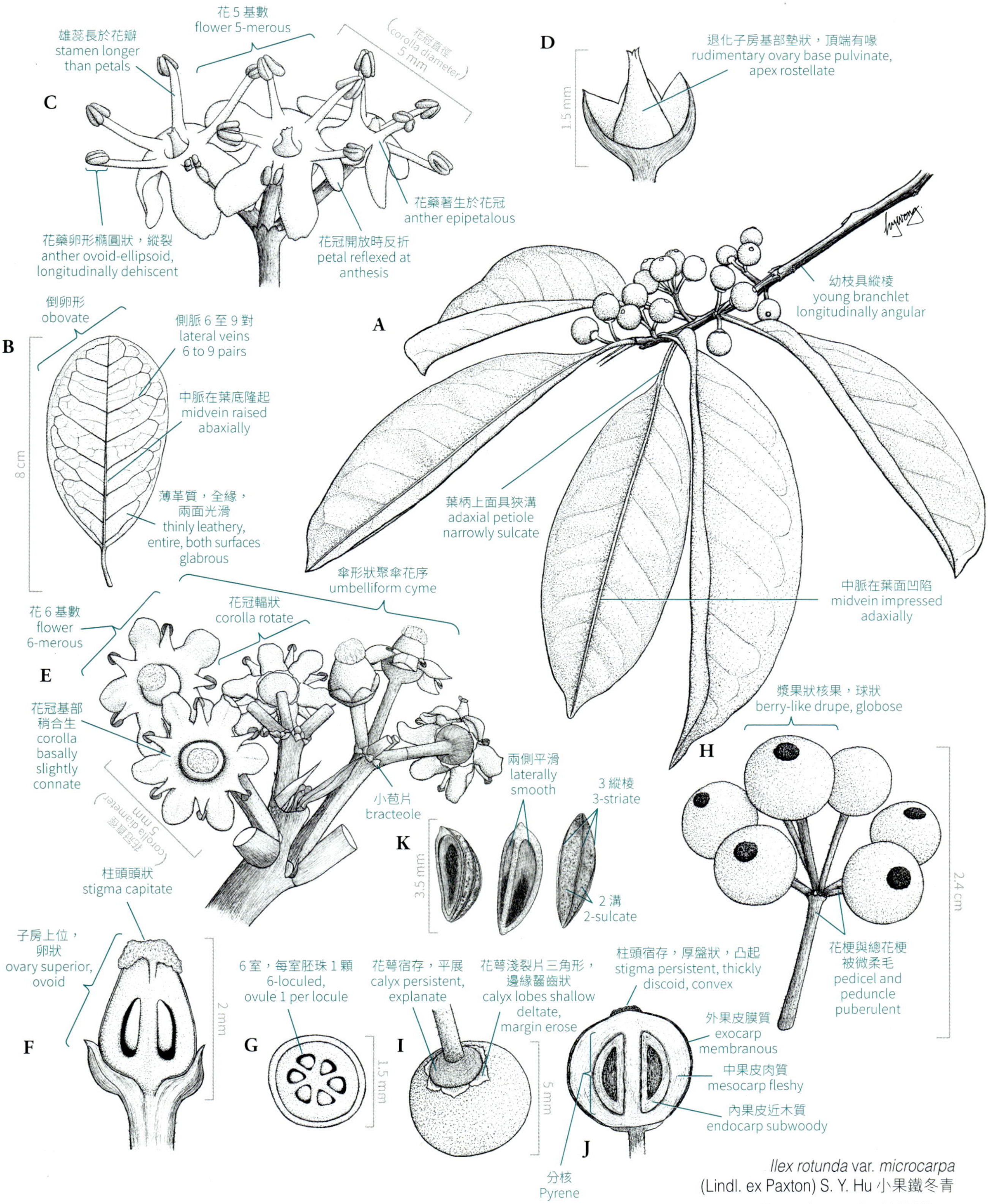

Ilex rotunda var. *microcarpa*
(Lindl. ex Paxton) S. Y. Hu 小果鐵冬青

Illustration 007 by H. Y. Wong
Authentication by D. T. W. Lau

植物結構 PLANT STRUCTURES:
A. 果枝 Fruiting branch **B.** 葉底 Leaf, abaxial **C.** 雄花花序 Male inflorescence **D.** 雄花解剖圖，展示退化子房 Dissected male flower, showing rudimentary ovary **E.** 雌花花序 Female inflorescence **F.** 子房，縱切面 Ovary, in L. S. **G.** 子房，橫切面 Ovary, in C. S. **H.** 果序 Infructescence **I.** 果實 Fruit **J.** 果實，縱切面 Fruit, in L. S. **K.** 分核 Pyrenes

參考標本 REFERENCE SPECIMENS: H.Y. Wong 001 (CUSLSH3000); H.Y. Wong 002 (CUSLSH3001); H. L. Wang 096 (CUSLSH3314); Shiu Ying Hu & Kung Hing Yung 315

參考文獻 REFERENCES:
Chen, S. K., & Feng, Y. X. (1999). Aquifoliaceae. In Delectis Florae Reipublicae Popularis Sinicae Agendae Academiae Sinicae Edita (Ed.), *Flora Reipublicae Popularis Sinicae* (Vol. 45(2), pp. 1–266). Science Press.
Chen, S. K., Ma, H. Y., Feng, Y. X., Barriera, G., & Loizeau, P. A. (2008). Aquifoliaceae. In Z. Y. Wu, P. H. Raven & D. Y. Hong (Eds.), *Flora of China* (Vol. 11, pp. 359–438). Science Press & Missouri Botanical Garden Press.
Hu, S. Y. (2008). Aquifoliaceae. In Hong Kong Herbarium, AFCD & South China Botanical Garden, CAS (Eds.), *Flora of Hong Kong* (Vol. 2, pp. 182–190). Agriculture, Fisheries and Conservation Department, HKSAR Government.
Stevens, P. F. (2021). Aquifoliales. Angiosperm Phylogeny Website. Retrieved July 4, 2021, from http://www.mobot.org/MOBOT/research/APweb/welcome.html
Xing, F. W., & Wang, F. G. (2010). Aquifoliaceae. In Fairy Lake Botanical Garden, Shenzhen & Chinese Academy of Sciences (Eds.), *Flora of Shenzhen* (Vol. 2, pp. 487–497). China Forestry Publishing House.
Ye, H. G. (2003). Aquifoliaceae. In South China Botanical Garden, the Chinese Academy of Sciences (Ed.), *Flora of Guangdong* (Vol. 5, pp. 381–411). Guangdong Science and Technology Press.

憑證標本
Voucher Specimen

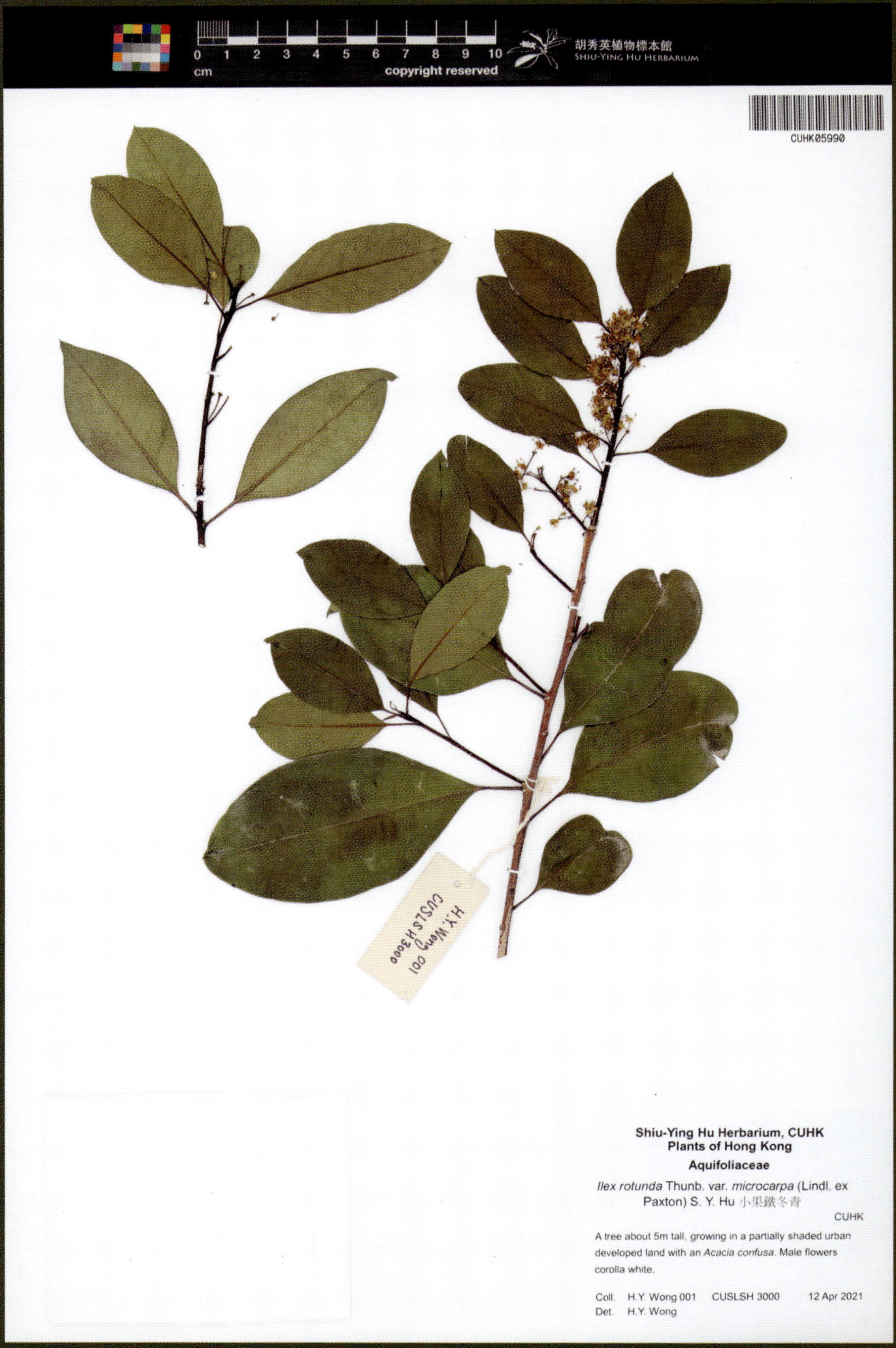

採集者編號 Collector no.
H. Y. Wong 001

採集年份 Year of collection
2021

憑證標本
Voucher Specimen

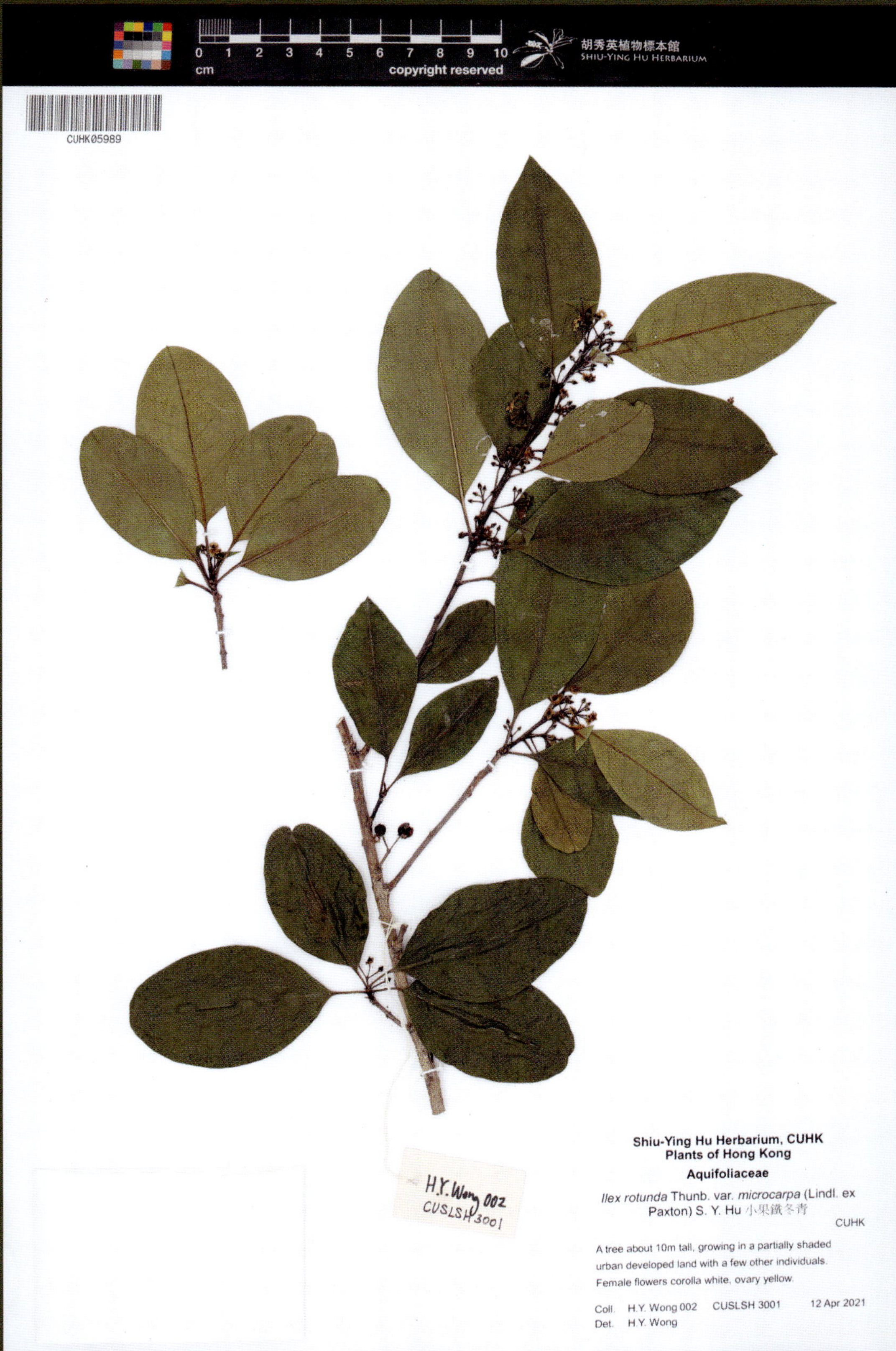

採集者編號 Collector no.
H. Y. Wong 002

採集年份 Year of collection
2021

憑證標本
Voucher Specimen

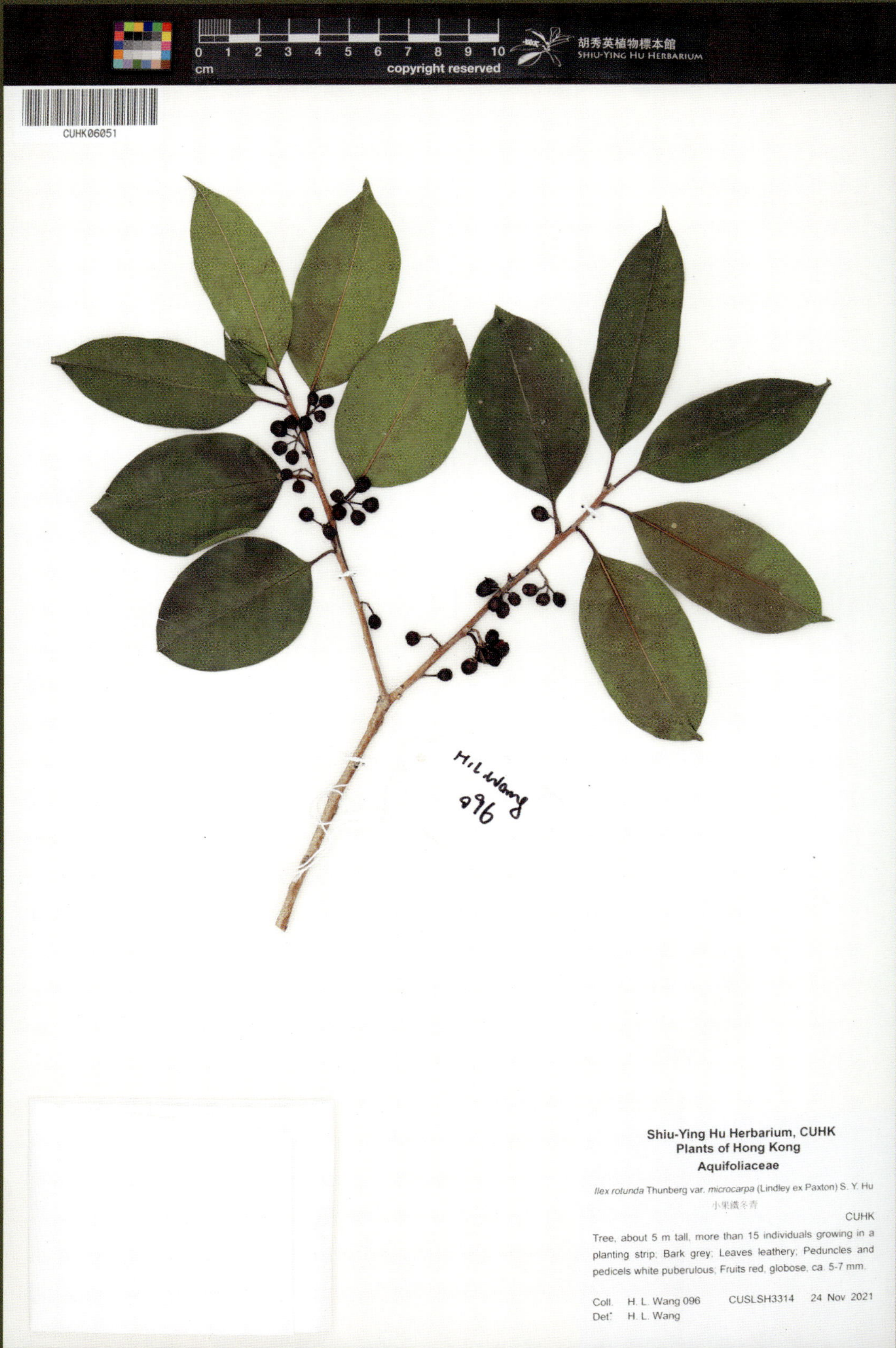

採集者編號 Collector no.
H. L. Wang 096

採集年份 Year of collection
2021

外觀 Overview

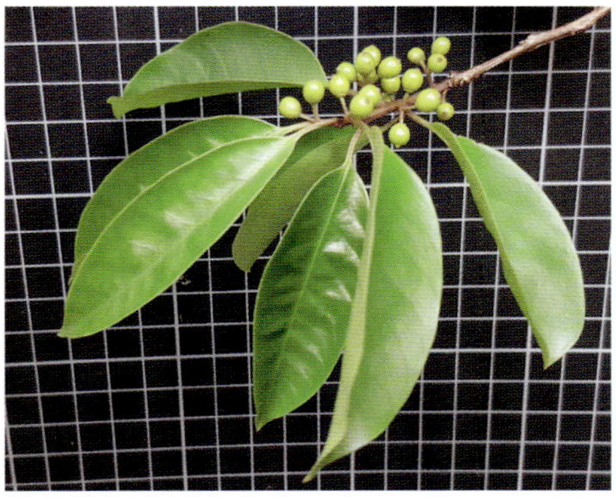

未成熟果枝
Fruiting branch with immature drupes

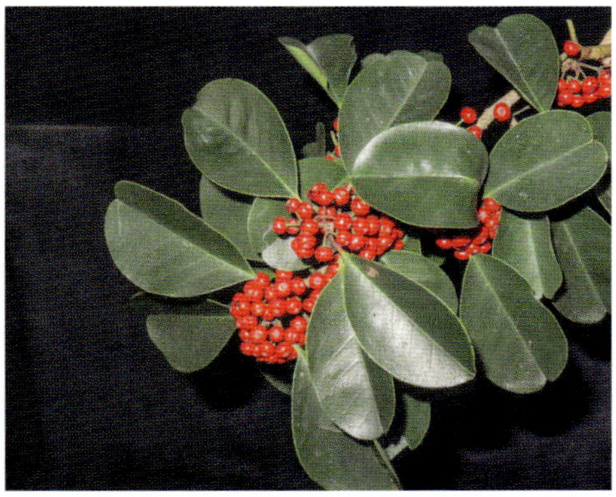

成熟果枝
Fruiting branch with mature drupes

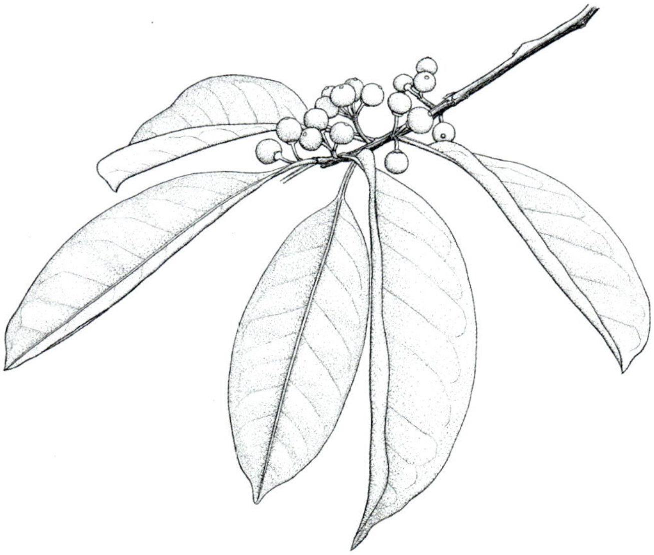

葉 Leaf

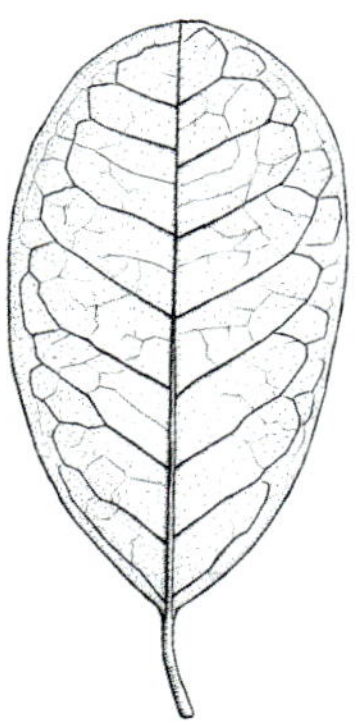

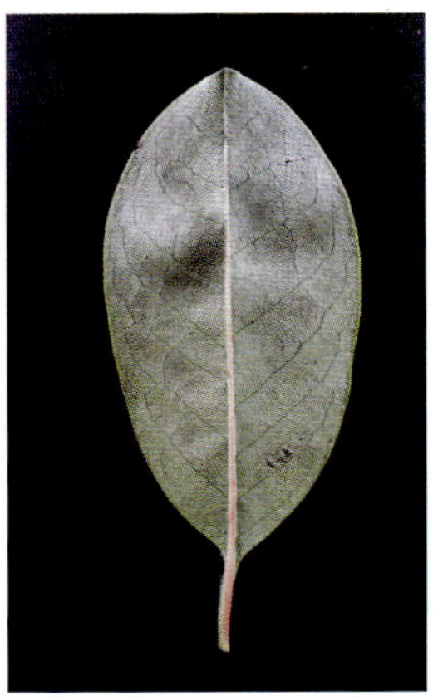

葉底，倒卵形的葉形
Leaf, abaxial, obovate

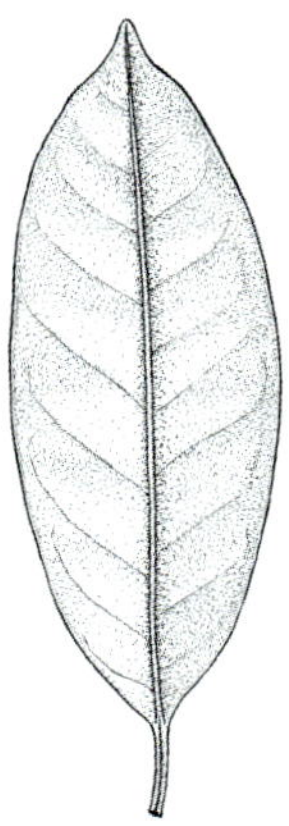

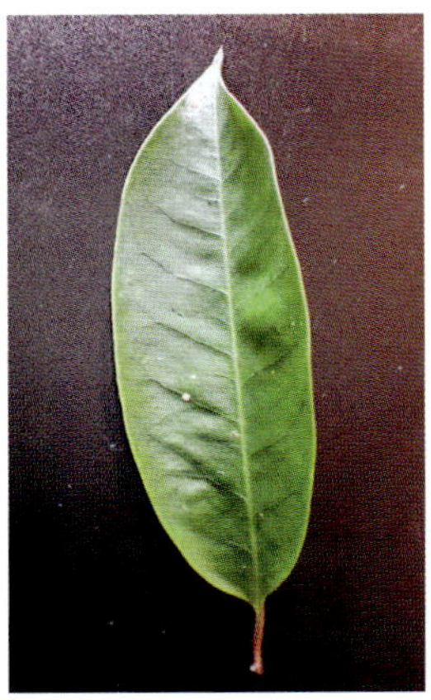

葉面，橢圓形的葉形
Leaf, adaxial, elliptic

花 Flower

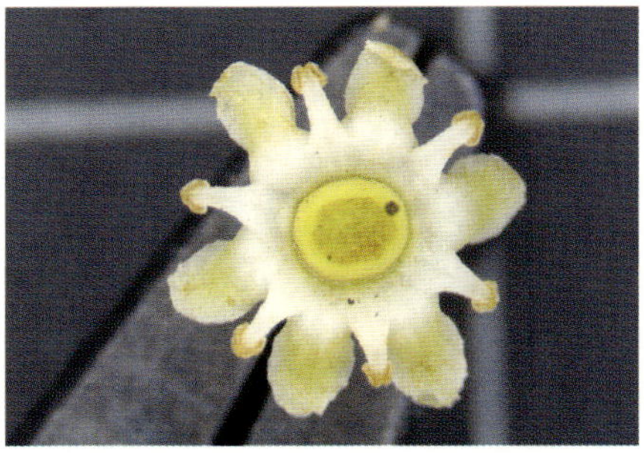
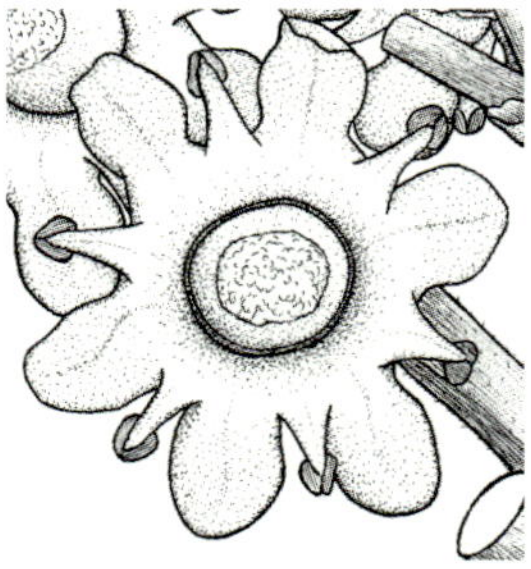

雌花，正面觀
Female flower, frontal view

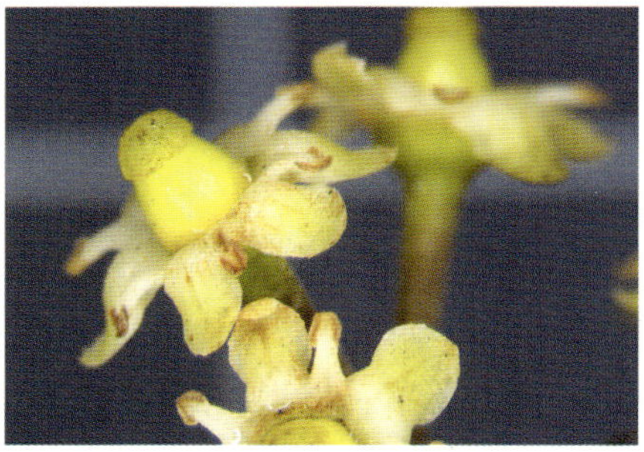
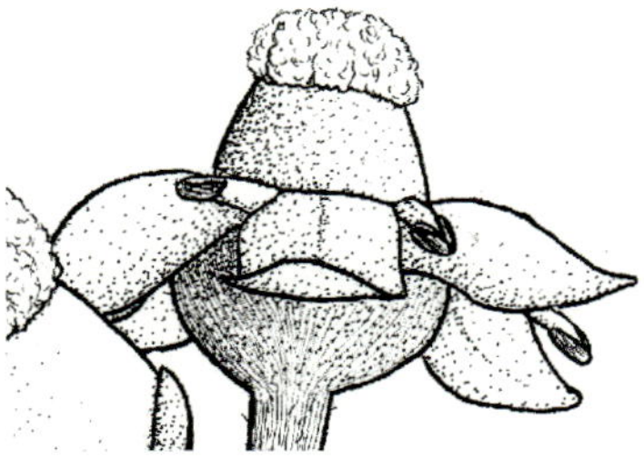

雌花，側面觀
Female flower, lateral view

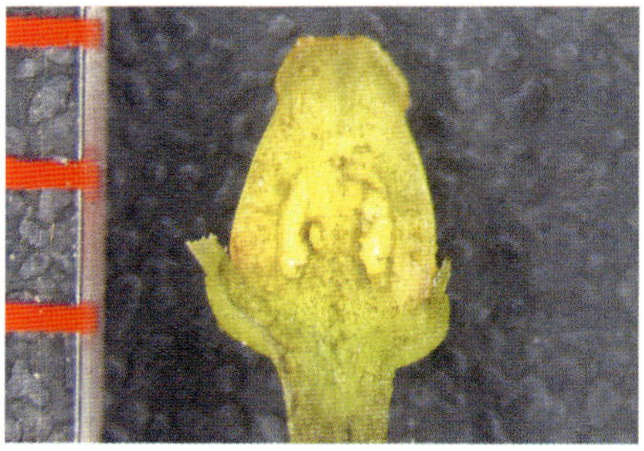
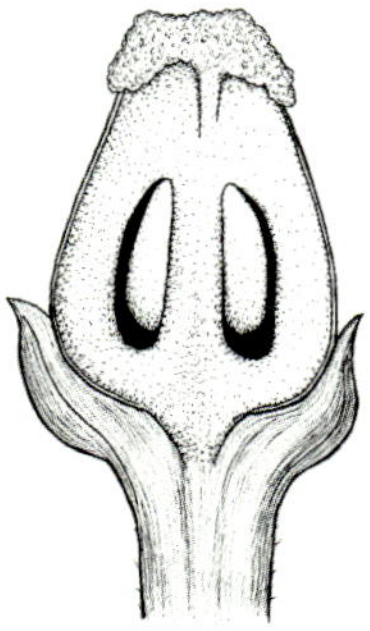

子房，縱切面
Ovary, in L. S.

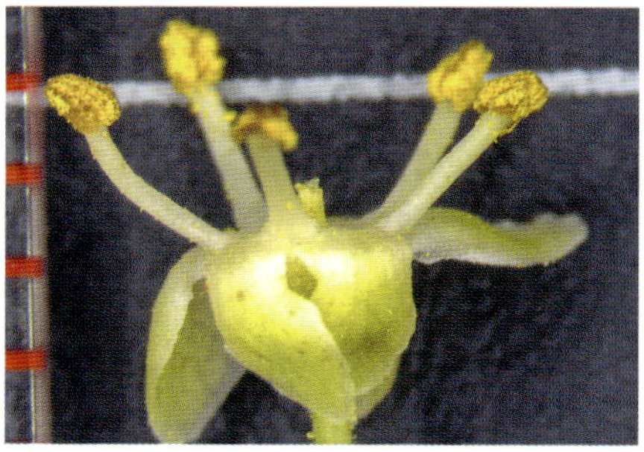
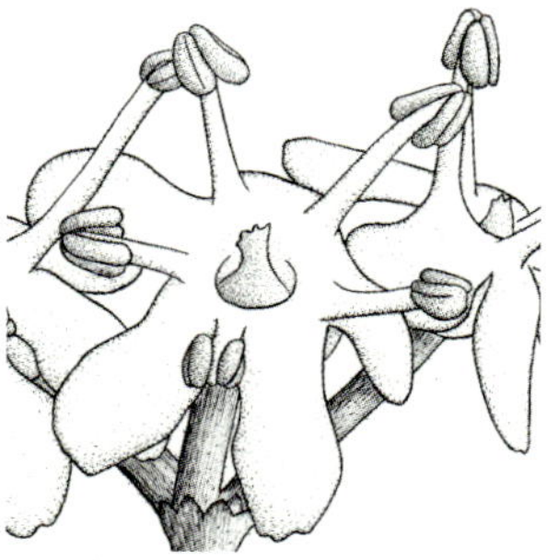

雄花，側面觀
Male flower, lateral view

果 Fruit

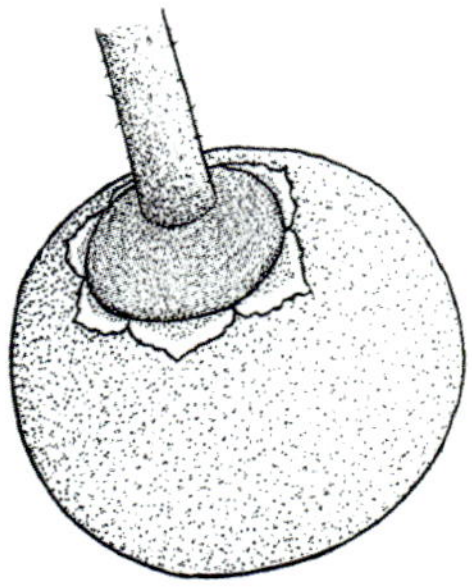
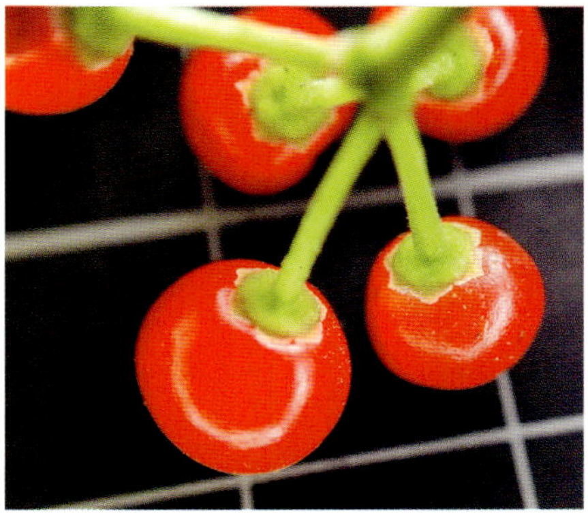

漿果狀核果 Berry-like drupes

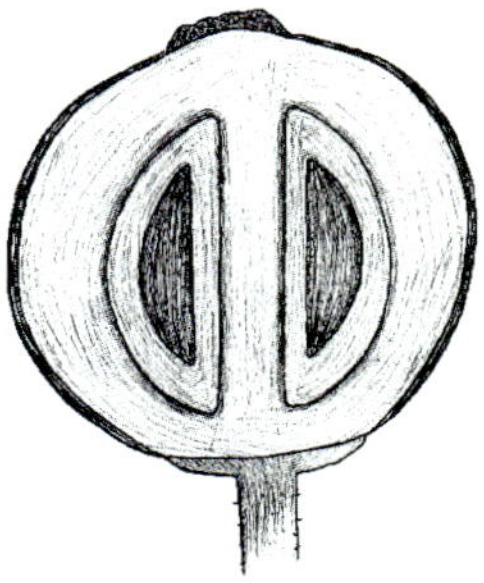
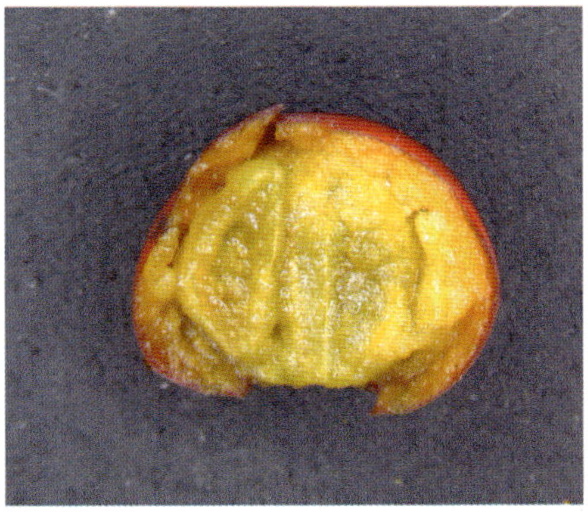

果實，縱切面 Fruit, in L. S.

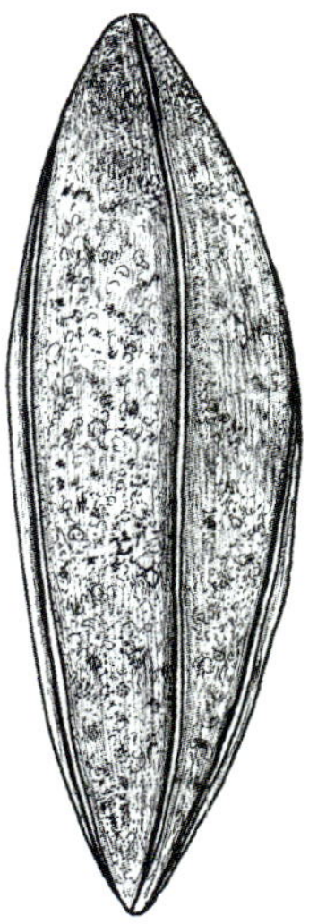
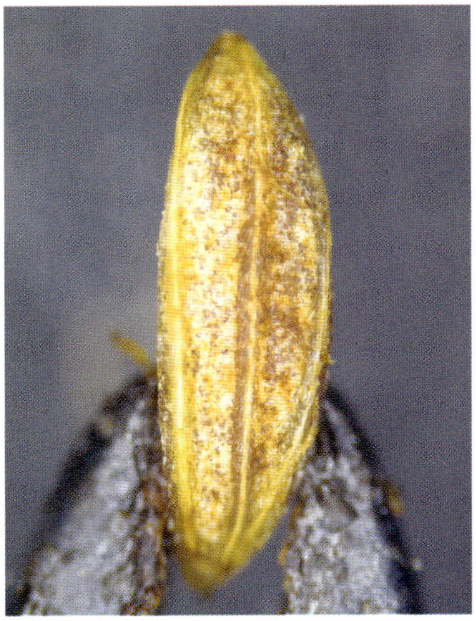

分核 Pyrene

Eleutherococcus trifoliatus (L.) S. Y. Hu
白簕

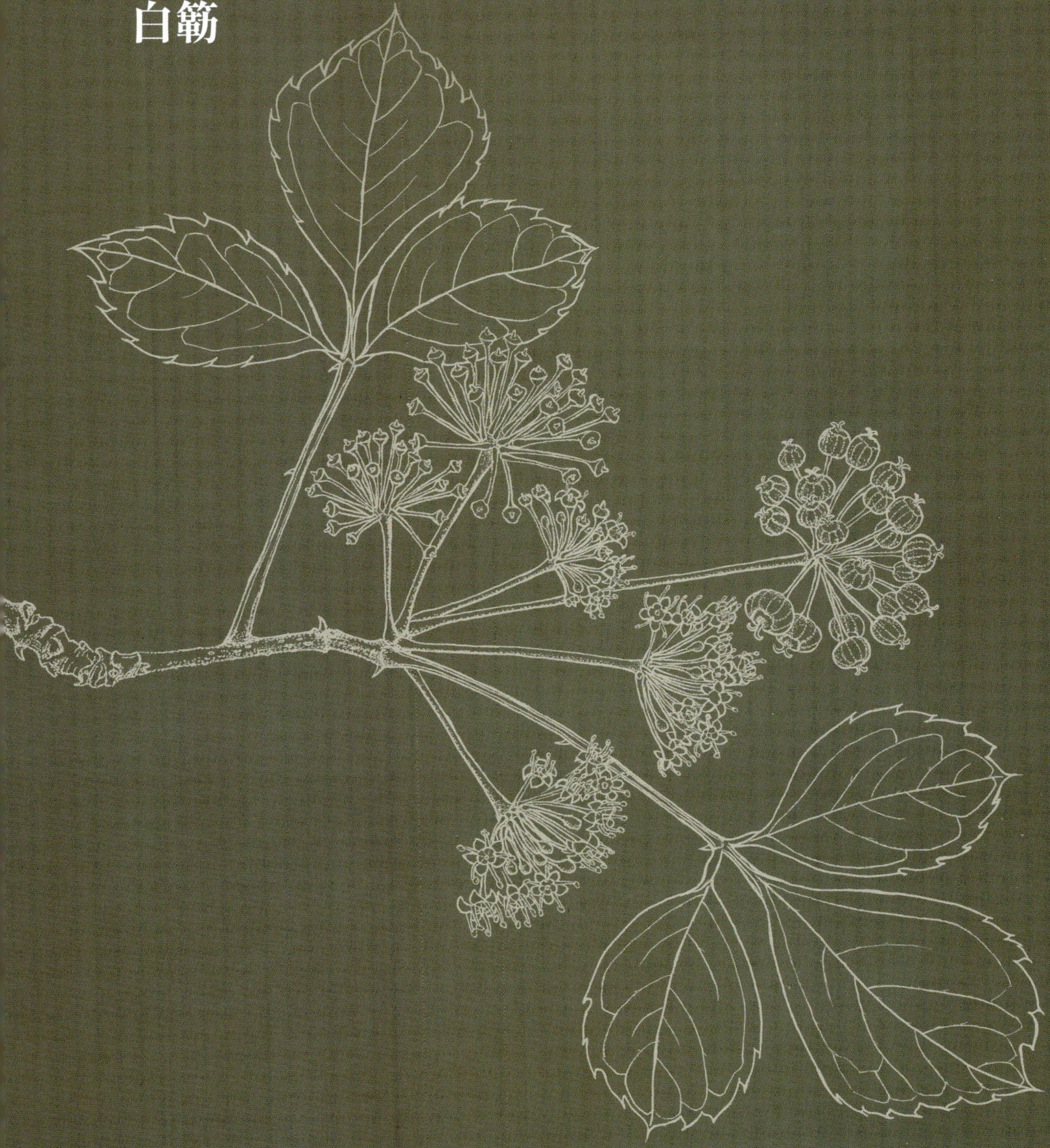

以五加屬植物入藥的歷史源遠流長，
其中白簕常用作治療多種痛症及炎症。

採集者編號
Collector no.
Shiu Ying Hu 5722

採集年份
Year of collection
1968

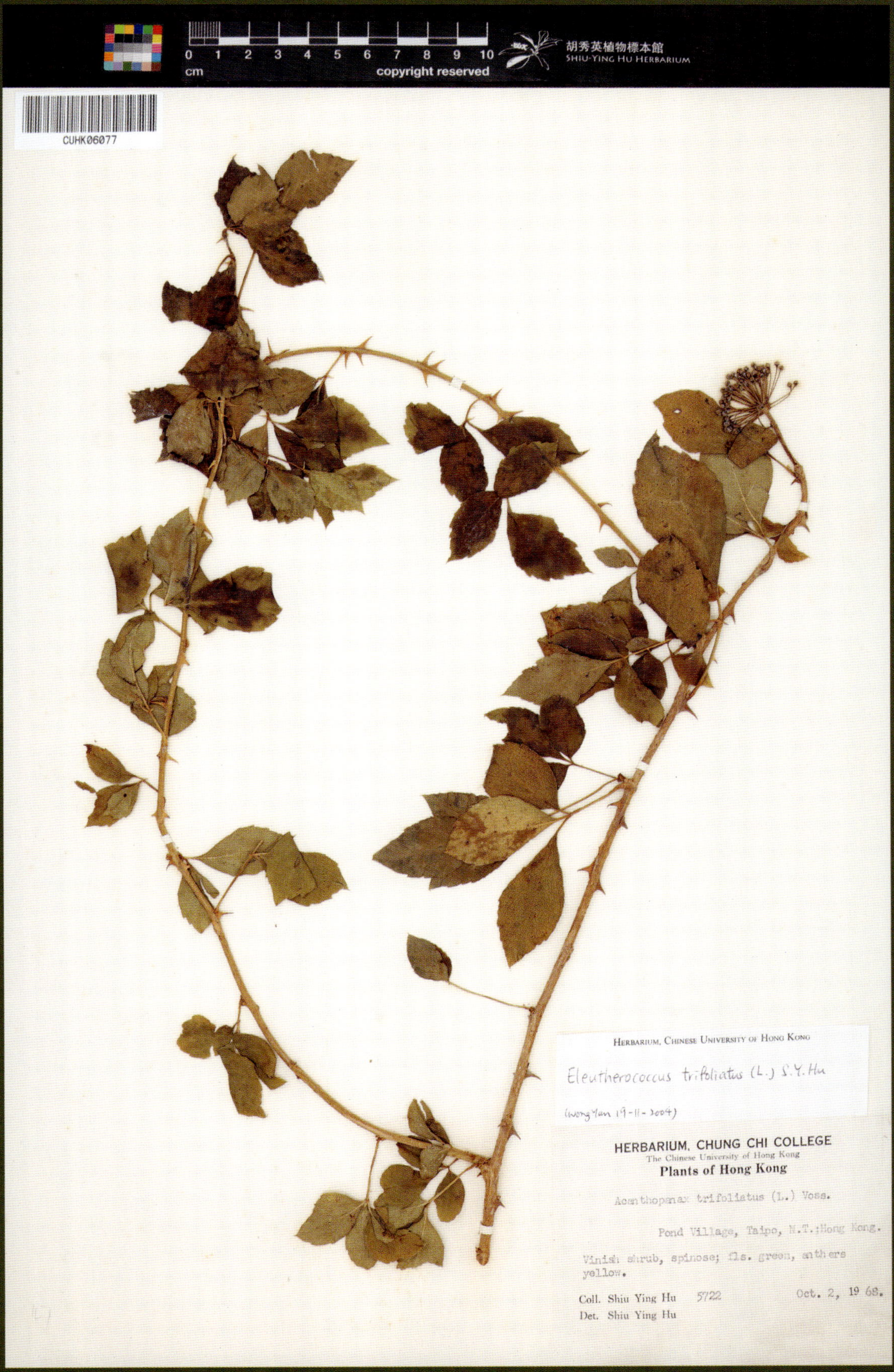
0 1 2 3 4 5 6 7 8 9 10
cm
copyright reserved
胡秀英植物標本館
SHIU-YING HU HERBARIUM
CUHK06077
HERBARIUM, CHINESE UNIVERSITY OF HONG KONG
Eleutherococcus trifoliatus (L.) S.Y. Hu
(wong Yan 19-11-2004)
HERBARIUM, CHUNG CHI COLLEGE
The Chinese University of Hong Kong
Plants of Hong Kong
Acanthopanax trifoliatus (L.) Voss.
Pond Village, Taipo, N.T.;Hong Kong.
Vinish shrub, spinose; fls. green, anthers yellow.
Coll. Shiu Ying Hu 5722 Oct. 2, 19 68.
Det. Shiu Ying Hu

白簕

學名	*Eleutherococcus trifoliatus* (L.) S. Y. Hu
中文名	白簕
科名	五加科
本地分布狀態	原生植物
繪圖編號	M. C. Li 005

花期	花色	果期	果色
8至9月	淺綠色	9至12月	黑色

白簕屬攀援狀灌木或藤本，常見於灌木叢和村邊。白簕含有豐富的鐵元素，在南中國的沿海地區被廣泛當作野菜食用。以五加屬植物入藥的歷史源遠流長，其中白簕常用作治療多種痛症及炎症。

《中國植物誌》(英文版) 的採用名稱
Eleutherococcus trifoliatus (Linnaeus) S. Y. Hu (科名：五加科)

《香港植物誌》的採用名稱
Eleutherococcus trifoliatus (L.) S. Y. Hu (科名：五加科)

參考文獻 References

1. Chen, S., Xu, Y., Liang, D., & Wang, R. (2020). The complete chloroplast genome of *Eleutherococcus trifoliatus* (Araliaceae): a wild edible plant in the coastal region of South China. *Mitochondrial DNA Part B 5*(1), 513–514. https://doi.org/10.1080/23802359.2019.1707131
2. Hong Kong Herbarium. (2022). *HK Herbarium-Eleutherococcus trifoliatus*. HK Plant Database. Retrieved March 17, 2022, from https://herbarium.gov.hk/en/hk-plant-database/plant-detail/index.html?pType=species&oID=9147

Eleutherococcus trifoliatus (L.) S. Y. Hu

Scientific name	*Eleutherococcus trifoliatus* (L.) S. Y. Hu
Chinese name	白簕
Family	Araliaceae Juss.
Local distribution status	Native species
Illustration number	M. C. Li 005

FLOWERING PERIOD	COLOUR OF FLOWER	FRUITING PERIOD	COLOUR OF FRUIT
August to September	Light green	September to December	Black

Eleutherococcus trifoliatus (L.) S. Y. Hu is a scandent shrub or climber, commonly found in shrublands and around villages.

Eleutherococcus trifoliatus (L.) S. Y. Hu is rich in iron and is regarded as a wild edible plant, growing along the coastal areas of South China. *Eleutherococcus* species have a long history of being used as traditional medicines. *Eleutherococcus trifoliatus* (L.) S. Y. Hu is especially useful for relieving pain and inflammation.

Name in *Flora of China*
Eleutherococcus trifoliatus (Linnaeus) S. Y. Hu (Family: Araliaceae)

Name in *Flora of Hong Kong*
Eleutherococcus trifoliatus (L.) S. Y. Hu (Family: Araliaceae)

3. Huang, Y. H., Li, J. T., Zan, K., Wang, J., & Fu, Q. (2021). The traditional uses, secondary metabolites, and pharmacology of *Eleutherococcus* species. *Phytochemistry Reviews,* 1–104. https://doi.org/10.1007/s11101-021-09775-z
4. Missouri Botanical Garden. (2022). *Name-Eleutherococcus trifoliatus.* Tropicos. Retrieved March 17, 2022, from https://www.tropicos.org/name/50008467
5. Xiang, Q. B., & Lowry II, P. P. (2007). Araliaceae. In Z. Y. Wu, P. H. Raven & D. Y. Hong (Eds.), *Flora of China* (Vol. 13, pp. 435–491). Science Press & Missouri Botanical Garden Press.

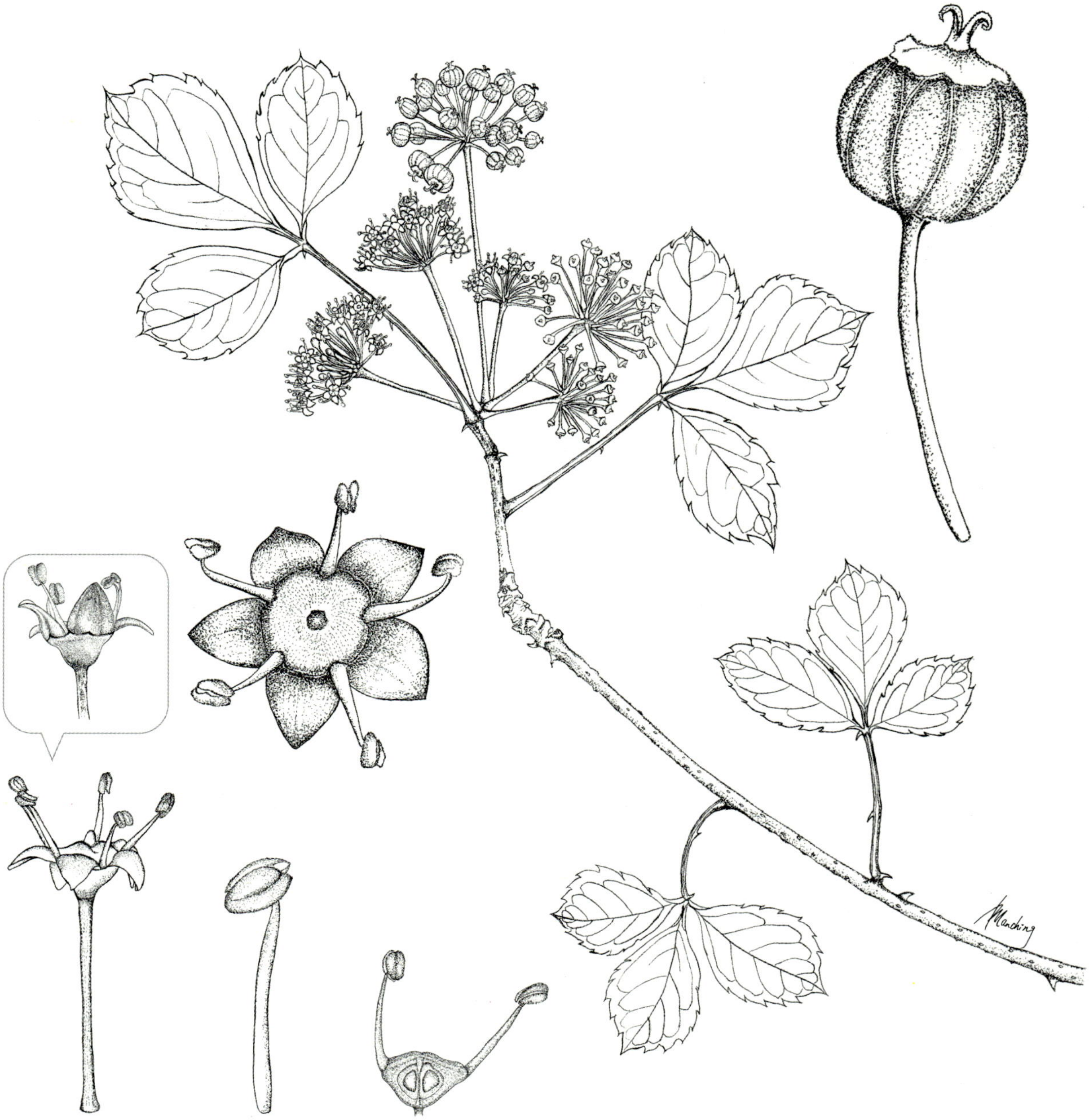

Eleutherococcus trifoliatus (L.) S. Y. Hu 白簕

Illustration 005 by M. C. Li
Authenticated by D. T. W. Lau

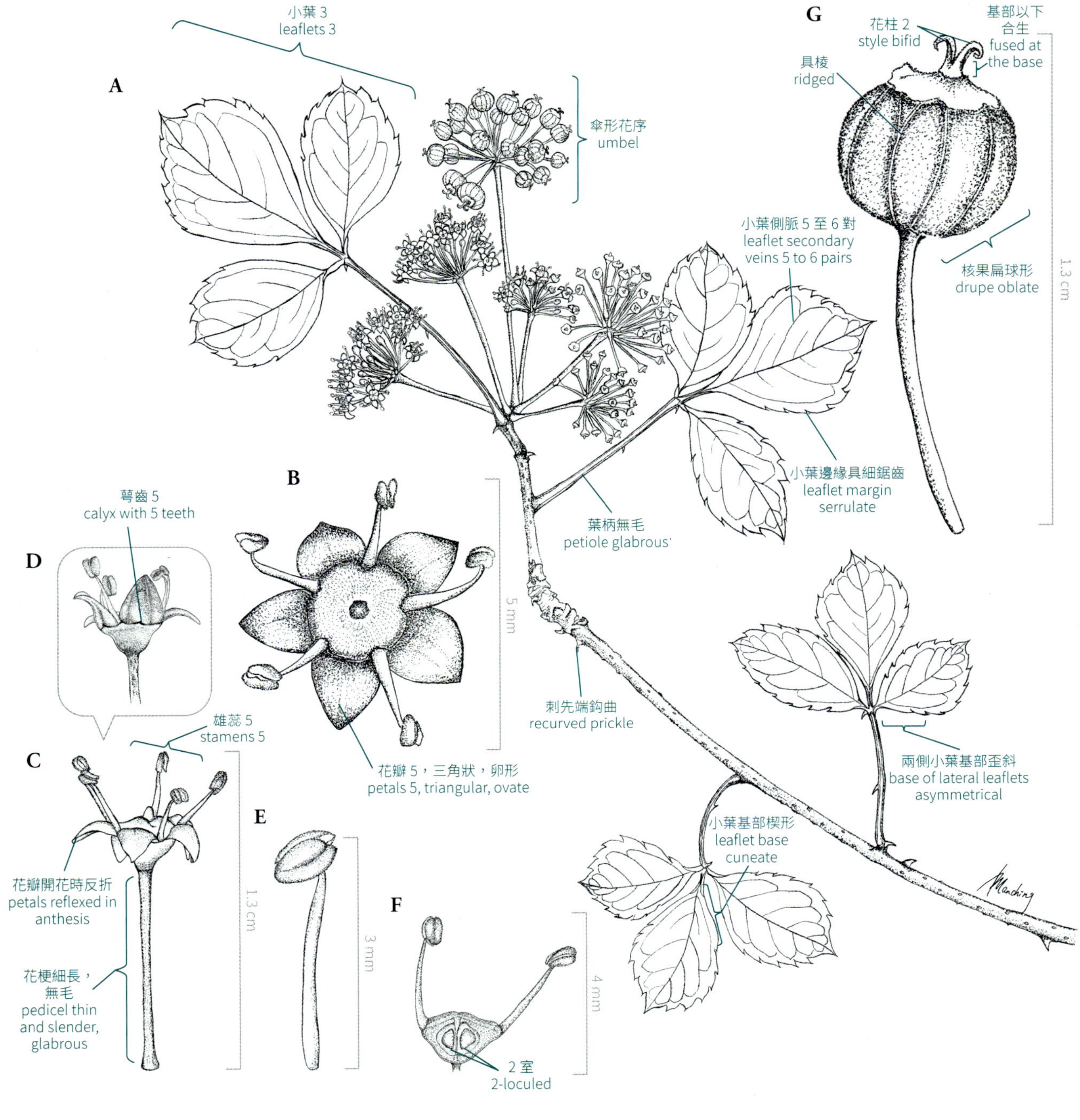

Eleutherococcus trifoliatus (L.) S. Y. Hu 白簕

Illustration 005 by M. C. Li

Authenticated by D. T. W. Lau

植物結構 PLANT STRUCTURES:

A. 花枝 Flowering branch **B.** 花，正面觀 Flower, frontal view **C.** 花，側面觀 Flower, lateral view **D.** 花，側面觀局部放大 Close-up of flower, lateral view **E.** 雄蕊 Stamen **F.** 子房，縱切面 Ovary, in L. S. **G.** 果實 Fruit

參考標本 REFERENCE SPECIMENS: M. C. Li 131 (CUSLSH3278); Shiu Ying Hu 5722

參考文獻 REFERENCES:

Chen, F. H. (1987). Araliaceae. In South China Botanical Garden, the Chinese Academy of Sciences (Ed.), *Flora of Guangdong* (Vol. 1, pp. 272–295). Guangdong Science and Technology Press.

Deng, Y. F. (2008). Araliaceae. In Hong Kong Herbarium, AFCD & South China Botanical Garden, CAS (Eds.), *Flora of Hong Kong* (Vol. 2, pp. 291–296). Agriculture, Fisheries and Conservation Department, HKSAR Government.

Hoo, G., & Tseng, C. J. (1978). Araliaceae. In Delectis Florae Reipublicae Popularis Sinicae Agendae Academiae Sinicae Edita (Ed.), *Flora Reipublicae Popularis Sinicae* (Vol. 54, pp. 1–190). Science Press.

Stevens, P. F. (2022). Apiales. Angiosperm Phylogeny Website. Retrieved March 17, 2022, from http://www.mobot.org/MOBOT/research/APweb/welcome.html

Xiang, Q. B., & Lowry II, P. P. (2007). Araliaceae. In Z. Y. Wu, P. H. Raven & D. Y. Hong (Eds.), *Flora of China* (Vol. 13, pp. 435–491). Science Press & Missouri Botanical Garden Press.

憑證標本
Voucher Specimen

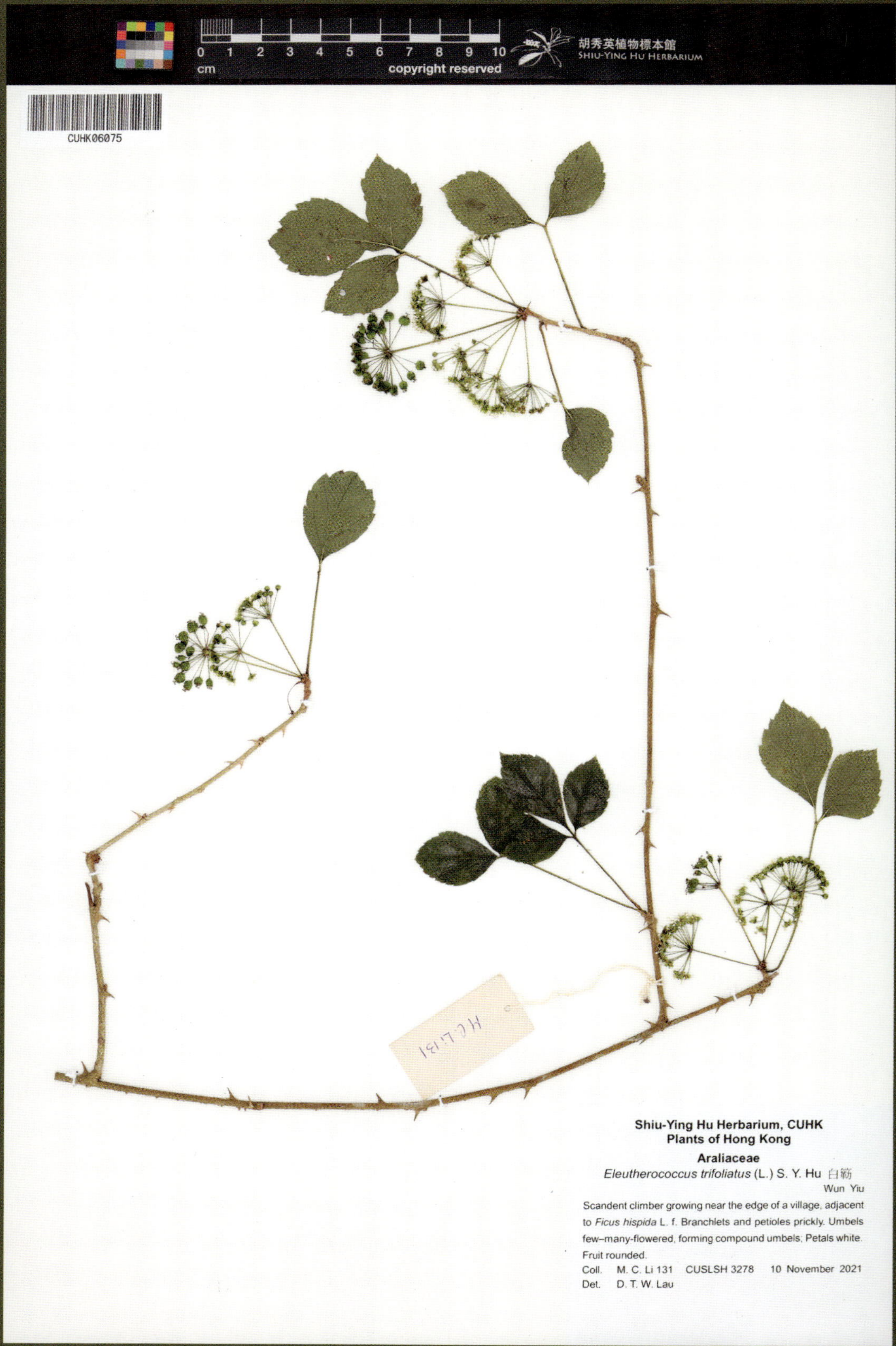

採集者編號 Collector no.
M. C. Li 131
採集年份 Year of collection
2021

外觀 Overview

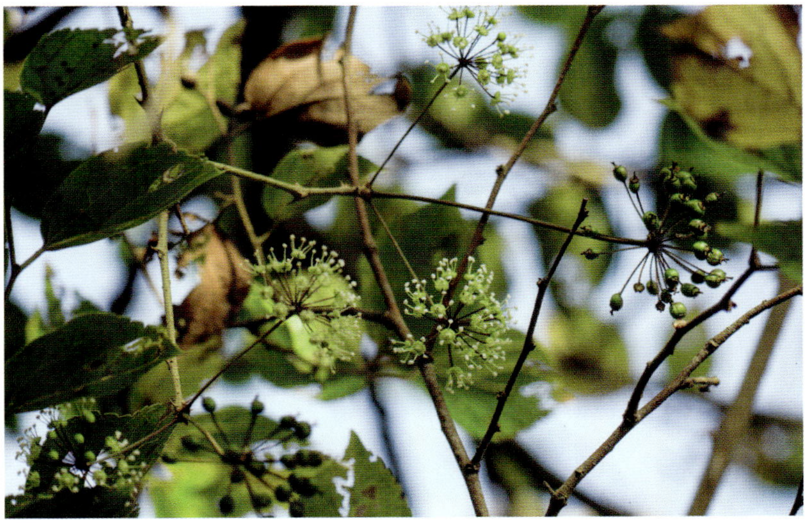

花枝 Flowering branch

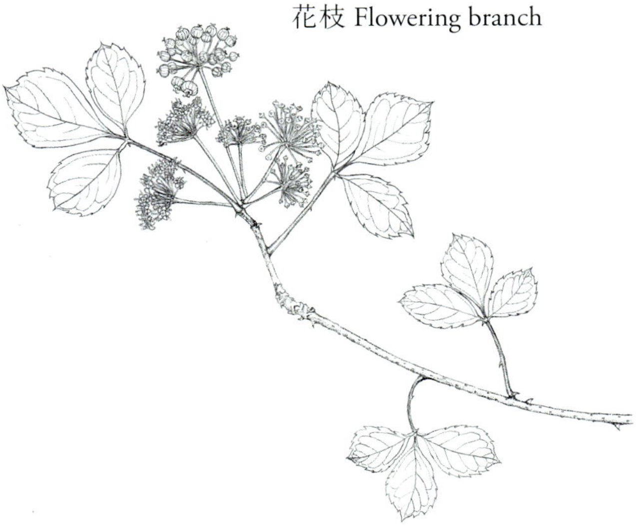

葉 Leaf

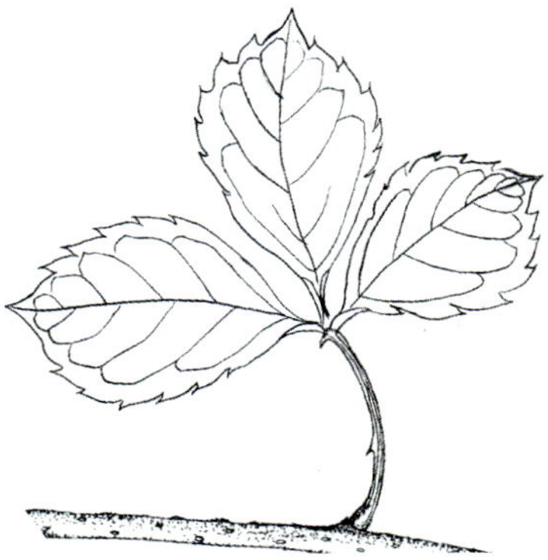

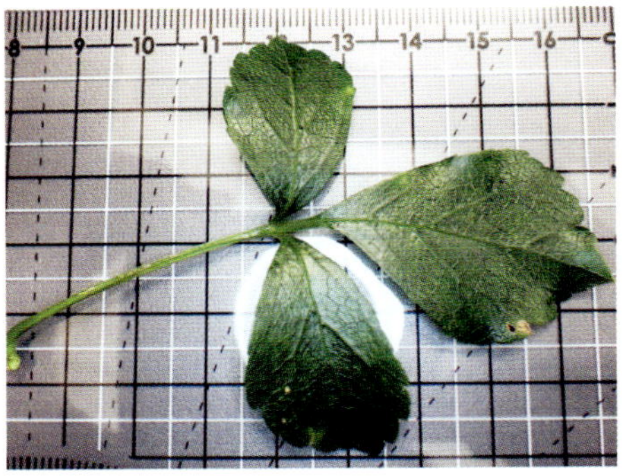

小葉葉面 Leaflets, adaxial

花 Flower

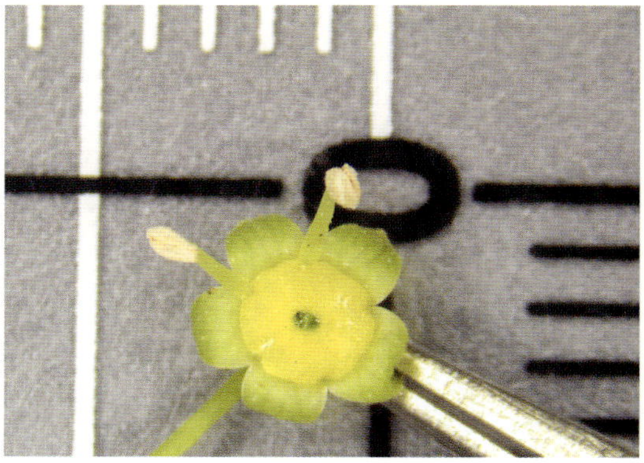

花，正面觀 Flower, frontal view

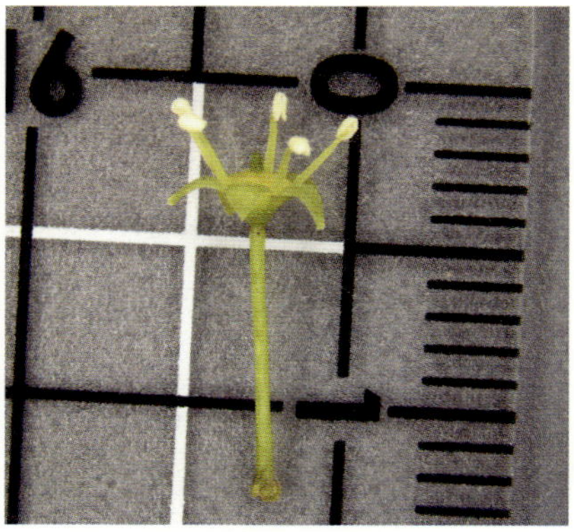

花，側面觀 Flower, lateral view

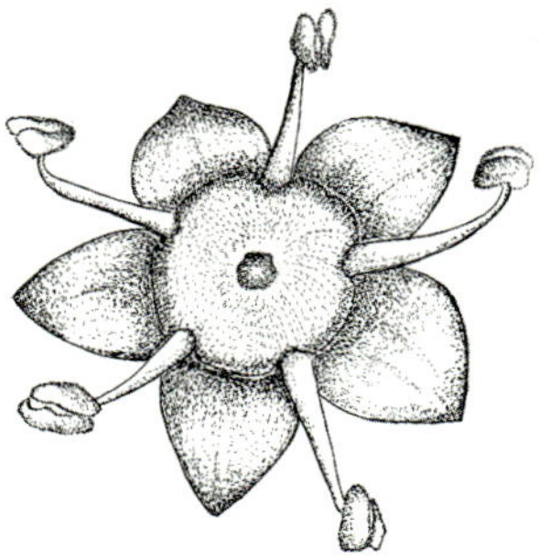

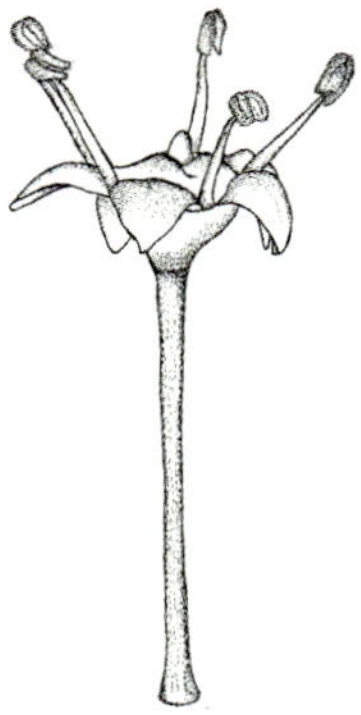

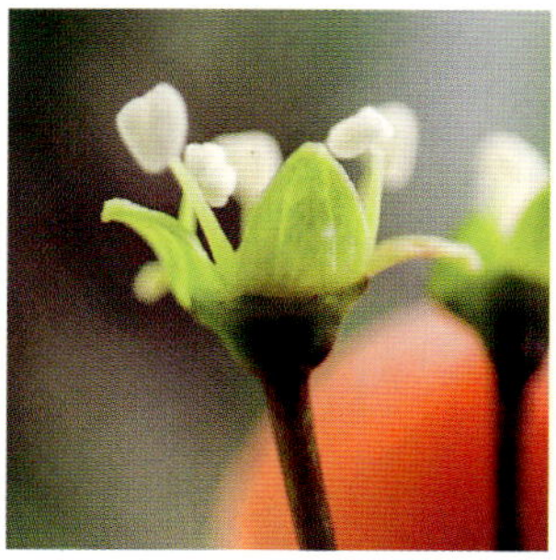

花，側面觀局部放大
Close-up of flower, lateral view

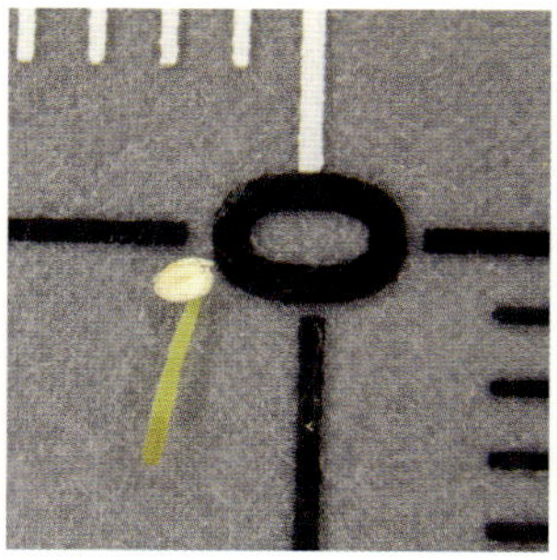

雄蕊 Stamen

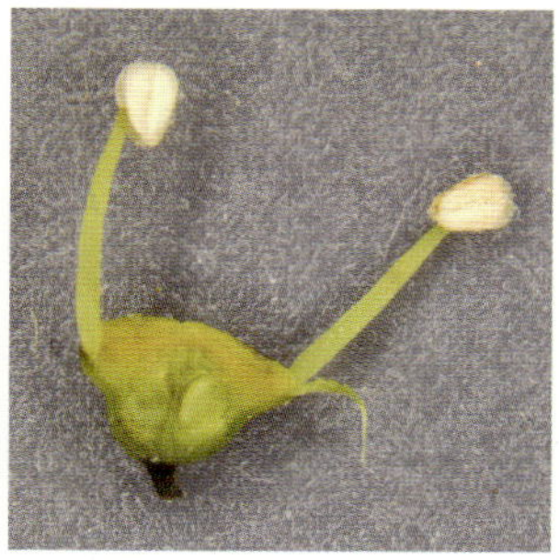

子房，縱切面 Ovary, in L. S.

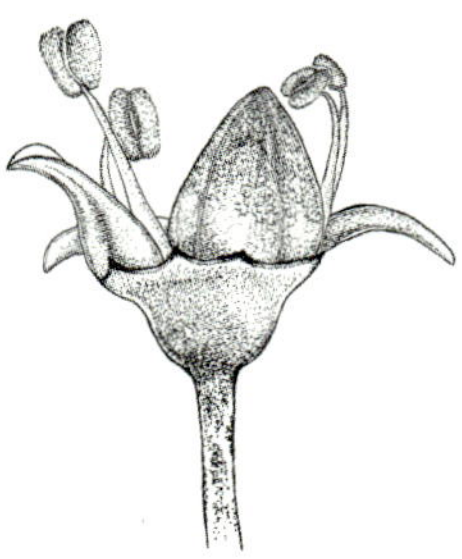

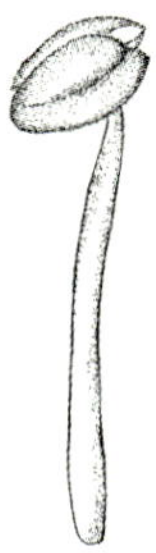

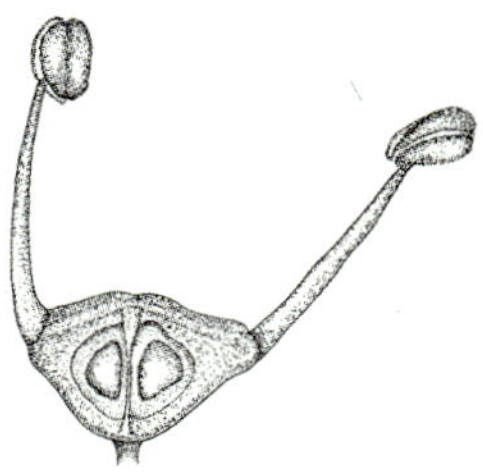

果 Fruit

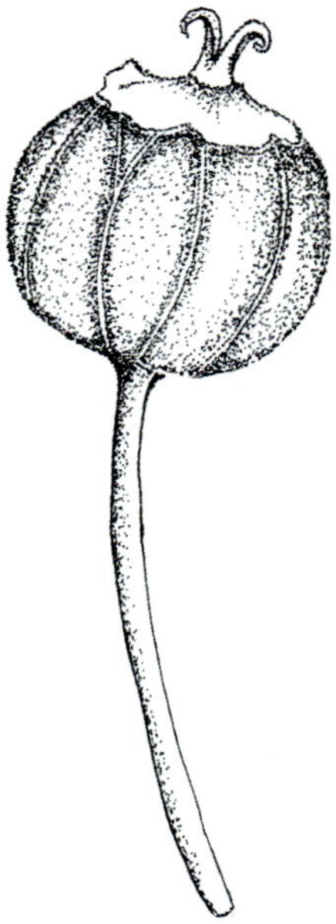

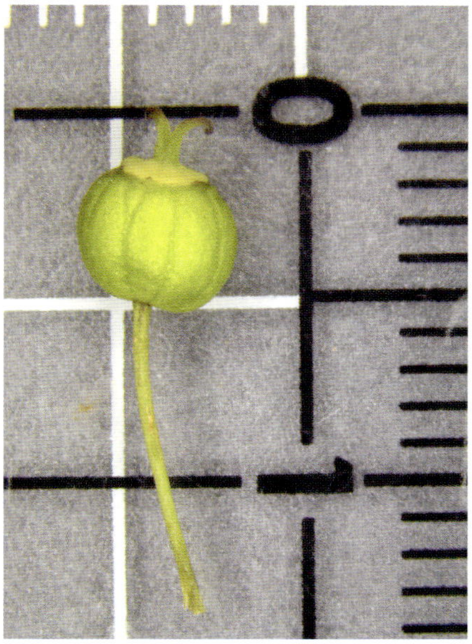

核果 Drupe

鳴謝
Acknowledgements

伍絜宜慈善基金

贊助出版

漁農自然護理署香港植物標本室

提供珍稀植物標本

陳耀文

胡秀英植物標本館技術員

撰寫〈從白簕的科學命名了解植物分類學家工作 —— 紀念胡秀英教授〉一文

Wu Jieh Yee Charitable Foundation

Publication Sponsor

Hong Kong Herbarium
Agriculture, Fisheries and Conservation Department
of the HKSAR Government

provided Voucher Specimens of precious, rare and endangered or protected species

Yiu-Man Chan

Technican, Shiu-Ying Hu Herbarium
Contributor of the essay
'In Remembrance of Prof Shiu-Ying Hu:
The Naming History of
Eleutherococcus trifoliatus (L.) S. Y. Hu'

本書收入的植物名錄
Checklist of Plants

學名 Scientific names	中文名 Chinese names	科 Families	香港稀有及瀕危的植物 Rare and Endangered Species in Hong Kong	頁 Page
		裸子植物 Gymnosperm(s)		
Alsophila spinulosa (Wall. ex Hook.) R. M. Tryon	刺桫欏	Cyatheaceae Kaulf. 桫欏科	✓	3
		被子植物 Angiosperm(s)		
Aquilaria sinensis (Lour.) Spreng.	土沉香	Thymelaeaceae Juss. 瑞香科	✓	143
Artabotrys hongkongensis Hance	香港鷹爪花	Annonaceae Juss. 番荔枝科	✓	37
Artocarpus hypargyreus Hance ex Benth.	白桂木	Moraceae Gaudich. 桑科	✓	117
Camellia oleifera C. Abel	油茶	Theaceae Mirb. 山茶科	✓	169
Dalbergia assamica Benth.	南嶺黃檀	Fabaceae Lindl. 豆科	✓	85
Dalbergia candenatensis (Dennst.) Prain	彎枝黃檀	Fabaceae Lindl. 豆科	✓	95
Eleutherococcus trifoliatus (L.) S. Y. Hu	白簕	Araliaceae Juss. 五加科		239
Ilex rotunda var. *microcarpa* (Lindl. ex Paxton) S. Y. Hu	小果鐵冬青	Aquifoliaceae Bercht. & J. Presl 冬青科		225
Illicium dunnianum Tutcher	紅花八角	Schisandraceae Blume 五味子科	✓	15
Iris speculatrix Hance	小花鳶尾	Iridaceae Juss. 鳶尾科	✓	61
Lysimachia alpestris Champ. ex Benth.	香港過路黃	Primulaceae Batsch 報春花科	✓	157
Michelia chapensis Dandy	樂昌含笑	Magnoliaceae Juss. 木蘭科	✓	27
Oroxylum indicum (L.) Benth. ex Kurz	木蝴蝶	Bignoniaceae Juss. 紫葳科		209
Paris polyphylla var. *chinensis* (Franch.) H. Hara	華重樓	Melanthiaceae Batsch 黑藥花科	✓	49
Pavetta hongkongensis Bremek.	香港大沙葉	Rubiaceae Juss. 茜草科	✓	193
Polygala hongkongensis Hemsl.	香港遠志	Polygalaceae Hoffmanns. & Link 遠志科		105
Rhododendron simsii Planch.	紅杜鵑	Ericaceae Juss. 杜鵑花科	✓	181
Rhodoleia championii Hook.	紅花荷	Hamamelidaceae R. Br. 金縷梅科	✓	73
Sapindus saponaria L.	無患子	Sapindaceae Juss. 無患子科		129

詞彙表
Glossary

2-loculed	2 室	a locule is a cavity with a carpel, anther or sporangium; 2-loculed means having two cavities
2-sulcate	2 溝	presence of two grooves
3-striate	3 縱棱	presence of three longitudinal ridges
abaxial	葉底 / 遠軸的	the lower surface of a leaf or a flattened organ, or the side facing away from the axis
actinomorphic	輻射對稱	a structure radially symmetric, usually from the top view
acuminate	漸尖	long tapering tip with slightly incurved sides and intersection angle < 45 degrees
adaxial	葉面 / 近軸的	the upper surface of a leaf or a flattened organ, or the side facing towards the axis
anatropous	倒生胚珠	the ovule is hanging in reversed form with the micropyle pointing downward
androecium	雄蕊群	a collective term of stamens, the male parts of a flower
anther dorsifixed	花藥背部著生	filament is attached to the middle portion of the anther's back
anthers epipetalous	花藥著生於花冠	stamens are attaching on the petals
anthesis	花冠開放時	duration when flowers open, usually at the optimal time of pollination
apex rostellate	頂端有喙	the tip ending in a small beak
apiculate	短尖頭	sharp but flexible apex with length to width ratio > 3:1
arcuate	弧曲	structure curved like a bow
base attenuate	基部漸狹	margin gradually narrowing towards the base; intersection angle < 45 degrees
base cuneate	基部楔形	margin tapering towards the base, with intersection angle 45–90 degrees; = cuneiform
berry	漿果	a fleshy and indehiscent fruit with one to many seeds
bract	苞片	a modified leaf-like structure usually present at the peduncle base of an inflorescence
bracteole	小苞片	a modified leaf-like structure present along the pedicel of a flower

branchlet	幼枝	small and young branches near the terminal end of the branching system
C. S.	橫切面	cross-section
ca.	約	approximately
calyx	花萼	a collective term of sepals, usually the outermost layers of a flower
calyx tube	花萼筒	sepals fused to form a tube
campanulate	鐘狀	a bell-shaped tube with nearly equal length and width
capitate	頭狀	structure with the broadest apex, but a smaller club size as compared to clavellate
capsule	蒴果	a dehiscent fruit developed from a syncarpous ovary; splits by valves or pores in maturity
carpel	心皮	a basic unit of gynoecium; includes ovary, style and stigma
caudate	尾尖	abruptly narrowing to a long tail-like structure with intersection angle < 45 degrees and incurved sides
clavate	棍棒狀	club-shaped; thickened towards the end, usually at the apex
connective	藥隔	the tissue layer separating two pollen sac (theca) of an anther
corolla	花冠	a collective term of petals; usually colourful and positioned in between stamens and calyx
cucullate	盔狀	hood-shaped
cuneate	楔形	tapering gradually near the base with an acute angle
decurrent	下延	base extension of tissue (e.g. leaf or stipule) downward to form wings or ridges
distichous	2列	structures arranged on two sides of an axis; seems on a same plane; = two-ranked
dorsifixed	背部著生	describes the filament attachment at the middle of the connective tissue
drupe	核果	an indehiscent fruit with exocarp, mesocarp and woody endocarp enclosing a seed
elliptic	橢圓形	ellipse-shaped, broadest in the middle; length to width ratio from 2:1 to 3:2
exocarp	外果皮	the outermost layer of the fruit wall (pericarp)
falcate	鐮刀形	sickle-shaped, like a short sword curved to one side
fasciculate	簇生	structure arranged in a tight bundle; seems to originate from one point; = fascicle
filament	花絲	the stalk of the stamen; usually an elongated structure originated from the corolla or other basal floral parts
filiform	線狀的	thread-like structure, cylindrical and very slender; long, thin and flexuous

fimbriate	流蘇狀	fringed margin with thread-like processes thicker than cilia
flower disk	花盤	a fleshy and circular structure outside or within the stamens; sometimes lobed and nectariferous
funicle	珠柄	an ovule stalk that connects to an ovary wall; same as seed stalk
fusiform	紡錘狀	a structure widest at the middle and tapering at both ends; = spindle-shaped
glabrous	無毛 / 光滑	smooth surface and without hair, trichomes or scales
gynoecium	雌蕊	the female parts of a flower; composed of one or more than one pistils
gynophore	子房柄 / 雌蕊群柄	ovary stalk; gynoecium stalk
head inflorescence	頭狀花序	a number of flowers (tubular and/or ligulate) on a nearly flattened receptacle, externally covered with involucral bracts
head infructescence	頭狀果序	a number of fruits on a nearly flattened receptacle, externally covered with involucral bracts
hirsute	粗毛	stiff and relatively long hair; not harsh in general, but coarse and rough
hirtellous	小粗毛	diminutive hirsute
hispid	硬毛	long and harshly stiff hair; more sharply bristly than hirsute; sometime fragile
imbricate	覆瓦狀排列	tissue layers overlapping, with each layer having one side external and the other side partially covered by a nearby one
indusium	囊群蓋	a protective layer covering the sporangia of ferns
inflorescence	花序	a group of flowers on one or a number of axes (flowering branches) with bracts; a simple one has only one flower
infructescence	果序	a group of fruits on one or a number of axes (fruiting branches) with bracts; a simple one has only one fruit
involucral bract	總苞片	a modified leafy structure at the base of an inflorescence peduncle
keel	龍骨瓣	two united petals with a ridge like a boat bottom; common in the flower of Fabaceae
L. S.	縱切面	longitudinal section
lanceolate	披針形	an elongated egg shape with a slender pointed apex; length to width ratio from 6:1 to 3:1
lateral veins	側脈	veins originated from a central main vein; = secondary vein
leaflet	小葉	a leafy structure that is a division unit of a compound leaf
legume	莢果	a dry and single-carpel fruit that dehisces along two sutures at maturity; common in Fabaceae (legume family)

lenticel	皮孔	a spongy area in the bark of stems that allows gaseous exchange; also used to describe similar tissue on roots and pericarps
linear	線形	a long and narrow shape with long parallel sides; length to width ratio from 6:1 to 12:1
lobes	裂片	an about halfway division of an organ; commonly used to describe corolla or calyx
loculicidal	室背開裂	a ripened capsule splitting into various units along the central midrib
longitudinally angular	縱棱	with an angle along the longitudinal edge
longitudinally dehiscent	縱裂	splitting longitudinally
margin erose	邊緣齧齒狀	margin jagged and uneven as if nibbled
margin sinuous	邊緣波狀	wavy margin with shallow lobes
mesocarp	中果皮	middle layer of pericarp (fruit wall); freshly and prominent in drupe
midrib / midvein	中脈	central major axis / vein of a leaf blade; seems to extend from the petiole
mucronate	凸尖	short stiff apical extension with length to width ratio from 3 to 1
oblanceolate	倒披針形	an elongated inverted egg shape with a tapering base; with length to width ratio from 6:1 to 3:1
oblate	扁圓球狀	globular but slightly compressed along the poles
oblong	矩圓形	a shape of short parallel sides and rounded ends; length to width ratio from 2:1 to 3:2
oblong-elliptic	長圓狀橢圓形	a 2-dimensional structure of elongated elliptic shape with parallel margins from side view
oblong-obovate	長圓狀倒卵形	inverted egg-shaped, with short parallel sides, broadest near the apex
obovate-elliptic	倒卵狀橢圓形	more or less an inverted egg-shaped, broadest between the apex and the middle; length to width ratio is > 2:1
ovary	子房	basal part of a pistil containing the ovules; usually swollen
ovary superior	子房上位	a flower structure with the sepals, petals and stamens originated below the ovary
ovate	卵形	egg-shaped, broadest near the base; with length to width ratio from 2:1 to 3:2
ovoid	卵球狀	3-dimensional egg-shaped, broadest at the base
ovoid-ellipsoid	卵形橢圓狀	a 3-dimensional structure with elliptic side view; broadest at the base
ovule	胚珠	an immature seed within the ovary

panicle	圓錐花序	the indeterminate main axis of an inflorescence; has several lateral branches which are raceme
paripinnate	偶數羽狀複葉	two opposite leaflets at the terminal end of a pinnate leaf
pedicel	花梗	the stalk of a single flower
peduncle	花序柄 / 總花梗	the stalk of an inflorescence
peltate	盾形	a disk-like shape with a stalk attachment pointing away from the margin
perianth	花被	a collective term for the calyx and the corolla
pericarp	果皮	the fruit wall; sometimes can be differentiated in three layers: exocarp, mesocarp and endocarp
petal	花瓣	a single piece of corolla; usually colourful and positioned in between stamens and calyx
petaloid	花瓣狀	a structure of form or color like a petal
petiolule	小葉柄	the stalk of a leaflet
pilose	長柔毛	long, soft and rather straight hairs; scattered, not dense but somewhat shaggy
pinna	羽片	a leaflet of a pinnate leaf, or the primary division of a compound leaf
pinnate	羽狀複葉	the main rachis have lateral leaflets arranged in opposition or alternately, like the pattern of a feather
pinnatipartite	羽狀深裂	the main rachis have lateral lobes in several rows, with the division of lobes more than half-way from the margin
pinnule	小羽片	a lateral segment of a pinna
pistil	雌蕊	the female organ of a flower, composed of ovary, style and stigma
placentation	胎座	arrangement of ovules inside an ovary
placentation axile	中軸胎座	ovules that arise from the placental tissue at the central axis of a syncarpous ovary
placentation free central	特立中軸胎座	ovules that arise from a central freestanding placental tissue of a unilocular ovary
placentation parietal	側膜胎座	ovules that arise from the inner wall of a syncarpous ovary
plant specimen	植物標本	A pressed and dried specimen made from flowering / fruiting branches. The freshly collected part is processed with a series of procedures to ensure the final specimen can be well authenticated and preserved intact in a herbarium for at least 100 years. The collector and determinator names, number, location, habitat and date are the key information of the specimens. The authentic specimens deposited in a herbarium can be accessible as a standard reference to verify the identity of unknown plant samples.

porous dehiscence	頂孔開裂	pollens are released via a pore opening on the apex of anthers
puberulent	微柔毛	rather dense, soft and extremely short hairs; hardly visible; minutely pubescent
puberulous	短柔毛	rather dense, soft and very short hairs; just visible
pubescent	柔毛	dense, soft and short hairs; easily visible
pyrene	分核	a seed with a hard layer of endocarp
racemose	總狀花序	the unbranched main axis of an indeterminate inflorescence with pedicellate flowers, the youngest on the top
rachis	葉軸	the main axis extending from a petiole; bearing pinnae or leaflets
reinform	腎形	kidney-shaped
retuse	微凹	slight apical incision
rhizome	根狀莖	underground stem with nodes and buds; usually prostrate and grow laterally
rosette	蓮座狀	organs, usually leaves, radiating from a point at or near ground level
salverform	高腳碟狀	a long corolla tube with abruptly widening lobes at the terminal end
schizocarp	果爿	a carpellary component, usually with one seed, splitting from a fruit
secondary veins	側脈	veins originated from a central main vein = lateral veins
sericeous	絹毛 / 絲綢狀毛	silky and shiny; soft straight hairs appressed on the surface
serrate	鋸齒邊緣	margin with sawlike teeth pointing upward
serrulate	細鋸齒	diminutive of serrate
sheath	鞘	a tubular tissue enclosing another organ; usually developed from a leaf blade or petiole
sorus	孢子囊群	a structure bearing or covering a group of sporangia
spatulate	匙形	spoon- or spatula-shaped; an oval to oblong structure with an extended basal stalk
sporangia receptacle	囊托	an outgrowth tissue from a leaf that bears a number of sporangia
sporangium	孢子囊	a capsule-like structure containing spores
stamen	雄蕊	a male organ of a flower, consisting of filament and anthers
stamens diadelphous	雄蕊二體	stamens fused in two groups or bundles; usually found in the flowers of Fabaceae
stamens monadelphous	單體雄蕊	stamens with filaments united into a tube; usually found in the flowers of Malvaceae
staminodes	不育雄蕊	infertile stamens that are usually smaller and not bearing pollen; sometimes petaloid in form and colour

stigma	柱頭	terminal enlarged apex of a pistil for receiving pollen
stipe / petiole	葉柄	leaf stalk connecting stem and leaf blade
stipule	托葉	a leaf-like structure at the base of a leaf stalk; usually in pairs
stolon	匍匐莖	a prostrate stem along the ground, with shoots or roots at the nodes
strigose	糙伏毛	sharp stiff hairs almost parallel to the surface
style	花柱	an extension part of an ovary with the terminal end of stigma
suborbicular	近圓形	close to a circular shape
subsessile	近無柄	the stalk is very short and inconspicuous
subulate	尖錐形	awl-shaped; stout needle-shaped; narrowly triangular
tepal	花被片	a laminate tissue or a unit of perianth; could be sepal or petal
thyrse	聚傘圓錐花序	an inflorescence of indeterminate main axis with lateral cymose branches
tomentose	絨毛	dense, soft, short and woolly hairs; matted and appressed to the surface
triangular-ovate	三角狀卵形	a triangular egg shape with one of the flat sides on bottom; length to width ratio from 2:1 to 3:2
tripinnatipartite	三回羽狀深裂	the main rachis have lateral bipinnatipartite leaves
umbel	傘形花序	a cluster of flowers with similar pedicels arising from the same point of a penduncle
umbelliform cyme	傘形狀 聚傘花序	a compound cyme that forms a similar shape as an umbel
villous	長絨毛	long, fine and soft hairs, slightly denser than pilose hairs; often curly and more or less erect
voucher plant specimen	植物憑證標本	a kind of plant specimen that is specially collected / selected as a verifiable and permanent reference of a plant individual or population, which is used for analysis of their population, ecology, DNA, chemicals or pharmacology, and is cited in scientific publication
wing	翼瓣	two lateral petals of the flowers in Fabaceae; enclosed by a larger petal (standard)
zygomorphic	兩側對稱	bilateral symmetry

團隊簡介

劉大偉（作者）

香港中文大學生命科學學院胡秀英植物標本館　館長

劉大偉博士是一位植物學家，曾參與多項有關植物分類學、草藥鑒定及藥理學的研究項目，現任香港中文大學生命科學學院胡秀英植物標本館館長，專責管理「香港植物及植被計劃」。劉博士的教研範圍包括植物分類學、生物多樣性及中藥鑒定。

李敏貞（作者及植物繪圖師）

香港中文大學生命科學學院胡秀英植物標本館　教研助理

畢業於香港珠海學院新聞及傳播學系，正修讀香港大學環境管理理學碩士課程。現職於香港中文大學生命科學學院胡秀英植物標本館，主要負責科學繪圖、植物學STEAM教育計劃及植物學研究課題。

出生於小鄉村，小時候受到家門前種下的一棵黃皮樹啟發，深深被大自然的生命力吸引，後來更開始嘗試以植物做素材畫畫，期望把眼前的美好記錄下來。加入標本館工作後，有幸接觸兩位教授級畫家——曾孝濂老師和馬平老師筆下的科學繪圖，開始意識到科學繪圖對植物鑒定的重要性，因此期望透過製作本圖鑑，把前輩們對植物分類的熱誠承傳下去。

王曉欣（作者及植物繪圖師）

香港中文大學生命科學學院胡秀英植物標本館　教研助理

畢業於香港中文大學環境科學系，在學期間曾參與自然教育的工作。現職於香港中文大學生命科學學院胡秀英植物標本館，工作範疇包括科學繪圖、植物學STEAM教育計劃及植物學研究項目等。自小在祖母的薰陶下，對民間草藥及食用野草產生了興趣。進入大學以後，有幸修讀了關於香港植物的課程，認識到不同的方式去欣賞和學習植物知識。後來以實習生身分在標本館工作，開始接觸到科學繪圖，並被兩位教授級畫家——曾孝濂老師和馬平老師留下的原稿深深震撼。其作品的細緻和精確度體現了超卓的工匠精神，是推動自己和團隊不斷調整各種繪圖、解剖和拍攝技巧，以及堅持來回校對畫稿的原因。希望透過本書，能夠把前輩的精神傳承下去。

王天行（編輯及教育顧問）

香港中文大學生命科學學院胡秀英植物標本館　教育經理

畢業於千禧年代的中文大學生物系，現為胡秀英植物標本館「植物學STEAM教育計劃」的成員。在STEM教育工作具有豐富經驗，曾參與建立香港植物及植被數據庫，十多年來製作或參與多個大型科普教育平台和教育計劃，有幸能協助本圖鑑的編輯。

About the Editorial Team

Dr David T. W. Lau (Author)

Curator of the Shiu-Ying Hu Herbarium, School of Life Sciences,
The Chinese University of Hong Kong

Dr David T. W. Lau is a botanist who has been researching plant taxonomy, herbal authentication and pharmacology. He is now a curator of the Shiu-Ying Hu Herbarium in the School of Life Sciences at The Chinese University of Hong Kong, and is responsible for managing the 'Hong Kong Flora and Vegetation' project. His research and education interests include plant taxonomy, biodiversity and Chinese medicine authentication.

Man-Ching Li (Author & Botanical Illustrator)

Education Assistant, Shiu-Ying Hu Herbarium, School of Life Sciences,
The Chinese University of Hong Kong

Man-Ching Li graduated from the Department of Journalism and Communication, Chu Hai College of Higher Education, and is pursuing a Master's programme in Environmental Management at the University of Hong Kong. She is currently working at the Shiu-Ying Hu Herbarium, School of Life Sciences, The Chinese University of Hong Kong, where her major roles include scientific illustration, the Botany STEAM Education Project, and botanical research studies.

Born in a small village, she was inspired by the vitality of the *Clausena lansium* (Lour.) Skeels in her front yard at a young age. Afterwards, she started to use plants as art themes to capture precious moments in nature. After joining the Herbarium team, she had the opportunity to encounter the scientific illustrations drawn by two top botanical artists, Mr. Xiao-Lian Zeng and Mr. Ping Ma, and realized the importance of scientific illustrations for plant identification. Through producing this illustrated book, she hopes to pass on their enthusiasm in the field of plant taxonomy.

Hiu-Yan Wong (Author & Botanical Illustrator)

Education Assistant, Shiu-Ying Hu Herbarium, School of Life Sciences,
The Chinese University of Hong Kong

After graduating as an Environmental Science major at The Chinese University of Hong Kong and participating in environmental education work while pursuing her studies, she is now working at the Shiu-Ying Hu Herbarium, School of Life Sciences, The Chinese University of Hong Kong. Her job duties include scientific illustration, the botany STEAM education programme, and botanical research projects. Nurtured by her grandmother, she developed an interest in folk medicinal and edible wild plants. After entering university, she enrolled in a course about plants in Hong Kong, where she realized there are different ways to appreciate and learn about plants. Later, she worked in a herbarium as an intern and encountered scientific illustrations. She was astonished by the manuscripts left behind by two top botanical artists, Mr. Xiao-Lian Zeng and Mr. Ping Ma. The spirit of these two professional artists was revealed in the details and precision of their work, and this motivated her to continuously improve her different illustrations, dissections and photography skills, and to verify the drafts repeatedly with the team. She hopes that the spirit of the former illustrators will be passed on via this illustrated book.

Tin-Hang Wong (Editor & Education Consultant)

Education Manager, Shiu-Ying Hu Herbarium, School of Life Sciences,
The Chinese University of Hong Kong

Tin-Hang Wong graduated in 2000 from the Biology Department of The Chinese University of Hong Kong, and works as a team member with the Botany STEAM Education Programme at the Shiu-Ying Hu Herbarium. Experienced in STEM education, he participated in establishing the Hong Kong Flora and Vegetation Web Database, and developed and participated in many Science Education platforms and schemes. Wong is honoured to help with editing this guide.